油田化学

(英汉对照)

OILFIELD CHEMISTRY

秦文龙 等◎编著

中国石化出版社

内容提要

本书在介绍胶体与表面化学、高分子化学基础知识的基础上，依据油气生产环节组成，主要介绍油气井钻井、完井、压裂、酸化、修井、堵水、调剖、三次采油、油气集输、腐蚀与防护等过程中存在问题的化学本质和解决这些问题的化学用剂及其作用原理。同时，对近年来油田化学领域的一些新产品、新技术等进行了简要介绍。

本书是石油类高等院校石油工程、储运和应用化学等专业的教学用书，也可供从事油田化学的专业研究人员和工程技术人员参考。

图书在版编目（CIP）数据

油田化学：英汉对照 / 秦文龙等编著 . —北京：中国石化出版社，2019.9
 ISBN 978-7-5114-5507-9

Ⅰ.①油… Ⅱ.①秦… Ⅲ.①油田化学–英、汉 Ⅳ.①TE39

中国版本图书馆 CIP 数据核字（2019）第 186433 号

未经本社书面授权，本书任何部分不得被复制、抄袭，或者以任何形式或任何方式传播。版权所有，侵权必究。

中国石化出版社出版发行
地址：北京市东城区安定门外大街 58 号
邮编：100011　电话：(010)57512500
发行部电话：(010)57512575
http://www.sinopec-press.com
E-mail: press@sinopec.com
北京科信印刷有限公司印刷
全国各地新华书店经销

*

787×1092 毫米 16 开本 21 印张 451 千字
2019 年 12 月第 1 版　2019 年 12 月第 1 次印刷
定价：88.00 元

Preface/ 前言

随着石油工业的发展以及非常规油气资源勘探开发水平的提高，油气田化学剂的需求越来越大，各种新产品、新技术不断涌现，油田化学知识及技术发展日益受到国内外石油工程科技人员的重视。近些年来，国际化教育已成为石油高等院校的发展趋势，编写油田化学中英文专业教材，将对国际化型石油工程师的培养具有促进意义。

油田化学是化学、地质、石油工程等多门类学科的结合，并在实践－理论－实践的过程中不断发展的一门新兴学科。该学科主要利用各门基础化学理论，研究钻井、采油和原油集输过程中出现的各种化学问题。油田化学的基础理论主要包括胶体/表面化学和高分子化学，涉及的应用领域包括钻完井、压裂酸化、堵水调剖、化学驱油、油气集输、腐蚀与预防等油田生产环节。本书在吸取目前同类教材优点的基础上，更加注意基本概念和基本原理的介绍及知识的循序渐进，文字浅显易懂，图文并茂，有利于读者对油田化学基础知识的理解和掌握。同时，本书在编写过程中，尽量展示了该学科的新思维和新技术，旨在使读者有所裨益。

本书第1章、第2章、第5章由秦文龙编写，第3章由张现斌编写，第4章由靳文博、秦文龙编写，第6章由肖曾利编写，第7章由何延龙编写，第8章、第9章由秦国伟编写，第10章由王洋编写，第11章、第12章、第13章由肖荣鸽编写，第14章由李冉编写。本书在编写过程中，得到了诸多同行专家和西安石油大学石油工程学院、国际教育学院有关领导的大力指导和支持，在此一并表示感谢。

本书采用中英双语方式编写，可作为石油院校石油工程、储运和应用化学等专业的教学用书，也可供从事油田化学的专业研究人员和工程技术人员参考。

油田化学涉及面十分广泛，由于编者水平有限，书中不妥之处在所难免，敬请各位同行和读者批评指正。

编著者

Contents/ 目录

第 1 章　Colloid and surface chemistry/ 胶体与表面化学

1.1	Colloid chemistry 1	1.1　胶体化学 1
1.2	Interfacial phenomena 19	1.2　界面现象 19
1.3	Surfactant 28	1.3　表面活性剂 28
1.4	Emulsion 39	1.4　乳状液 39

第 2 章　Polymer chemistry/ 高分子化学

2.1	Polymeric compounds 47	2.1　高分子化合物 47
2.2	Solution of polymer 54	2.2　聚合物溶液 54

第 3 章　Chemistry and chemicals of drilling fluids/ 钻井液化学

3.1	Basic knowledge of clay minerals 64	3.1　黏土矿物基础 64
3.2	Functions and compositions of drilling fluids 72	3.2　钻井液功能与组成 72
3.3	Drilling fluids systems 76	3.3　常见钻井液体系 76
3.4	Special additives for drilling fluids 83	3.4　钻井液处理剂 83

第 4 章　Cement slurry chemistry/ 水泥浆化学

4.1	Basic composition of portland cement 105	4.1　油井水泥及组成 105
4.2	Hydration of the cement 108	4.2　水泥的水化 108
4.3	Important properties of cement slurry 114	4.3　水泥浆的重要性能 114
4.4	Cement additive 118	4.4　油井水泥外加剂 118
4.5	Cement slurry system and its application ... 132	4.5　水泥浆体系及应用 132

第 5 章 Fracturing Fluid Chemistry / 压裂液化学

5.1 Definition and performance requirements of fracturing fluid 136
5.2 Classification of fracturing fluid system 139
5.3 Addition agents of fracturing liquid 147

5.1 压裂液的定义及性能要求 136
5.2 压裂液体系分类 139
5.3 压裂液添加剂 147

第 6 章 Acidification and acid additives / 酸化及酸液添加剂

6.1 Acid treatment process 154
6.2 Acid type and additives 157
6.3 Acidification technology with retarded acid 170

6.1 酸处理工艺 154
6.2 酸液类型及添加剂 157
6.3 缓速酸酸化技术 170

第 7 章 Chemical water plugging and profile control technology / 化学堵水与调剖技术

7.1 Water plugging of oil production well 177
7.2 Profile control of water injection well 192

7.1 油井堵水 177
7.2 注水井调剖 192

第 8 章 Oil well chemistry paraffin (wax) inhibition and removal technology / 油井化学清防蜡技术

8.1 Chemical structure features of paraffin (wax) 207
8.2 Paraffin (wax) deposition in oil wells and factors affecting the deposition 212
8.3 Oil well chemistry paraffin (wax) inhibition and removal technology 214

8.1 蜡的化学结构特征组成 207
8.2 油井结蜡现象和影响结蜡的因素 212
8.3 油井化学清防蜡技术 214

第 9 章 Chemical sand control / 化学防砂

9.1 Causes and hazards of sand production in oil and gas wells 228
9.2 Classification of chemical sand control 231

9.1 油气井出砂的原因及危害 228
9.2 化学防砂分类 231

第 10 章　Chemical flooding/ 化学驱

10.1	Polymer flooding 239	10.1	聚合物驱 239
10.2	Alkaline flooding 247	10.2	碱驱 247
10.3	Surfactant flooding 254	10.3	表面活性剂驱 254
10.4	Combination flooding 265	10.4	复合驱 265

第 11 章　Crude oil demulsification and defoaming of foamed crude oil/ 原油破乳与起泡原油的消泡

11.1	Crude oil emulsion 270	11.1	原油乳状液 270
11.2	Crude oil demulsifier 271	11.2	原油破乳剂 271
11.3	Defoaming of foamed crude oil 279	11.3	起泡原油的消泡 279

第 12 章　Pour point depression and drag reduction transportation of crude oil / 原油降凝与减阻输送

12.1	Pour point depression transportation of crude oil 284	12.1	原油的降凝输送 284
12.2	Drag reduction transportation of crude oil 289	12.2	原油的减阻输送 289

第 13 章　Metal corrosion and chemical protection/ 金属的腐蚀与化学防护

13.1	Metal corrosion 294	13.1	金属腐蚀 294
13.2	Prevention method of metal corrosion 295	13.2	金属防腐蚀方法 295

第 14 章　Oilfield wastewater treatment/ 油田含油污水处理

14.1	Overview of oilfield wastewater 305	14.1	油田含油污水概述 305
14.2	Technology and treatment agent of oilfield wastewater treatment 309	14.2	油田含油污水处理技术及处理剂 309

References/ 参考文献 323

Colloid and surface chemistry
胶体与表面化学 第1章

1.1 Colloid chemistry

The science studying colloids and surface phenomena associated with colloids is called colloid chemistry. Systems and processes of oilfield chemistry researches are very complex, and most systems belong to or involve with colloid dispersion systems. For example, crude oil is actually an emulsion of oil and water, which can be classified as a colloid, while a reservoir is a large and complex highly dispersed system. During the process of oilfield development, there are many oilfield chemistry working fluids used, such as drilling fluids, completion fluids, profile control and water plugging fluids, acidizing fluids, fracturing fluids, displacement fluids that used to improve oil recovery and other fluids. All these fluids are related to colloid system.

1.1.1 Basic concepts of colloid

1.1.1.1 Basic concepts

Research object of colloid chemistry is mainly the multiphase dispersion system. In order to illustrate the classification and characteristics of the multiphase dispersion system, several basic concepts shall be introduced first:

1. Phase and phase interface

Phase refers to the uniform part of a substance that with the same physical and chemical properties. For

1.1 胶体化学

研究胶体及其相关的表面现象的科学被称为胶体化学。油田化学研究的体系和过程十分复杂，绝大多数体系属于或涉及胶体分散体系。如原油其实是油和水的乳状液，可归为胶体，而油藏则是巨大而复杂的高度分散体系。油田开发过程中涉及的钻井液、完井液、调剖堵水液、酸化液、压裂液及提高采收率的驱替液等油田化学的工作液无不与胶体体系有关。

1.1.1 胶体的基本概念

1.1.1.1 基本概念

胶体化学的研究对象主要是多相分散体系。为了说明多相分散体系的分类及特性，先介绍几个基本概念。

1.相和相界面

相，是指那些物质的物理性质和化学性质都完全相同的均匀部分。例如，0℃时冰水体系中的水和冰，是一种物质的两种存在形式，所以称冰和水为两个不

example, water and ice in an ice-water system are two forms of one substance at 0℃, so ice and water could be identified as two different phases. Moreover, phase marks difference in form and property of a substance. For instance, in an oil-water system, although both water and oil are liquid, property of the two is quite different. Therefore, oil and water shall also be two different phases. If there are two or more phases in one system, the system is called a multiphase system.

Phase systems are generally classified into homogeneous systems (such as water-ethanol mixture) and heterogeneous systems (such as oil-water mixture). There is no interface between dispersion phase and dispersion medium in homogeneous system, but there is interface in heterogeneous system.

The contact surface between two phases is called interface. If one of the two phases in contact is gas, the interface could also be called a surface. Contact surface between a liquid phase and a solid phase is called an "interface". At present, the two terms of "surface" and "interface" are interchangeable, so all physical and chemical phenomena occurring on interfaces are called surface (interface) phenomena.

2. Dispersion phase and dispersion medium

Dispersion system is a system in which one or several substances are dispersed in another substance. In a multiphase dispersion system, the dispersed substance is called dispersion phase, and the substance that disperses it is called dispersion medium (Figure 1-1). The former is discontinuous phase and the latter is

同的相。它标志着物质的存在形式和性质的差别。又如，油水体系中的水和油虽然都是以液体形式存在，但两者的性质不同，所以，油与水也是两个不同的相。体系中有两个或两个以上的相，称为多相体系。

相体系一般分为均相体系（如水-乙醇）和非均相体系（如油-水），均相体系的分散相和分散介质之间没有界面，而非均相体系存在界面。

相与相之间的接触面称为相界面。如果相互接触的两相中，一相为气体，这样的相界面又称为表面。液相与固相之间的分界面称为"界面"。目前，"表面"与"界面"两个名词通用。凡发生在相界面上的物理化学现象都称为表(界)面现象。

2.分散相和分散介质

分散体系是一种或几种物质分散在另一种物质中所形成的体系。在多相分散体系中，被分散的物质称为分散相，另一种分散它的物质称为分散介质（图1-1）。前者为不连续相，后者为连续相。例如，钻井液中，黏土颗粒分散在水中，黏土为分散相，水为分散介质。水分散在原

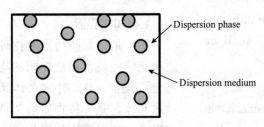

Figure 1-1 Schematic of dispersion phase and dispersion medium

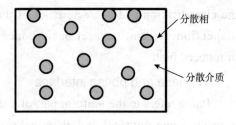

图1-1 分散相和分散介质示意图

continuous phase. For example, in a drilling fluid, clay particles are dispersed in water, clay is dispersion phase, and water is dispersion medium. In case that water dispersed in crude oil, water is dispersion phase, and crude oil is dispersion medium.

3. Dispersity and specific surface area

Dispersion degree of a dispersion system can be represented by particle size d of the dispersion phase, or $1/d$, i.e. dispersity D. Besides, the specific surface area could also be used. Specific surface area is the surface area per unit volume (or per unit mass) of substance.

It should be particularly noted that the concepts of dissolution and particle dispersion are completely different. The dispersion system formed by dissolution is called a true solution (or molecular dispersion system), and the dispersion system formed by particle dispersion is a multiphase dispersion system. In thermodynamics, multiphase system is considered as unstable system and has a spontaneous coalescence tendency. It is also a multiphase heterogeneous system with a physical interface between dispersion phase and dispersion medium.

1.1.1.2 Dispersion system and colloid

According to dispersion degrees, the dispersion system can be divided into fine dispersion system and coarse dispersion system. Colloid actually belongs to fine dispersion system, which consist of finely divided particles or macromolecules (such as glue, gelatin, proteins, etc.). A dispersion system with particle size ranging from 1 to 100 nm was designated to be a colloid. In coarse dispersion systems, particle size is larger than 100 nm. Suspensions, foams and the like belong to coarse dispersion systems. Usually, if in a system that dispersion phase and dispersion medium are mixed in form of molecular or ionic, there is no interface formed because the system is a uniform single phase. The system in which molecular radius is less than

油中，水是分散相，原油是分散介质。

3. 分散度和比表面积

分散体系的分散程度的量度，既可用分散相的粒子大小d表示，又可以用$1/d$，即分散度D表示，还有一种方法就是用比表面积表示。所谓比表面积，就是单位体积(或单位质量)物质所具有的表面积。

应该特别指出的是，溶解与颗粒分散是完全不同的两个概念。通过溶解而形成的分散体系称为真溶液(或分子分散体系)，而通过颗粒分散而形成的分散体系是多相分散体系。它在热力学上是不稳定系统，有自动聚结的倾向；它还是多相不均匀系统，分散相与分散介质之间存在着物理界面。

1.1.1.2 分散体系和胶体

分散体系的含义比胶体更广泛。按分散体系分散度的不同，可将分散体系分为细分散体系与粗分散体系。胶体实际上是细分散体系，是指分散相颗粒大小在$1\sim100$nm的分散体系。分散相颗粒大于100nm的为粗分散体系，悬浮体、泡沫等则属于粗分散体系。分散相与分散介质以分子或离子形式彼此混溶，没有界面，是均匀的单相，分子半径大小在1nm以下，通常把这种体系称为真溶液(也称分子分散体系)，如$CuSO_4$溶液。

必须指出：上面所规定的胶体粒子大小的界限，完全是人为

1nm can be considered as a true solution (or a molecular dispersion system), such as $CuSO_4$ solution.

It must be pointed out that the colloidal particle size range specified above is not a constant standard but a relative concept. It is quite common that particle sizes of many systems (such as some foams, emulsions, and suspensions, etc.) are not in this range but showing many common properties as colloids. In this case although belonging to the coarse dispersion system, they are also studied as colloid systems. In some books, the size range of colloidal particle is defined as dimension range from 1 to 1000 nm.

Many properties of dispersion system are related to the size or dispersity of dispersion phase. As listed in the Table 1-1, property differences of dispersion systems depend essentially on the size of dispersion phase particles and their hydrophilic properties.

的大致划分。往往在此界限以外的物系，例如，有一些属于粗分散体系的泡沫、乳状液、悬浮液也具有许多与胶体共同的性质，也作为胶体系统来研究。也有些书中把胶体颗粒尺寸范围划定为1～1000nm。

分散体系的许多性质均与分散相的大小或分散度有关。表1-1列出了分散体系的一些性质比较。表中所列性质的差异基本上取决于分散相粒子的大小及其亲水性能。

Table 1-1 Properties of some dispersion systems

Type	Size of dispersion phase /nm	State of dispersion phase	Example	Comparison of properties				
				Homogeneous or heterogeneous	Thermodynamic stability	Diffusion speed	Through semi-permeable membrane or not	Type of solution formed
Molecular dispersion system	<1	Atom, molecule and ion	Solution	Homogeneous	Stable	Fast	Through	True solution
Colloid dispersion system	1-100	Colloidal particle	Sol	Heterogeneous	Not stable	Slow	Not through	Gas, liquid or solid sol
Coarse dispersion system	>100	Coarse particle	Emulsion Suspension	Heterogeneous	Not stable	Slow	Not through	Emulsion Suspension

表1-1 分散体系的一些性质比较

类 型	分散相尺寸/nm	分散相状态	举例	性质比较				
				均相或多相	热力学稳定性	扩散速度	透过半透膜情况	形成何种溶液
分子分散体系	<1	原子、分子、离子	溶液	均相	稳定	快	透过	真溶液
胶体分散体系	1～100	胶粒	溶胶	多相	不稳定	慢	不透过	气、液或固溶胶
粗分散体系	>100	粗粒	乳状液 悬浮液	多相	不稳定	慢	不透过	乳状液 悬浮液

1.1.1.3 Classification of colloid system

Since one of the important features of the colloid system is based on size of dispersion phase particles, it is very clear that as long as the particle size of dispersion phase in different aggregation states is in the range from 1 to 100 nm, a colloid system can be formed in dispersion medium of different states, except that dispersion phase and dispersion medium are both gases. Therefore, the colloids can be classified into 8 types according to aggregation state of dispersion phase and dispersion medium (Table 1-2). Conventionally, if dispersion medium in a colloid system is liquid, the system will be called as lyosol or sol. If the medium is water, the system will be called hydrosol. When the medium is solid, the system will be called solid sol. If the medium is gas, it will be called aerosol.

1.1.1.3 胶体体系分类

由于胶体体系的重要特征之一是以分散相粒子的大小为依据的，显然，除了分散相与分散介质都是气体而不能形成胶体体系外，只要不同聚集状态分散相的颗粒大小在1~100nm，则在不同状态的分散介质中均可形成胶体体系。因此，按分散相和分散介质的聚集状态可以将胶体分成8类（表1-2）。习惯上，把分散介质为液体的胶体体系称为液溶胶或溶胶，如介质为水的称为水溶胶。当介质为固体时，称为固溶胶；当介质为气体时，称为气溶胶。

Table 1-2　Classifications according to aggregation state of dispersion phase and dispersion medium

Dispersion phase	Dispersion medium	Alternative names	Examples
Gas Liquid Solid	Gas	Aerosol	— Fog Dust
Gas Liquid Solid	Liquid	Lyosol / Sol	Foam Emulsions(such as milk , emulsified crude oil) Sol, suspension and gel (such as drilling fluid)
Gas Liquid Solid	Solid	Solid sol	Foamed plastics, bread Pearl, hydrogel Alloy, ruby, tinted glass

表1-2　按分散相和分散介质的聚集状态分类

分散相	分散介质	名　称	实　例
气 液 固	气	气溶胶	— 云雾 烟尘
气 液 固	液	液溶胶	泡沫 乳状液，如牛奶、乳化原油 溶胶、悬浮液与凝胶，如钻井液
气 液 固	固	固溶胶	泡沫塑料、面包 珍珠、水凝胶 合金、红宝石、有色玻璃

1.1.2 Preparation and properties of colloid

In colloid chemistry, lyosol (sol) is the most common colloid in which dispersion medium is liquid. In this section, we will focus on issues related to sol.

1.1.2.1 Preparation and purification of sol

To prepare a sol, the necessary condition is that particle size of dispersion phase should be in the range from 1 to 100 nm, and a proper stabilizer (such as electrolyte or surfactant) shall be also added into the system to make it stable enough. In principle, there are two methods for preparing a sol: one is dispersion method, which is mainly making larger solid particles smaller; The other is coacervation method, in which molecules or ions will be polymerized into colloidal particles, as shown in Figure 1-2.

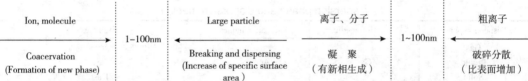

Figure 1-2 Formation methods of colloid

Sols obtained by coacervation method are all with feature of polydisperse, i.e. the systems contain various types of particles with different sizes, and some of which may exceed the colloidal particle range. However, chemical method for sol preparation usually results in a lot of electrolytes. Although proper amount of electrolyte could work as stabilizer for the sol, too much electrolytes will reduce its stability. Therefore, in order to obtain relatively pure and stable sol, the prepared sol must be purified. Coarse particles in sol could be removed by filtration (colloidal particles are small enough to pass through pores of ordinary filter paper), sedimentation or centrifugation. The excessive electrolytes must be removed by dialysis (also known as dialyse).

1.1.2.2 Kinetic properties of sol

This section mainly introduces diffusion and sedimentation of colloidal particles, which are attributed to kinetic properties of sol.

(1) Diffusion.

Like solute molecules in solution, small particles in sols are always in endless disorderly movement (Brownian motion). According to molecular motion theory, essence of Brownian motion is thermal motion of the particles. Therefore, compare with dilute solution, a sol has no other essential difference except that its particles are much larger than molecules or ions in true solution and its concentration is much lower than usual dilute solution. Therefore, some properties of dilute solution, such as diffusion, osmotic pressure or other properties can also be found on sols, only in different degrees. When there is a concentration difference, particles of the sol will definitely diffuse from high concentration side to low (Figure 1-3). Difference between diffusion of a true solution and a sol is that the former is the result of molecules (or ions) thermal motion, while the latter is caused by Brownian motion of colloidal particles.

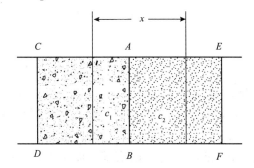

Figure 1-3 Schematic representation of diffusion

(2) Sedimentation and sedimentation equilibrium.

Particles dispersed in gas or liquid medium are subjected to two opposite forces:

① Gravity. If density of particles is larger than that of medium, the particles will sink due to gravity. This phenomenon is called sedimentation.

② Diffusion force (caused by Brownian motion). On the contrary of sedimentation, diffusion force can promote upward movement of particles with higher concentration in the lower part of system, and then concentration of the system tends to be uniform.

When the two forces are equal, the equilibrium state is reached, which is called "sedimentation equilibrium". At equilibrium, particle concentration in each horizontal plane remains the same, but a concentration gradient will be formed from bottom to the top (Figure 1-4). The situation is similar to that of atmosphere on the ground. The farther away from the ground, the atmosphere with lower atmospheric pressure becomes thinner.

②扩散力（由布朗运动引起）。与沉降作用相反，扩散力能促进体系中下部较浓的粒子向上运动，使体系浓度趋于均匀。

当这两种作用力相等时，就达到平衡状态，谓之"沉降平衡"。平衡时，各水平面内粒子浓度保持不变，但从下向上会形成浓度梯度（图1-4），这种情况正如地面上大气分布的情况一样，离地面越远，大气越稀薄，大气压越低。

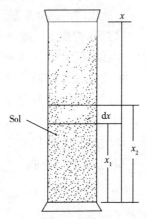

Figure 1-4 Schematic representation of sedimentation equilibrium

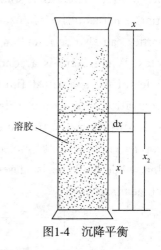

图1-4 沉降平衡

According to Stokes's Law, size of particles will significantly influence sedimentation speed. Generally, clay particles equal to or larger than 0.1μm will sink to bottom of container after being placed for a certain period of time. Whereas particles smaller than 0.1μm will be difficult to sink. Many external conditions, such as convection or mechanical vibration, will influence sedimentation of particles. Especially when the particles are smaller than 0.1μm, diffusion against sedimentation should also be considered. This is why it will take days or even years for many sols to reach sedimentation

根据Stocks定律可知，粒子的大小对沉降速度影响很大。一般等于或大于0.1μm的黏土粒子，放置一段时间以后，将会下沉到容器的底部，而小于0.1μm的粒子就难下沉。由于外界条件的影响，如对流、机械振动等，都可能阻止粒子沉降。特别是粒子小于0.1μm时，还应考虑与沉降作用相对抗的扩散作用。也正因为如此，许多溶胶往往需要几天甚至几年才能达到沉降平衡。这个

equilibrium. This fact not only explains why sols can remain stable for a long period of time without sedimentation, but also explains fundamentally why sols are unbalanced systems.

1.1.2.3 Optical properties of sols

Many colloids show bright colors while others not; some colloids are opalescent, while some others are quite clear; when viewing from different angles, some colloids will show beautiful color reflections; the colors of some colloids will change continuously with standing time. In summary, the reason why colloids could show such colorful optical properties is related to scattering and absorption of light by colloids. The intensity of scattering and reflection of light is related to size of particles. When particle size is within range of colloid, light scattering is more obvious; when particle diameter is larger than wavelength of light, reflection takes domination, which lead to a turbid system. Therefore, when a light beam passes through a colloidal solution and is viewed from the side that is perpendicular to beam travel direction, a turbid but bright light column can be seen. This opalescence phenomenon is called Tyndall effect. This effect is an important feature of most colloid systems and the easiest way to discriminate sol and molecular solution.

1.1.2.4 Electrical properties of sol

(1) Electrokinetic phenomena.

Electrokinetic phenomena of colloids include electrophoresis, electroosmosis, as well as generation of streaming potential and sedimentation potential. Electrokinetic phenomena can be summarized into the following categories Figure1-5:

(2) Causes for sol particles being charged.

The existence of electrokinetic phenomena indicates that surface of colloidal particles is always charged, some with positive charges and some with negative charges. In fact, not only sols, all interfaces

事实不仅说明了溶胶在相当长的时间内能保持稳定而不沉降的原因，而且也从根本上说明了为什么溶胶是不平衡体系。

1.1.2.3 溶胶的光学性质

许多胶体具有鲜艳的颜色，而有些胶体则不带色；有的胶体呈现乳光，有些胶体外观却很清亮；有的胶体从不同的角度观看，显示出不同的绚丽色带；有的胶体颜色还会随着放置时间而不断地变化。胶体之所以会呈现出如此丰富多彩的光学性质，简单说来，和胶体对光的散射与吸收有关。由于散射与反射的强弱与质点大小有关。当质点大小属于胶体范畴时，散射明显；当质点直径大于光的波长时，主要发生反射，使体系呈现浑浊。因此，当一束光线通过胶体溶液，在与光束前进方向相垂直的侧向上观察，可看到一个浑浊发亮的光柱。这种乳光现象称为丁达尔效应（Tyndall effect）。此效应是大多数胶体体系的一个重要特征，也是判别溶胶与分子溶液的最简便的方法。

1.1.2.4 溶胶的电学性质

（1）电动现象。

胶体的电动现象包括电泳、电渗、流动电位与沉降电位的产生。电动现象可以归纳如下几类（图1-5）：

（2）溶胶粒子带电的原因。

电动现象的存在，说明了胶粒表面总带有电荷，有的带正电荷，有的带负电荷。其实，不仅

that are in contact with polar medium (such as water) are always charged. The presence of interface charges will affect distribution of ions in solution. Hence ions with oppositely charges will be attracted to the interface, and ions with same charges are rejected from the interface. Due to their thermal motion, the ions will establish a diffuse electric double layer with certain distribution pattern at the interface. This pattern will determine electrical properties and other physicochemical properties of the sol.

是溶胶，凡是与极性介质（如水）相接触的界面总是带电的。界面电荷的存在影响到溶液中离子在介质中的分布，带相反电荷的离子被吸引到界面附近，带相同电荷的离子则从界面上被排斥。由于离子的热运动，离子在界面上建立起具有一定分布规律的扩散双电层。这种分布状态决定了溶胶的电性以及其他的物理化学性质。

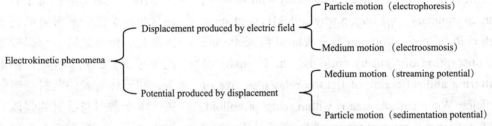

Figure 1-5　Electrokinetic phenomena

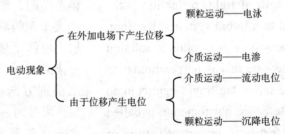

图1-5　电动现象

Principal reasons for sol particles being charged are:

① Ionization.

Clay particles and glass are all silicates, which can be ionized in water, so their surfaces are with negative charges, and liquid phase in contact with them are with positive charges. Silica sol with negative charges in weak acidic and alkaline medium is also attributed to silica ionization on surface of particles. Type and quantity of charges generated by ionization are closely related to pH value of the medium.

② Lattice displacement.

In aluminum-oxide octahedron and silicon-oxide tetrahedron of clay minerals, when Al^{3+} (or Si^{4+}) is

溶胶粒子带电的主要原因有：
①电离作用。

黏土颗粒、玻璃等皆属硅酸盐，在水中能电离，故其表面带负电，而与其接触的液相带正电。硅溶胶在弱酸性和碱性介质中带负电荷，也是因为质点表面上硅酸电离的结果。电离产生的电荷种类、数量与介质pH值有密切关系。

②晶格取代作用。

黏土矿物的铝氧八面体和硅氧四面体中，当 Al^{3+}（或 Si^{4+}）被部分低价的 Mg^{2+} 和 Ca^{2+} 所取代，

partially replaced by Mg^{2+} and Ca^{2+}, the clay crystals will be with negative charges. In order to maintain electrical neutrality, clay surface adsorbs some cations, forming a diffuse electric double layer in water. Lattice displacement is a special case of clay particles with charges, which is rare in other sols.

③ Ion adsorption.

Sol particles have a large specific surface area and strong adsorption capacity. It adsorbs ions from medium and follows the selective adsorption law, so that sol particles are always charged, as shown in Figure 1-6.

④ Unsaturated linkage.

Large crystals that composed of numerous small crystal cells are often with broken bonds, forming unsaturated bonds with a certain amount of charges. For example, unsaturated bonds are often found at fractures on edge of clay minerals, which result in clay with charges.

1.1.3 Electrical double layer (EDL) and micelle structure

1.1.3.1 Diffuse electrical double layer theory

(1) Formation and structure of the EDL.

According to electrical neutrality principle and theory of diffuse electrical double layer proposed by Stern in 1924, if colloidal particles are charged, there must be counter-ions around them. And number of counter-ions is the same with number of charges. An EDL is then formed at the solid-liquid interface. On the one hand, counter-ions in EDL are attracted by the solid surface charges and getting close to the solid surface; on the other hand, due to thermal motion of counter-ions, they could diffuse into the liquid phase. As a result of these two opposite effects, the counter-ions are diffused around colloidal particles and forming a diffuse electrical double layer.

As shown in Figure 1-6, diffuse electrical double

会使黏土晶体带负电，为了维持电中性，黏土表面就吸附了一些阳离子，在水中形成扩散双电层。晶格取代是黏土颗粒带电的一种特殊情况，在其他溶胶中很少见。

③离子吸附作用。

溶胶粒子具有巨大的比表面积，具有很强的吸附能力。它从介质中吸附离子，并遵循选择性吸附规律，使溶胶粒子带有一定性质的电荷，如图1-6所示。

④未饱和键。

由无数小晶胞组成的大晶体，常常有破开的断键，形成了带有一定电荷的未饱和价键。如黏土矿物边缘上断裂处常有未饱和的价键，使黏土带电。

1.1.3 扩散双电层与胶团结构

1.1.3.1 扩散双电层理论

（1）扩散双电层的形成与结构。

由于电中性原因，根据1924年Stern提出的扩散双电层模型理论，胶体粒子带电，那么在它周围必然分布着电荷数相等的反离子，于是，在固液界面形成双电层。双电层中的反离子，一方面受到固体表面电荷的吸引，靠近固体表面；另一方面，由于反离子的热运动，又有扩散到液相内部去的能力。这两种相反作用的结果，使得反离子扩散分布在胶粒周围，形成扩散双电层。

如图1-6所示，扩散双电层

layer is the part that positive charges from excess on solid surface to none, consisting of an adsorption layer (also called a compact layer) and a diffusion layer. The adsorption layer refers to the part that counter-ions and water molecules are closely adsorbed on the solid surface; and the diffusion layer refers to the part that adsorption of counter-ions and water molecule become weak. Interface between adsorption layer and diffusion layer is called sliding surface. When solid moves, adsorption layer moves with it.

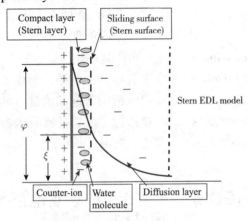

Figure 1-6　Schematic of EDL

Potential fall from surface of solid to homogeneous liquid phase is called surface potential φ_0. Potential fall from sliding surface to homogeneous liquid phase is called ξ potential (also called zeta potential).

According to physical image of the EDL, an electromotive force equation can be derived according to distribution of charges:

$$\xi = \frac{\mu u}{\varepsilon E} \qquad (1-1)$$

where ξ is the zeta potential of the particle;
μ is the viscosity of the liquid;
u is the electrophoresis speed of the particle;
ε is the dielectric constant of the liquid;
E is electric field strength.

Value of zeta potential indicates degree of colloidal particles charged and stability of the colloid system. The higher the value of zeta potential demonstrates the

是从固体表面到过剩正电荷为零的部分，由吸附层（也称紧密层）和扩散层两部分组成。其中，吸附层是固体表面紧密吸附的反离子和水分子所组成的部分；而扩散层则是反离子和水分子吸附的力较弱的部分。吸附层和扩散层的分界面称为滑动面，固体运动时带着吸附层一起运动。

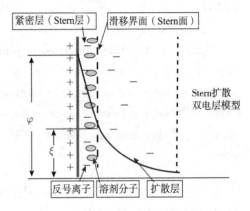

图1-6　扩散双电层示意图

从固体表面到均匀液相的电势降称为表面电势 φ_0，从滑动面到均匀液相的电势降称为 ξ 电位（也称 Zeta 电位）。

从扩散双电层的物理图像出发，根据电荷的分布规律，可以导出电动方程：

$$\xi = \frac{\mu u}{\varepsilon E} \qquad （1-1）$$

式中　ξ——从滑动面到均匀液相的电势降；
　　　μ——介质黏度；
　　　u——电泳速度；
　　　ε——介电常数；
　　　E——电场强度。

ξ 电位的大小表示了胶粒带电的多少和胶体体系稳定性的高

more the colloidal particles charged and the higher the stability of colloid system; conversely, the lower the zeta potential is, the less colloidal particles are charged. When zeta potential equals to zero, it indicates that colloidal particles are not charged, and stability of the colloidal system is low.

(2) Influence of electrolytes on EDL.

It is found that addition of electrolytes could cause decrease in zeta potential of the colloid, which is the compression effect of electrolyte on EDL. Degree of compression is related to counter-ions concentration and valence: The higher the valence of counter-ions is, the stronger the compression will be; the larger the counter-ion concentration is, the greater the compression enhancement will be shown. When electrolyte concentration increases to a certain value, zeta potential is zero, which is called isoelectric point.

1.1.3.2 Micelle structure

Since size of colloidal particles is normally range from 1 to 100 nm, each colloidal particle must be composed of many molecules or atoms. The core part of the colloidal particle is called the nucleus, which is an insoluble aggregate formed by a large number of molecules or atoms. Nucleus of colloidal particle is surrounded by EDL consisting of an adsorption layer and a diffusion layer. The overall structure is called a micelle, in which nucleus and adsorption layer forms a "colloidal particle".

Fajans' law shows that the particles with the same chemical elements with colloidal particles will be preferentially adsorbed by the colloidal particles. According to this rule, property of charges on colloidal particles is determined by ions associated with the nucleus. For example, during preparation of AgI sol by dilute $AgNO_3$ solution and KI solution, if KI is excessive, according to Fajans' law, the nucleus is tending to adsorb I^- from the solution, leading to the colloidal particles negatively charged (Figure 1-7). When micelles are dispersed in a liquid medium, the liquid is commonly known as a sol.

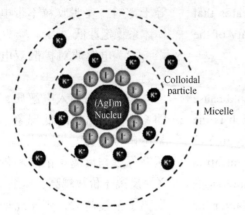

Figure 1-7 Sol structure of AgI with negative charges

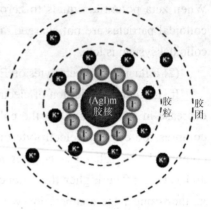

图1-7 AgI负电溶胶结构示意图

1.1.4 Stability and coagulation of sol

Colloids often appear in industrial, agricultural production and scientific researches. When it is needed, it shall be prepared and stabilized, such as preparation of drilling mud, etc; and sometimes it is not desirable, such as demulsification of crude oil and so on. Therefore, only by understanding stability factors of the sol, conditions could be properly selected for stabilizing or destroying the sols.

1.1.4.1 Stability of sols

If the sol dispersion system can maintain its dispersive state for a long time, and the particles are in uniform suspension without breaking, the sol system is stable. There are two different concepts in colloidal stability: dynamic (sedimentation) stability and coalescence stability.

Sedimentation stability refers to the property that whether dispersion phase particles are easy to sink or not under the gravity. Generally, sink speed of the dispersion phase particles is used to measure stability of dynamic stability.

Coalescence stability refers to a particular property

1.1.4 溶胶的稳定和聚沉

在工农业生产和科学研究中常常遇到胶体问题，有时需要胶体，就得制备它并使它稳定，例如钻井泥浆的配制等。但有时也不希望胶体生成或需要破坏胶体，例如原油破乳等。因此，只有了解溶胶的稳定因素，才能适当选择条件，使其稳定或破坏。

1.1.4.1 溶胶的稳定性

溶胶分散体系若能长久保持其分散性状态，各微粒处于均匀悬浮状态而不破坏，就称为具有稳定性。所谓胶体的稳定性有两种不同的概念：动力（沉降）稳定性和聚结稳定性。

沉降稳定性是指在重力作用下分散相粒子是否容易下沉的性质。一般用分散相下沉速度的快慢来衡量动力稳定性的好坏。

聚结稳定性是指分散相粒子是否容易自动地聚结变大的性质。

of dispersion phase particles that whether they will easily coagulated and become larger or not. Regardless of their sedimentation velocity, as long as dispersion phase particles do not reduce dispersity spontaneously and result in larger coalescence degree, the colloid system is with good coalescence stability.

It must be pointed out that dynamic stability and coalescence stability are two different concepts, but there is a connection between them. If dispersion phase particles become larger spontaneously due to coalescence, their weight will increase and then sink inevitably. Therefore, if coalescence stability is not good enough, dynamic stability will be destroyed eventually. Hence coalescence stability is the fundamental stability property of a colloid system.

In order to explain stability of sols, the DLVO (*Derjaguin-Landau-Verwey-Overbeek*) theory was proposed in the 1940s. According to the theory, under certain conditions, whether a sol is stable or coagulates is depending on the attraction force and repulsion force between the particles. If repulsive force is greater than attraction, the sol is stable, otherwise it is unstable.

1.1.4.2 Coagulation of sol

Coagulation of sol refers to the process in which the colloidal particles are coalesced and sunk. During this process, the sol will be destroyed by separation of dispersion phase and dispersion medium.

All factors that weaken or eliminate sol stability may cause coagulation of the sol, in which the most effective factors are charge and water removing. If charges on the surface of colloid particles are neutralized, the zeta potential will become lower, and the hydration film will become thinner. During this process, the possibility of coagulation after particles collision caused by Brownian motion becomes larger, transforming small particles into big ones. The particles will then sink under gravity and the colloid system is

分散相粒子，不管其沉降速度如何，只要它们不自动降低分散度，聚结度变大，该胶体就是聚结稳定性好的体系。

必须指出，动力稳定性与聚结稳定性是两种不同的概念，但是它们之间又有联系。如果分散相粒子自动聚结变大，那么，由于重量加大，必然引起下沉。因此，失去聚结稳定性，最终必然失去动力稳定性。由此可见，在上述两种稳定性中，聚结稳定性是最根本的。

为了解释溶胶的稳定性，20世纪40年代形成了DLVO理论。该理论认为，溶胶在一定条件下是稳定存在还是聚沉，取决于粒子间的相互吸引力和静电斥力。若斥力大于吸力则溶胶稳定，反之则不稳定。

1.1.4.2 溶胶的聚沉

溶胶的聚沉指的是溶胶颗粒聚结下沉，使分散相和分散介质分离而使溶胶破坏的过程。

一切削弱或消除溶胶稳定性的因素都能使溶胶聚沉。但重要的是去电、去水作用。如果能中和掉胶粒表面所带的电荷，使ξ电位降低，使水化膜变薄，布朗运动引起胶粒碰撞后聚并的可能性增大，从而使小颗粒变为大颗粒，以致在重力作用下使之沉降，溶胶体系破坏。

生产上常用的聚沉方法很多，如长时间放置使之老化、加热、

destroyed eventually.

In production, there are many commonly used methods for coagulation, such as aging, heating, concentrating, adding electrolyte (or polymer compound), or mixing two kinds of sols with opposite charges.

1.1.5　Gel

Gel refers to a system in which under certain conditions, dispersion phase particles (colloid particles or macromolecules in a sol or solution) are connected to each other, forming a grid structure filled by the dispersion medium. If water is dispersion medium in gel system, it is usually called a hydrogel. All hydrogels are similar in appearance (semi-solid, no fluidity), such as tofu and jelly. Gel systems are widely applied in oilfield chemistry, such as drilling fluids, fracturing fluids, profile controlling/water plugging agents, etc.

1.1.5.1　Characteristics of gel

Gel is with the following characteristics: Firstly, gel is very different with sol (or solution). Sol is with good fluidity, but gel systems are not only lose fluidity, but also show some mechanical properties of solids, such as certain strength, elasticity, and yield value. Secondly, a gel is not exactly the same as real solid. It consists of solid-liquid phases and belongs to colloid dispersion system. Therefore, the structural strength of the gel system is often limited and subjected to change. The changes of any condition（such as temperature, medium composition, external force applied, etc.）, will cause its structural damage, irreversible deformation, and resulting in fluidity. It can be shown that gel is a special form of dispersion system because their properties are between solid and liquid.

1.1.5.2　Classification of gel

According to whether dispersion phase is

浓缩、外加电解质（或高分子化合物）及两种异性溶胶的混合都会使之发生聚沉。

1.1.5　凝胶

凝胶是指分散相粒子（溶胶或溶液中的胶体粒子或高分子）在一定条件下相互连接，形成空间网架结构，分散介质充填于网架结构的空隙中形成的体系。通常把分散介质为水的凝胶称为水凝胶。所有水凝胶的外表很相似，呈半固体状，无流动性，如豆腐、果冻等。凝胶在油田化学中应用广泛，如钻井液、压裂液、调剖堵水剂等使用时均涉及该类体系。

1.1.5.1　凝胶的特征

凝胶具有如下特征：其一，凝胶与溶胶（或溶液）有很大的不同。溶胶具有良好的流动性，而凝胶体系不仅失去流动性，并且显示出固体的力学性质，如具有一定的强度、弹性和屈服值等。其二，凝胶和真正的固体又不完全一样，它由固液两相组成，属于胶体分散体系，故结构强度往往有限，易于遭受变化。改变条件，如改变温度、介质成分或外加作用力等，往往能使结构破坏，发生不可逆变形，结果产生流动。由此可见，凝胶是分散体系的一种特殊形式，故性质介于固体和液体之间。

1.1.5.2　凝胶的分类

通常按照分散相是大分子物

macromolecular substance or low molecular weight substance, gels can be divided into two types: elastic gel and rigid gel. The formation mechanisms of their network structure are quite different.

（1）Non-elastic gel.

When electrolytes are added into rigid particle (such as SiO_2, V_2O_5 or Fe_2O_3, etc.) sols, dispersion phase particles are compressed by diffusion layer of the electric double layer, causing zeta potential and stability of the sol to decrease gradually. Due to colliding by Brownian motion, the particles are connected to each other before their stability is completely lost, which form a three-dimensional network structure (Figure 1-8) by Van der Waals forces and chemical bonds. The structure is fully filled with medium and the system change into a non-elastic gel (or rigid gel). If concentration of electrolyte is further increasing and reaching to the critical flocculation concentration, the flocculation will appear in the sol and sink at bottom of vessel. Generally, mixing dried dispersion phase of the rigid gel with medium (such as water) and heating, it will neither swell and return to original gel state, nor return to original sol state. That's why rigid gel is also called irreversible gel.

质还是低分子物质，可将凝胶分为弹性凝胶和刚性凝胶两类，它们的网状结构形成的机理不同。

（1）刚性凝胶。

一些刚性质点（如SiO_2、V_2O_5、Fe_2O_3等）溶胶中加入电解质时，分散相粒了因双电层的扩散层被压缩而引起ξ电位下降，使溶胶稳定性逐渐降低。粒子因布朗运动而碰撞，在完全失去稳定性之前，粒子间靠范德华力及化学键相互连接而形成三维网络结构（图1-8），结构中充满介质，体系"固化"成为刚性凝胶（或非弹性凝胶）。如果电解质浓度进一步增加达到临界絮凝浓度时，溶胶聚沉，絮凝体沉降于容器底部与介质完全分离。刚性凝胶干燥后的分散相与介质（如水）混合加热，一般不会发生溶胀作用再返回原来的凝胶状态，也不会重新形成原来的溶胶状态。故这种凝胶又称为不可逆凝胶。

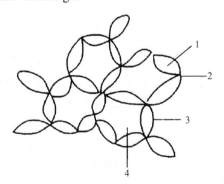

1—Dispersion phase particle；2—Unstable surface region；
3—Stable surface region；4—"Crystal cell" filled with medium

Figure 1-8 Schematic of flocculation structure

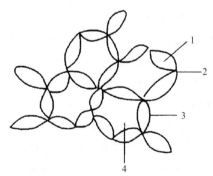

1—分散相粒子；2—失去稳定性的表面区域；
3—仍保持稳定的表面；4—充满介质的"晶胞"

图1-8 絮凝结构示意图

（2）Elastic gel.

During process of changing temperature,

（2）弹性凝胶。

柔性的线型大分子溶液在改

concentrating solution or adding small amount of electrolyte, sometimes even in process of standing, the flexible linear macromolecular solution will form a network structure (Figure 1-9) by hydrogen bonding and other interaction forces. This gel is called elastic gel. The most obvious feature of this kind of gel, such as gelatin (which is a protein), is that the dried gel could be heated and dissolved in a medium (such as water) to form a sol, and the sol could coagulate into jelly-like gel during cooling by gelation. The jelly-like gel could back to dried gel state through syneresis and drying. This operation can be repeated, so this type of gel is also called reversible gel.

变温度、浓缩溶液、加入少量电解质，甚至在静置过程中，大分子间靠氢键等作用力会形成网络结构（图1-9），所形成的凝胶称为弹性凝胶。这种凝胶的特点是，诸如明胶（是一种蛋白质）这样的干凝胶在水中加热溶解所得到的溶液，在冷却过程中发生胶凝作用成为胶冻状的凝胶，而凝胶经脱水收缩再干燥重新变为干凝胶，这一操作可以反复进行，故这类凝胶又称可逆凝胶。

Figure 1-9   Schematic of jelly structure

图1-9   柔性大分子凝胶结构示意图

1.1.5.3   Difference between gel and jelly

Traditionally, gels composed of flexible macromolecular substances are often referred to as jelly or soft gel, while the term "gel" is often used to express rigid gels narrowly. Specifically, there are two differences between gel and jelly:

(1) Difference on chemical structure.

Gel is formed by crosslinking of chemical bonds (for example, SiO_2 gel particles are connected by Si—O—Si covalent bonds), in which the macromolecules are gelatinized by crosslinking under the effect of chemical agents, oxygen or high temperature. It is impossible to revert a gel into a flowable solution without chemical bond damage, thus it is irreversible. Jelly is a network structure formed by non-chemical bond force (such as

1.1.5.3   凝胶和冻胶的区别

按照传统习惯，凡由柔性大分子物质构成的凝胶多称为冻胶或软胶，而凝胶一词常狭义地指刚性凝胶。具体来看，凝胶与冻胶有两个方面的区别：

（1）化学结构上的区别。

凝胶是化学键交联（例如SiO_2凝胶质点间形成了Si—O—Si共价键），在化学剂、氧或高温作用下，使大分子间交联而凝胶化。不可能在不发生化学键破坏的情况下重新恢复为可流动的溶液，为不可逆凝胶。而冻胶是由次价力（如氢键等）缔合而成的

hydrogen bonds). Under condition of temperature rise, mechanical agitation, oscillation or large shear force, the structure will break and jelly will become a flowable solution, hence it is reversible.

(2) Difference of liquid content in network structures.

Gel contains moderate amount liquid, while jelly contains high amount liquid. Usually, liquid content in jelly is greater than 90% (volume fraction).

1.2 Interfacial phenomena

One of the important characteristics of colloid system is its large surface area. In general, any surface could actually be called interface. Complex physical or chemical phenomena can occur on any two-phase interface, and they are collectively referred to as interface phenomena. Surface chemistry (also known as interface chemistry) is a science that studies surface phenomena. There are two main interface phenomena to be studied in oilfield chemistry. One is how highly dispersed clay will influence performance of the drilling fluid system and how to eliminate the adverse effects. The other is physical and chemical phenomena of the oil-water-rock interface during oil recovery process.

1.2.1 Surface tension and surface energy

As shown in Figure 1-10, since molecular attraction of gas to liquid is less than molecular attraction of the liquid itself, the M_2 (liquid molecule) at interface is subjected to a resultant force directed into the liquid. This resultant force is called net attractive force. Broadly, all liquid molecules in surface layer are subjected to attractive force that perpendicular from the interface and toward interior direction of the liquid. Therefore, all surfaces of liquid have spontaneous shrinkage tendency. If the volume is constant, sphere is with the smallest surface area in all geometries. This is why some small water droplets and mercury beads are often spherical (the droplet gravity does not play a major role in this condition). If a liquid molecule is needed to move from interior to liquid surface for increasing surface area, the molecular attractive force must be overcame. The work required to do so will be converted into energy of the surface layer i.e. surface energy (surface free energy or the interfacial free energy). The consumed work W is proportional to the increased surface area A:

$$W = \sigma A \quad (1-2)$$

where the proportional coefficient σ is called surface tension or interfacial tension (IFT), which is equal to the work done by increasing per unit surface area, then surface tension σ is:

$$\sigma = \frac{W}{A} \quad (1-3)$$

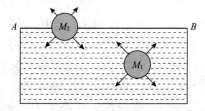

Figure 1-10　The force of molecule at interface

It can be shown that under constant temperature conditions, surface tension is equal to free energy per unit area (specific surface free energy) in thermodynamics. For planes, direction of surface tension is parallel to the gas-liquid interface; for curved faces, direction of surface tension is the tangential direction of work done along the gas-liquid interface. When surface area is needed to be increased, it is necessary to overcome this force to do work, and that is why σ is called surface tension. Conversion relationships of various units of surface tension are:

$1\text{erg/cm}^2 = 1\text{dyn}\cdot\text{cm/cm}^2 = 1\text{dyn /cm} = 10^{-3}\text{N/m} = 1\text{mN/m}$

为最小。这就是一些小水滴、汞珠常呈球形的原因（此时液滴重力不起主要作用）。如果我们要把一个液体分子从液体内部移到液体表面以增大表面积，就必须克服液体内部自身的分子引力而做功，所消耗的功将转化为表面层的能量——表面能。所消耗的功W与所增加的表面积A成正比：

$$W = \sigma A \quad (1-2)$$

式中，比例系数σ称为表面张力，它等于增加单位表面积所做的功，其表达式为：

$$\sigma = \frac{W}{A} \quad (1-3)$$

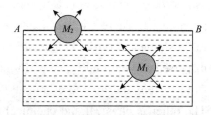

图1-10　界面分子受力示意图

在热力学上可以证明，在恒温条件下，表面张力等于单位面积上的自由能（比表面自由能）。对于平面，表面张力的方向平行于气-液界面；对于弯曲面，表面张力的方向为沿气-液界面所做功的切线方向。增加表面积时，必须克服这个力而做功，这就是将σ称为表面张力的原因。表面张力各种单位的换算关系为：

$1\text{erg/cm}^2 = 1\text{dyn}\cdot\text{cm/cm}^2 = 1\text{dyn /cm} = 10^{-3}\text{N/m} = 1\text{mN/m}$

根据表面张力的量纲，也可

Depending on dimension of surface tension, it can also be regarded as the force per unit length of the surface. Surface tension exists in liquid-gas, liquid-liquid, liquid-solid, gas-solid and solid-solid interfaces. Since surface tension on solid-solid interface could not be measured experimentally, only liquid-gas, liquid-liquid, liquid-solid and gas-solid are studied for surface tension. Conventionally, surface tension on liquid-liquid and liquid-solid interface is referred to as interfacial tension (IFT), and interfacial tension on liquid-gas and gas-solid interface is called surface tension.

Attractive force between molecules is the fundamental source of surface tension. Due to different molecules, the molecular attractive force is different, resulting in different surface tension between different substances. Table 1-3 shows surface tension values of some pure liquids at 20℃. It can be seen that surface tension of metallic mercury is the largest, and surface tension of water is second.

以将表面张力看作是作用于界面边界线单位长度上的力。在液－气、液－液、液－固、气－固、固－固界面上都存在表面张力。固－固界面上的表面张力无法用实验方法测量，所以一般只研究液－气、液－液、液－固、气－固间的表面张力。习惯上将存在于液－液、液－固、固－固界面上的表面张力称为界面张力，而将物质与本身蒸汽（或空气）接触时的界面张力称为表面张力。

表面张力产生的根本原因是分子间的引力。由于相同分子或不同分子之间分子引力的差异，不同物质之间具有不同的表面张力。表1-3为某些纯液体在常温下的表面张力，可见金属汞的表面张力最大，水的表面张力次之。

Table 1-3 Surface tension values of some pure liquids at 20℃

Liquid	Surface tension/(mN/m)	Liquid	Surface tension/(mN/m)
Ether	17	Benzene	29
Carbon tetrachloride	26.9	Trichloromethane	28.5
Toluene	28.5	Carbon disulfide	32.3
Mercury	484	Water	72.8
n-octane	21.8	n-hexane	18.4

表1-3 某些纯液体在20℃的表面张力

液　体	表面张力/（mN/m）	液　体	表面张力/（mN/m）
乙醚	17	苯	29
四氯化碳	26.9	三氯甲烷	28.5
甲苯	28.5	二硫化碳	32.3
汞	484	水	72.8
n-辛烷	21.8	n-己烷	18.4

In summary, intermolecular forces can cause net attractive force, while net attractive force causes surface tension. Surface tension is always tangent to liquid surface and perpendicular to net attractive force.

综上所述，可以得出结论：分子间力可以引起净吸力，而净吸力引起表面张力。表面张力永远和液体表面相切，而和净吸力相互垂直。

1.2.2 Adsorption

Adsorption refers to the phenomenon that a substance is concentrated at a phase interface. It is an important surface phenomenon of substances. Adsorption occurs at surface of substance or at interface of two phases. It will increase as area of adsorption surface increases. Since reservoir rock is a porous medium with large specific surface area, adsorption phenomenon in oil reservoir cannot be ignored.

Due to asymmetry of molecular attractive force, there is excess surface free energy on surface of substances. And the surface free energy, i.e. surface tension of the substance is always trying to reduce. This trend of reduction could be reached by reducing contact area, or surface free energy by adsorbing molecules of adjacent substances, which is the basic principle of adsorption. Adsorption can occur either on solid surfaces or liquid surfaces.

1.2.2.1 Adsorption on solid surfaces

Generally, adsorption on solid surfaces is with the following characteristics:

（1）Adsorption amount of adsorbate on solid surface will increase as the adsorption area increases.

（2）Solid surface is not only uneven but also with not uniform composition. Since different substances are with different adsorption properties, solid adsorption is selective.

（3）All adsorption are exothermic, thus adsorption capacity will decrease while temperature increases.

（4）Adsorption capacity is proportional to concentration of adsorbate. Gases are with large compressibility, thus adsorption capacity of solid surface to gas will increase as pressure increase (concentration increase).

Structure of liquids is more complicated than that of gases. In fact, liquid is often impure. For example,

1.2.2 吸附

所谓吸附，指一种物质在相界面上浓集的现象。吸附是物质的一种重要表面现象。吸附现象发生在物质表面或两相界面，吸附作用随着吸附面面积的增大而增大。储油岩石是一种比表面积极大的孔隙介质，因此油层中的吸附现象是不容忽视的。

在物质表面，由于分子引力的不对称而存在一种过剩的表面自由能，而且物质总是力图缩小其表面自由能，即表面张力。这种缩小的趋势可以通过减小接触面积来满足，也可以通过吸附相邻的物质分子来减小其本身的表面自由能，这就是吸附的基本原理。吸附作用既可以发生在固体表面，也可以发生在液体表面。

1.2.2.1 固体表面的吸附作用

一般来讲，固体的表面吸附具有以下特点：

（1）固体表面对被吸附物质的吸附量随吸附面积的增大而增大。

（2）固体表面凹凸不平，其表面物质成分也不均一，不同物质具有不同的吸附性能，所以固体吸附具有选择性。

（3）吸附作用都是放热的，所以吸附量随温度的升高而减小。

（4）吸附量与被吸附物质的浓度成正比。气体具有较大的压缩性，因而固体表面对于气体的吸附量随压力的升高（浓度增大）而增大。

液体的结构比气体复杂。实

water often dissolves various salts or other substances, and petroleum itself is a complex mixture. Solids could adsorb both solvents and solutes. Therefore, on solid surface, the adsorption of liquid is also more complicated than that of gas.

In general, polar adsorbents tend to adsorb polar substances, and non-polar adsorbents tend to adsorb non-polar substances. Water is a strong polar substance, and rocks that are consisted of minerals also belong to polar substance, that is why water is easily adsorbed by rocks. The various hydrocarbons in petroleum are non-polar substances, which indicating it will not be easily adsorbed by rocks. However, hydrocarbon compounds containing oxygen, sulfur and nitrogen in petroleum, such as naphthenic acid, colloids and asphaltenes are with polar structure which could be adsorbed by rocks. Therefore, the degree of petroleum adsorption on rock surface often depends on the type and content of polar substances in petroleum.

1.2.2.2 Adsorption on liquid surface

The interface properties of pure liquids are relatively simple. In practice, however, liquids are often with other substances that will change original interface properties of the liquids greatly. For example, dissolved organics (such as alcohols, acids, etc) will reduce surface tension of water. If inorganic salts are dissolved in water, its surface tension will be increased. Further studies have also found that distribution of dissolved substances in solutions is not uniform also. For some dissolved substances, concentration at interface is greater than that in bulk solution, while others are opposite. This concentration change phenomenon at interface is called adsorption on liquid surface. When concentration of dissolved substances at interface is greater than that in bulk solution, it will be referred to as positive adsorption, on the contrary, if it less than that in bulk solution, it is referred to as negative adsorption. After being dissolved, a substance that shows positive adsorption is called a surface active substance, and substance that undergoes negative adsorption is called a surface non-active substance. It is evidently that the phenomenon of surface activity is still related with adsorption.

Adsorption on liquid surface can also be explained by surface free energy. For pure liquids, when temperature is constant, surface tension will remain a certain value. In order to reduce surface free energy, the only way is to reduce liquid surface area. When other substances are dissolved into the liquid, the following two situations will occur: In case that dissolved substance at interface reduced surface tension of the liquid, in order to achieve purpose of reducing surface free energy, concentration at the interface must be increased. Then positive adsorption is triggered. In case that dissolved substance at interface increased surface tension of the liquid, the dissolved substance will move toward to the bulk solution as much as possible, and the concentration at interface will be reduced, which lead to a negative adsorption.

1.2.3 Wetting

If a liquid is dropped on a solid surface, the bead may immediately spread along solid surface or remain shape of droplets adhered to the surface. The former is referred that the liquid is wetting to the solid surface, and the latter is non-wetting. It can be said that wetting describes a liquid property of whether a liquid is easily spread on a solid surface. It is actually a process in which the gas on solid surface is replaced by the liquid.

Degree of liquid wetting on solid surface is expressed by the contact angle. Contact angle is the angle between solid-liquid surface or liquid-liquid interface. It is specified to measure from the larger density side as the θ shown in Figure 1-11. When θ is less than 90°, the solid surface is wetting (such as water on glass);θ is equal to 0°, which indicates a complete

液体后发生正吸附的物质称为表面活性物质，发生负吸附的物质称为非表面活性物质。可见物质具有表面活性的现象仍然与吸附作用有关。

液体表面上的吸附作用仍可用表面自由能来解释。对于纯液体，当温度不变时，其表面张力为一定值。要降低表面自由能，只有减小液体的表面积。当液体中溶解有其他物质时，就会出现以下两种情况：如果溶解物质在界面上存在的结果是降低液体的表面张力，为了达到降低表面自由能的目的，溶解物质就必须增大在界面上的浓度，于是发生正吸附。如果溶解物质在界面上存在的结果是增大液体的表面张力，则溶解物将极力跑向液体内部，减小其在界面上的浓度，于是发生负吸附。

1.2.3 润湿

如果在固体表面滴一滴液体，这液珠可能立即沿固体表面扩散开来，也可能仍以液滴形式附着于固体表面。我们将前一种情况称为液体对固体表面润湿，将后一种情况称为不润湿。可以这么讲，润湿是指液体在固体表面是否容易铺开的性质，它实际上就是固体表面上的气体被液体取代的过程。

液体对固体表面的润湿程度用润湿接触角表示。润湿接触角是固体表面与液体表面或液体与液体界面之间的夹角，并规定从密度较大的一方算起，如图1-11

wetting. When θ is greater than 90°, the solid surface is non-wetting (such as water on teflon); θ is equal to 180°, which indicates a complete non-wetting. The solids that could be wetted by liquid is called lyophilic solid, such as hydrophilic solid or hydrophobic solid.

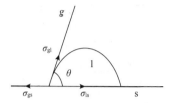

Figure 1-11 Contact angle of a liquid drop on a solid surface

Wetting characteristics of solid surfaces is mainly depending on molecular properties of the solid and liquid. Therefore, different solid and liquid shall have different wetting properties. That is to say the same solid may show different wetting properties when it is combined with different liquids. Similarly, when it is with different solid surfaces, the same liquid may display different wetting properties.

In oilfield chemistry, the most research on wetting of solid surfaces is focusing on wetting of reservoir rocks. The wetting of reservoir rocks is an important factor in distribution control of oil and water in pores, and closely related with oil recovery. Surface wetting of reservoir rocks can be divided into hydrophilic wetting, hydrophobic wetting and neutral wetting.

1.2.4 Curved interface phenomenon

In daily life, some common phenomena are related with bending of liquid surface or interface, such as towel can absorb water, or cracks will appear when a wet clod was dried. In oilfield chemistry, the most important curved interface phenomenon is the oil-water interface formed by oil and water in reservoir capillary pores. It is always resulting in difficulties of oil displacement in capillary pores.

中的 θ。当 θ < 90° 时，液体润湿固体；θ = 0° 为完全润湿。当 θ > 90° 时，液体不润湿固体，θ = 180° 表示完全不润湿。能够被液体润湿的固体，称为亲液体固体，如亲水固体、亲油固体。

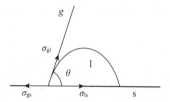

图1-11 液体对固体的润湿

固体表面的润湿性主要取决于固体和液体的分子性质。因此，不同的固体和液体应具有不同的润湿性质。也就是说，同一种固体，对于不同的液体组合可以有不同的润湿性质。同一种液体组合在不同的固体表面可以表现出不同的润湿性质。

油田化学中对固体表面润湿性研究最多的是储油岩石的润湿性，它是控制孔隙中油水分布的重要因素，油层润湿性与石油采收率有密切关系。油层岩石表面的润湿性可分为水湿、油湿和中性润湿。

1.2.4 弯曲界面现象

日常生活中常见的毛巾会吸水、湿土块干燥时会裂缝等现象都与液面或界面弯曲有关。油田化学中，重要的界面弯曲现象是油与水在储层毛细管道中形成的弯曲油水界面，从而使得毛细孔道中的油难以被驱替出来。

1.2.4.1 Capillary pressure

For water surface in a cup, pressure on both sides of the interface is balanced and equal. However, it is known from ordinary physics that due to presence of surface tension, there is an additional pressure on any curved interface (usually defined as concave or convex surfaces), as shown in Figures 1-12.

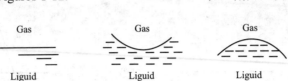

Figure 1-12 Concave and convex surfaces

For a spherical-shaped interface, direction of pressure is same with concave direction of the liquid surface. As the arrows indicate in Figure 1-13, magnitude of pressure is determined by the Laplace equation:

$$p_c = \frac{2\sigma}{R} \quad (1-4)$$

where R is curvature radius of a spherical-shaped interface;
p_c is additional pressure on the curved surface;
σ is interfacial tension.

The additional force p_c generated in capillary is also called capillary force p_c. Assuming the capillary radius is r and the contact angle of a liquid is θ, it can be seen that the curvature radius R is equal to $r/\cos\theta$ (Figure 1-13). When R is substituted by $r/\cos\theta$ in formula (1-4), the equation will become:

$$p_c = \frac{2\sigma}{r}\cos\theta \quad (1-5)$$

1.2.4.1 毛细管压力

在一杯水平界面层处，界面内外两侧的压力是平衡、相等的。但从普通物理学中知道，由于表面张力的存在，任何弯曲界面（通常规定为凹面和凸面两类）都存在一附加压力，如图1-12所示。

图1-12 凹面和凸面

对于球形弯曲液面，该压力的方向与液面的凹向一致，如图1-13中箭头所示。压力的大小由Laplace方程确定：

$$p_c = \frac{2\sigma}{R} \quad (1-4)$$

式中 R——球面曲率半径；
p_c——曲面上的附加压力；
σ——两相界面张力。

毛细管中所产生的附加力p_c就是我们通常所说的毛细管力p_c。假定毛细管半径为r，液体润湿接触角为θ，由图1-13可以看出，曲率半径$R=r/\cos\theta$，代入式(1-4)得：

$$p_c = \frac{2\sigma}{r}\cos\theta \quad (1-5)$$

式(1-5)表明，毛细管力与

Formula (1-5) shows that capillary force is inversely proportional to capillary radius, that is, the capillary force will increase as the capillary gets finer. Direction of the capillary force is always same with the non-wetting phase (Figure 1-13).

毛细管半径成反比，亦即毛细管越细，毛细管力越大。毛细管力的方向总是指向非润湿相一方。

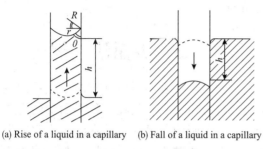
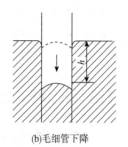

(a) Rise of a liquid in a capillary　　(b) Fall of a liquid in a capillary

Figure 1-13　Capillary

(a)毛细管上升　　(b)毛细管下降

图1-13　毛细现象

1.2.4.2　Rise and fall of liquid in capillary

Rise or fall of a liquid in a capillary (arising from the capillary force) is depending on the wetting characteristics. As shown in Figure 1-13(a), if the liquid wets the capillary (like water and glass or water and metal), liquid surface in the capillary will be concave. Since the pressure of liquid phase below concave surface is lower than that of the liquid that is with a flat surface at the same height, the liquid will be pressed into the capillary, and the liquid column will rise by a certain height h.

Similarly, if the liquid non-wetting (such as Hg in glass), liquid surface in the capillary will be convex as shown in Figure 1-13(b). Since the pressure of liquid phase below convex surface is higher than that of the liquid that is with flat surface at the same height, i.e., the gas pressure above the liquid surface is larger, thus the liquid column in capillary is lowered.

Reservoir rocks are with very complex pore system, which could be regarded as a network of irregular capillaries. Thus capillary forces will play an important factor in reservoir systems where liquids of oil and water are present in rock porous. According to the

1.2.4.2　毛细管上升和下降现象

如图1-13(a)所示，若液体能很好地润湿毛细管壁，则毛细管内的液面呈凹面。因为凹液面下方液相的压力比同样高度具有平面的液体中的压力低，因此液体将被压入毛细管内使液柱上升一定高度h。

同样，若液体不能润湿管壁，则毛细管内的液面呈凸面，如图1-13(b)所示。因凸液面下方液相的压力比同高度具有平面的液体中的压力高，亦即比液面上方气相压力大，所以管内液柱反而下降。

油岩石具有十分复杂的孔隙系统，这种孔隙系统可以看成是由一套不规则毛细管的毛细管网络组成。因此，油、水及岩石作为一个整体，其毛细管现象十分突出，对于原油开采有重要影响。由上述一般毛细管性质的讨论可知，对于亲水地层来说，毛细管

general capillary properties described above, it could be seen that for a hydrophilic formation, the capillary force is a power, but for the hydrophobic formation it is a resistance. In this case, an external force must be applied to overcome capillary pressure, in which the crude oil in rock porous could be displaced. Of course, reduction of capillary resistance can also be achieved by reducing the surface tension.

1.3 Surfactant

There are two important properties for surfactants, one is oriented adsorption on various interfaces, and the other is formation of micelles inside the solution. The former property is the basis that it could be used as emulsifiers, foaming agents or wetting agents, and the latter property is the reason why surfactants often show solubilization. Due to their unique physicochemical properties, surfactants have a wide range of applications in oilfields and have shown excellent effects in development, drilling, gathering and transportation.

1.3.1 Definition and structure of surfactant

Surfactant refers to a substance that can significantly change the interface properties (such as interfacial tension) of a system. In a narrow sense, surfactants are substances that could significantly reduce surface tension of water at very low concentrations (few grams per liter or 1~100 mmol/L, as shown in Tables 1-4). For example, soap and washing powder that you are familiar with are surfactants.

力是动力，而对亲油地层是阻力，必须施加外力克服毛细管压力才能将原油驱出。当然，毛细管阻力的减小也可以通过降低表面张力来实现。

1.3 表面活性剂

表面活性剂有两个重要的性质，一个是在各种界面上的定向吸附，另一个是在溶液内部能形成胶束。前一种性质是许多表面活性剂用作乳化剂、起泡剂、润湿剂的根据，后一种性质是表面活性剂常有增溶作用的原因。表面活性剂以其特殊的物理和化学性质在油田上应用十分广泛，在开发、钻采、集输等方面都显示出优异的作用。

1.3.1 表面活性剂的定义和结构

表面活性剂是指能显著改变物体界面性质（如界面张力）的物质。狭义地讲，表面活性剂是低浓度条件下（1~100mmol/L）就能显著降低水的表面张力的物质（表1-4）。例如，大家熟悉的肥皂、洗衣粉即是表面活性剂。

Table 1-4 Surface tension values of different aqueous solutions

Surface tension/（mN/m）	Surfactant $C_{12}H_{25}SO_4Na$/（mmol/L）	Ethanol /%
72	0	0
50	0.8	10
40	3	20
30	8	40
22	—	100

表1-4 不同水溶液的表面张力值

表面张力/(mN/m)	表面活性剂 $C_{12}H_{25}SO_4Na$/(mmol/L)	乙醇/%
72	0	0
50	0.8	10
40	3	20
30	8	40
22	—	100

Surfactant molecules are composed of a hydrophobic group (such as various alkyl groups) and a hydrophilic group (such as carboxyl groups, sulfonic groups and various polar groups), which means its molecular structure is "amphiphilic" (Figure 1-14). For example, molecular structure of soap is CH_3—$(CH_2)_n$—$COONa$, in which the hydrocarbon chain CH_3—$(CH_2)_n$— is a hydrophobic group (non-polar group), and the carboxyl group—$COONa$ is a hydrophilic group (polar group). The reason why surfactant can adsorb liquid surface and reduce liquid surface tension is that its special molecular structure. These two groups not only could prevent oil and water from rejecting each other, but also could connect oil and water phases together. That's why when surfactant molecules with amphiphilic structure are at the interface, net attractive force and interfacial tension will be reduced.

It must be pointed out that although molecular structure of surfactant is characterized by amphiphilic molecules, not all amphiphilic molecules are surfactants. Only when hydrophobic group of amphiphilic substance is with sufficient length, the substance could be called surfactant. For example, although in the family of fatty acid sodium salt, compounds with small number of carbon atoms (such as sodium formate, sodium acetate, sodium propionate, sodium butyrate, etc.) are all with

表面活性剂分子均由亲油基（如各种烃基）和亲水基（如羧基、磺酸基等各种极性基）等组成，即分子结构特点具有"两亲性"（图1-14）。例如，肥皂的分子结构为CH_3—$(CH_2)_n$—$COONa$，其中碳氢链CH_3—$(CH_2)_n$—是亲油基（非极性基团），羧基—$COONa$是亲水基（极性基团）。表面活性物质之所以能够吸附于液体表面，降低液体的表面张力，就是由这类物质特殊的分子结构决定的。这两种基团不仅具有防止油水两相互相排斥的功能，而且还具有把油水两相联结起来不使其分离的特殊功能，因此当具有两亲结构的表面活性剂分子处于界面上时，就能降低界面上的净吸力和界面张力。

必须指出，虽然表面活性剂分子结构的特点是两亲性分子，但并不是所有两亲性分子都是表面活性剂，只有亲油部分有足够长度的两亲性物质才是表面活性剂。例如，在脂肪酸钠盐系列中，碳原子数少的化合物（甲、乙、丙、

Figure 1-14 Molecular structure of surfactant

图1-14 表面活性剂的两亲结构

hydrophobic groups and hydrophilic groups, they could not act as soap, and could not be regarded as surfactants. Only when the number of carbon atoms increased to a certain extent, fatty acid sodium will show obvious surface activity and general soap properties.

1.3.2 Hydrophilic–hydrophobic balance (*HLB*) of surfactant

In surfactant molecule, relative strength on hydrophilicity of the hydrophilic group and hydrophobicity of the hydrophobic group are also a basic characteristic of the surfactant. The balance between hydrophilic–hydrophobic is called hydrophilic–hydrophobic balance (*HLB*). *HLB* value equals to 7 indicating hydrophilicity and hydrophobicity of the surfactant are similar. *HLB* value greater than 7 indicating hydrophilicity of the surfactant is greater than hydrophobicity, the surfactant belongs to hydrophilic surfactant. The more the value greater than 7, the stronger hydrophilic features will be shown. Surfactants with *HLB* value less than 7 shows stronger hydrophobicity than hydrophilicity. Most surfactants are with a *HLB* value between 1 and 20. The application range of surfactants with different *HLB* values is shown in Table 1-5.

1.3.2 表面活性剂的亲水亲油平衡值（*HLB*）

表面活性剂分子中的亲水基的亲水性和亲油基的憎水性的相对强弱，也是表面活性剂的一个基本特性，其数值标度称为亲憎平衡值（常用符号*HLB*值表示）。*HLB*值等于7，表示活性剂的亲水性和憎水性相近；*HLB*值大于7，表示活性剂的亲水性大于憎水性，属亲水性活性剂，而且比7大得越多其亲水性越强；*HLB*值小于7的活性剂，其憎水性比亲水性强。大多数表面活性剂的*HLB*值在1~20。不同*HLB*值的表面活性剂应用范围见表1-5。

Table 1-5　*HLB* values of surfactants and their applications

HLB value	Application	*HLB* value	Application
1-3	Defoamer	12-15	Wetting agent
3-5	Water-in-oil emulsifier	13-15	Detergent
8-18	Oil-in-water emulsifier	15-18	Solubilizer

表1-5　活性剂的*HLB*值及其应用范围

*HLB*值	应用领域	*HLB*值	应用领域
1~3	消泡剂	12~15	润湿剂
3~5	油包水型乳化剂	13~15	洗涤剂
8~18	水包油型乳化剂	15~18	增溶剂

Standard *HLB* values of surfactants are based on paraffin (*HLB* value is equal to zero), oleic acid (*HLB* value is equal to 1, potassium oleate (*HLB* value is equal to 20, and sodium dodecyl sulfate (*HLB* value is equal to 40). *HLB* value of other surfactants could be determined by emulsification effect comparison through emulsification test, or could be calculated by empirical formula. Obviously, *HLB* value of surfactants plays an important guiding role in surfactant application of different fields.

1.3.3 Classification of surfactants

1.3.3.1 Classification by solubility

According to their solubility in water, surfactants can be divided into water-soluble surfactants and oil-soluble surfactants. At present, most of surfactants are water-soluble, although oil-soluble surfactants are increasingly important.

1.3.3.2 Classification by molecular weight

Surfactant with molecular weight more than 10,000 is called polymeric surfactant; Surfactant with molecular weight between 1000 and 10000 is called moderate molecular weight surfactant; and surfactant with molecular weight between 100 and 1000 is called low molecular weight surfactant.

Most commonly used surfactants are low molecular weight surfactants. In moderate molecular weight surfactants, there is a polyether surfactant used widely in industry. It is prepared by condensation reaction of polyoxypropylene and polyoxyethylene. Although surface activity of polymeric surfactants is not outstanding, it has unique features in emulsification, solubilization, especially dispersion or flocculation performance. Thus it is a material with promising future.

1.3.3.3 Classification by application

According to its application, surfactants can

表面活性剂的 *HLB* 值均以石蜡的 *HLB*=0、油酸的 *HLB*=1、油酸钾的 *HLB*=20、十二烷基硫酸钠的 *HLB* = 40 作为标准，其他表面活性剂的 *HLB* 值可用乳化实验对比其乳化效果来确定，也可用经验公式计算出来。显而易见，掌握表面活性剂的 *HLB* 值，对于表面活性剂在不同方面的应用具有重要的指导作用。

1.3.3 表面活性剂的分类

1.3.3.1 按溶解性分类

按在水中的溶解性，表面活性剂可分为水溶性表面活性剂和油溶性表面活性剂两类，前者占绝大多数，油溶性表面活性剂日显重要，但其品种仍较少。

1.3.3.2 按分子量分类

相对分子质量大于10000者称为高分子表面活性剂；相对分子质量在1000～10000的称为中分子表面活性剂；相对分子质量在100～1000的称为低分子表面活性剂。

常用的表面活性剂大都是低分子表面活性剂。中分子表面活性剂有一种是聚醚型的，即聚氧丙烯与聚氧乙烯缩合的表面活性剂，在工业上占有特殊的地位。高分子表面活性剂的表面活性并不突出，但在乳化、增溶特别是分散或絮凝性能上有独特之处，很有发展前途。

1.3.3.3 按用途分类

表面活性剂按用途可分为表

be classified into surface tension reducing agents, penetrating agents, wetting agents, emulsifiers, solubilizers, dispersing agents, flocculants, foaming agents, defoamers, bactericides, antistatic agents, corrosion inhibitors, softeners, waterproofing agents, fabric finishing agents, leveling agents, etc. In addition, there are organometallic surfactants, silicon-containing surfactants, fluorine-containing surfactant and special reactivity surfactants.

1.3.3.4 Classification by type of ions

Ionic classification method is a commonly used classification method, which is actually a chemical structure classification method. It is desirable to combine mechanism of a surfactant with its chemical composition for finding the laws of property and function.

According to dissociation or non-dissociation, surfactant is divided into ionic surfactant and nonionic surfactant after it is dissolved in water. According to property of the charges, ionic surfactant can be further classified into anionic type (such as petroleum sulfonate $RAr-SO_3M$), cationic type (such as dodecyl ammonium chloride $[C_{12}H_{25}-NH_3]Cl$) and amphoteric surfactants (such as sodium lauraminopropionate $C_{12}H_{25}NHCH_2CH_2COONa$).

1.3.4 Surfactant solution

1.3.4.1 Critical micelle concentration

When soap is dissolved in water, its surfactant molecules can be stabilized in two ways: One is adsorbs on liquid surface, the polar group faces polar water, the non-polar group faces non-polar air, so that the polarity difference at water-air interface is reduced and the surface tension is lowered. The other is to stay in water and gather together. The hydrophobic non-polar tail groups are put inward and close to each other while the hydrophilic groups are put outward, then micelles will

面张力降低剂、渗透剂、润湿剂、乳化剂、增溶剂、分散剂、絮凝剂、起泡剂、消泡剂、杀菌剂、抗静电剂、缓蚀剂、柔软剂、防水剂、织物整理剂、匀染剂等类。此外，还有有机金属表面活性剂、含硅表面活性剂、含氟表面活性剂和反应性特种表面活性剂。

1.3.3.4 按离子型分类

离子型分类法是常用的分类法，它实际上是化学结构分类法。人们希望将表面活性剂的作用机理与其化学组成联系在一起，借以寻找性质与作用的规律。

表面活性剂溶于水后，按离解或不离解分为离子型表面活性剂和非离子型表面活性剂。离子型表面活性剂又可按产生电荷的性质分为阴离子型（如石油磺酸盐 $RAr-SO_3M$ 等）、阳离子型（如氯化十二烷基铵 $[C_{12}H_{25}-NH_3]Cl$ 等）和两性型表面活性剂（如十二烷基氨基丙酸钠 $C_{12}H_{25}NHCH_2CH_2COONa$ 等）。

1.3.4 表面活性剂溶液

1.3.4.1 临界胶束浓度

当肥皂溶解于水时，表面活性剂分子可以通过两种方式达到稳定：一是被吸附于液体表面，极性端朝向极性的水，非极性端朝向非极性的空气，从而使水-空气界面上的极性差减小，表面张力降低。二是分子留在水中，聚集在一起，憎水的非极性端向内互相靠拢，亲水端向外，形成

be formed as shown in Figure 1-15. As a result, the non-polar hydrophobic group is completely surrounded by the polar hydrophilic group and is isolated from water. Therefore, the micelles are hydrophilic and can be stably stay in water.

Figure 1-15 shows the distribution of surfactant molecules in water as a function of concentration. Initially, surfactant concentration is small and affects little on water surface tension. As the concentration increases, the concentration of surfactant in surface layer increases also, causing surface tension decreasing sharply. At the same time, a few small micelles are formed in water. Finally, the surface layer is entirely occupied by surfactant molecules. When the concentration is further increasing, only micelles can be formed, and the surface tension will no longer decrease with the increase of concentration, which is called critical micelle concentration(*CMC*). Generally, the *CMC* value of surfactant is in the range from 10^{-5} to 10^{-4} mol/L. Conventionally, solution with surfactant concentration lower than *CMC* is called active water, and solution with surfactant concentration higher than *CMC* is called micelle solution.

Factors affecting *CMC* are mainly salts, molecular structure of surfactant, alcohols, macromolecules, etc. *CMC* decreases with salt addition. *CMC* will decrease

胶束，如图 1-15 所示。其结果是非极性的憎水基完全被极性的亲水基包围在内部，与水脱离接触。所以，胶束表现为亲水性质，可以稳定地存在于水中。

图 1-15 表示表面活性剂在水中的分布随浓度的变化关系。一开始，表面活性剂浓度很小，对水的表面张力几乎没有影响。随着浓度的增加，表面层中表面活性剂浓度增大，使表面张力急剧降低，同时，在水中也形成少量的小胶束。最后，表面层全部被表面活性剂分子占据。再增大浓度时，就只能形成胶束，这时表面张力再也不会随着浓度的增加而降低了。我们将这时候的浓度称为临界胶束浓度，以 *CMC* 表示。一般表面活性剂的 *CMC* 值在 $10^{-5} \sim 10^{-4}$ mol/L。通常将表面活性剂浓度低于 *CMC* 的溶液称为活性水，表面活性剂浓度高于 *CMC* 的溶液称为胶束溶液。

影响表面活性剂 *CMC* 的因素主要有盐、表面活性剂的分子结构、醇、高分子等。*CMC* 随盐

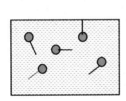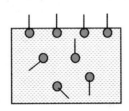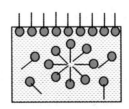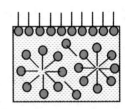

Figure 1-15 Distribution of surfactant molecules in water as a function of concentration

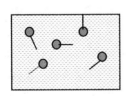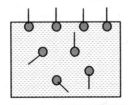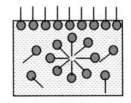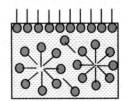

图1-15 表面活性剂在水中的分布与浓度的关系

along with increase of surfactant alkyl carbon atoms; and long chain fatty alcohol could reduce *CMC* of anionic surfactant. Low concentration of short chain fatty alcohol will lead to *CMC* decrease, but high concentration of short chain fatty alcohol could result in *CMC* increase. For dilute surfactant solutions, *CMC* of nonionic surfactants will be increased and that of ionic surfactants will be reduced in the presence of macromolecules.

When surfactant concentration exceeds *CMC*, the surfactant spherical micelles could continue to grow under appropriate conditions (temperature, electrolyte, pH, etc.), and convert into rod-like micelles (worm-like micelles), vesicles, etc. This will result in properties of surfactant similar with polymer solution, such as high viscoelasticity, shear thinning and so on.

1.3.4.2　Surfactant adsorption at interfaces

Orientation of the surfactant molecules in aqueous solution at liquid-gas or liquid-liquid interface is determined by molecular structure of surfactant (Figure 1-15). Because of the orientation of surfactant molecule at interface, a concentration difference between interface and bulk phase appears, this phenomenon is known as surface excess or adsorption.

1.3.5　Some important functional principles of surfactants

1.3.5.1　Wetting

It is known that water can wet surface of clean glass, but could not wet surface of paraffin. If a suitable surfactant is added to water, amphiphilic surfactant molecules will concentrate onto solid surface and will be aligned spontaneously: The polar group is directed to polar glass surface, and the non-polar group is directed to water. Due to adsorption of surfactant, an original hydrophilic polar solid surface becomes a hydrophobic non-polar surface, or an original hydrophobic solid surface becomes hydrophilic solid surface, this phenomenon is called wetting reversal. A surfactant that is capable of triggering wetting reversal on solid surface is particularly called a wetting agent.

There are various applications for wetting reversal phenomenon in drilling and oil recovery. For example, surfactant in mud can cause a hydrophobic steel surface, which will greatly reduce or prevent probability of bit balling and pipe sticking. When oil soaking method is used to release sticking, a surfactant resulting in a hydrophobic surface of drill pipe and mud cake shall be added. With the help of surfactant, oil film can penetrate quickly between drill pipe and mud cake, which is able to improve speed of sticking release.

While water flooding with active water, the surfactant could adsorb on surface of rocks to reverse rock surface from hydrophobic to hydrophilic, thereby improving oil displacement efficiency (Figure 1-16). At the same time, the relative permeability of oil and water will be changed to produce a favorable water-oil mobility ratio, and sweep efficiency of water flooding is improved.

面，非极性基指向水，由于表面活性剂的吸附，使原来亲水性极性固体表面变成亲油性非极性表面，或者原来亲油性固体表面变成亲水性表面的现象，称为润湿反转。能使固体表面产生润湿反转的活性剂，特称润湿剂。

润湿反转现象在钻井、采油中有各种应用。例如，泥浆中的活性剂能使钢铁表面亲油化时，则可大大减少或防止泥包钻头和黏附卡钻。发生黏卡后用油泡解卡时，加入能使钻杆和泥饼表面亲油化的活性剂，使油膜很快渗入钻杆与泥饼之间，则能大大提高解卡速度。

使用活性水驱油时，表面活性剂在地层岩石表面吸附，可使亲油的岩石表面反转为亲水表面，从而提高洗油效率（图1-16）。同时，还可以使油水相对渗透率发生变化，形成有利的水油流度比，提高注水波及系数。

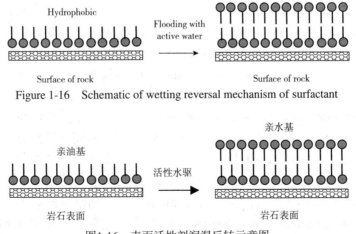

Figure 1-16　Schematic of wetting reversal mechanism of surfactant

图1-16　表面活性剂润湿反转示意图

1.3.5.2 Solubilization

When surfactant concentration is greater than CMC, the micelles will be formed. Core of these micelles are aggregated non-polar groups, which is with properties of hydrocarbon. Therefore, adding crude oil into micelle solution, the surfactant will adsorb on oil-water interface. Since oil-water interfacial tension is reduced, oil is able to be dissolved in hydrophobic core of the micelles. This phenomenon is called solubilization. After solubilization, volume of micelles increases, which is called swelling micelles. Solubilization refers to a phenomenon that when concentration of surfactant in an aqueous solution reaches to CMC, the water solubility of originally insoluble oil has significantly increased. Solubilization is a characteristic attribute of surfactant semidilute solutions (micelle solutions, microemulsions), while dilute surfactant solutions are not with the property. Why solubilization effect of surfactant solution could be found after micelles formed? According to the principle of similar dissolve mutually, when substances that being solubilized (such as oil) was added into micelle solution, the molecules of solubilized substances will be "hidden" into the non-polar region that is completely isolated from water. Then the whole solution becomes a homogeneous, stable system without any interface.

The solubilization of surfactant solutions are mainly with the following forms: Non-polar organic molecules "dissolved" in core of micelles; solubilized molecules interspersed in the "fence" structure of micelles; solubilized molecules are adsorbed on shell of micelles; solubilized molecules are adsorbed to interface between micelles and water (Figure 1-17).

Solubilizing effect of surfactant on oil is related with molecular structure of the surfactant. Nonionic surfactant is with the best alkane solubilization capacity among different surfactants with same alkyl chain, cationic surfactant and anionic surfactant follows. Polar

1.3.5.2 增溶作用

表面活性剂浓度大于 CMC 后，所形成胶束的内核是非极性基的聚集体，具有烃的性质。因此，如果在胶束溶液中加入原油，表面活性剂在油－水界面吸附，由于降低了油－水界面张力，油便溶解于胶束的憎水核心中，这种现象称为增溶。增溶后的胶束体积增大，称为溶胀胶束。增溶作用是指水溶液中表面活性剂的浓度达到临界胶束浓度之后，原来难溶于水的油溶解度显著增大。增溶是表面活性剂浓溶液（胶束溶液、微乳液）的特有属性，而表面活性剂稀溶液则无此属性。表面活性剂形成胶束后之所以具有增溶作用，是由于被增溶的物质（如油）加入胶束溶液中之后，依据相似相溶原理，油的分子"包藏"到胶束内部的非极性区域内与水完全隔离，并成为一个不存在任何界面的均相、稳定体系。

表面活性剂溶液的增溶主要有以下几种形式：非极性有机分子"溶解"于胶束内核；增溶的分子穿插于胶束的"栅栏"结构中；增溶的分子被吸附在胶束的外壳；被增溶的分子吸附于胶束与水的界面（图 1-17）。

表面活性剂对油的增溶作用与表面活性剂分子结构有关。具有相同烷基链的不同表面活性剂对烷烃的增溶顺序为：非离子型表面活性剂大于阳离子型表面活性剂大于阴离子型表面活性剂。极性有机物质（如醇）可以穿插于胶束的"栅栏"结构中，有利

organic materials (such as alcohols) can be interspersed into the "fence" structure of micelles, which is helpful for solubilization of alkanes. Inorganic salts could reduce *CMC* value of surfactant while increase micelle aggregation number, micelle volume and solubilization. Solubilization of ionic surfactants will increase along with temperature. When temperature is below the cloud point temperature, solubilization of nonionic surfactant will increase along with the increase of temperature.

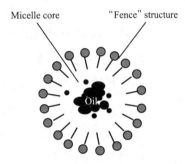

Figure 1-17 Schematic of solubilization

Solubilization is different from general dissolution because the latter refers to uniform dispersion of solute molecules in solvent molecules, while solubilization is aggregation of solute inside the micelles. Compare to emulsification, solubilization refers the process that solubilized substance with aggregation state dissolve into micelle. Solubilization is only a process of "dissolution" of solubilized substance in micelles and will not increase interfacial area of the system. Hence it is a thermodynamically stable system. However emulsification is a dispersing process that the phase interface area is increased, so that surface free energy of the system will be greatly increased, which indicates it is a thermodynamically unstable system.

1.3.5.3 Foaming

Foam is aggregation of large number of bubbles. It is a dispersing system that formed by gases dispersed into liquids. Bubbles generated from water stirring will disappear quickly, but those generated by stirring soap or washing powder solution could be stable for a long

于对烷烃的增溶。无机盐可以使表面活性剂的 *CMC* 值降低，使胶束聚集数增加，胶团增大，增溶量增加。升高温度，可以增大离子型表面活性剂的增溶作用。在低于浊点温度条件下，非离子型表面活性剂的增溶作用随温度上升而增大。

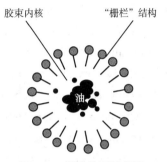

图1-17 增溶作用示意图

增溶作用不同于一般的溶解作用，因为一般的溶解作用是指溶质分子在溶剂分子中的均匀分散，而增溶作用则是溶质集中在胶束内部。增溶作用与乳化作用也有所不同，增溶过程是被增溶物以整团的形式溶入胶束区域内，它仅仅是被增溶物在胶束中"溶解"，不增加体系的界面面积，所以是一个热力学稳定体系。而乳化作用是增加相界面面积的分散过程，从而使体系的表面自由能大为增加，是热力学不稳定体系。

1.3.5.3 起泡作用

泡沫是大量气泡的密集体，是气体分散于液体中的分散体系。搅动普通水时生成的气泡很快就消失了，但搅动加有肥皂或洗衣粉的水时，生成的气泡却能稳定

time. Therefore, pure liquid could not form stable foams. To obtain stable foams, a stabilizer (or a foaming agent) must be added. Most of foaming agents are surfactants.

Surfactants can produce foaming effect because: It could reduce gas-liquid surface tension, thereby reducing the bubbles coalescence tendency; It could form an adsorption protective membrane with certain mechanical strength around bubbles (Figure 1-18), which is the main factor for foam stability. In addition, proper surface viscosity is necessary: This is that liquid membrane between the bubbles will flow away and become thin due to the gravity, and then the bubbles will break. If viscosity of liquid membrane is high, it is not easy to flow away, and mechanical strength of the membrane will become strong. However, if bubbles are too large, the membrane tends to become brittle, and foaming agent molecules inside of it are not easy to move freely. When the membrane is partially damaged, the "wound" cannot be quickly compensated by free foaming agent molecules, and bubbles become easily broken. When ionic surfactant is added, the hydrophilic group of surfactant molecule will be ionized in water. The repulsion of same ion charges of the surfactant will prevent liquid membrane thinning and bubbles aggregating.

The main function of defoaming agent is to destroy strength of the adsorption layer of bubble liquid membrane. For example, if the surfactant added is with high surface activity (easy to replace the foaming

相当长的时间。所以,纯液体不能形成稳定的泡沫,要得到稳定的泡沫必须加入稳定剂(或起泡剂),多数起泡剂是表面活性物质。

活性剂的起泡作用是：降低气－液表面张力,从而降低气泡的聚并趋势；在气泡周围形成具有一定机械强度的吸附保护膜(图1-18),这是泡沫稳定的主要因素；要有适当的表面黏度：气泡间液膜受地心引力等的作用流走,液膜变薄,易使气泡破裂。膜中液体黏度大则不易流走,还会使膜的机械强度增加。但太大则膜易变脆,而且膜中起泡剂分子不易自由移动,膜局部受损时,不能迅速弥补"伤口",反而使泡易破；对于离子型活性剂,亲水端在水中电离,两层活性剂离子的相同电荷的排斥作用阻碍液膜变薄和气泡聚并。

消泡剂的主要作用是破坏气泡液膜吸附层的强度,例如,加入表面活性较高(易于顶替起泡剂)但形成的吸附膜强度很差的

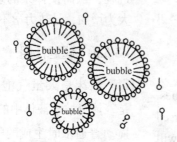 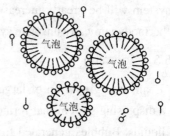

Figure 1-18　Adsorption of foam agent at gas-liquid interface　　图1-18　起泡剂在气－液界面上的吸附

agent) but strength of adsorption membrane formed by this surfactant is poor, stability of foam will be greatly reduced.

1.3.5.4 Washing

A surfactant could remove a substance (such as oil) from solid surface by a liquid (such as water). This function of surfactant is called washing. Surfactant with washing function is called cleaning agent or detergent. Washing is result of a synthetic effect, including wetting reversal, emulsification and solubilization. For example, during the process of washing oil film from sandstone surface by surfactant solution, the sandstone surface is turned to hydrophilic surface by the surfactant. When oil film is washed off, the surfactant will emulsify it, leading to an emulsification of oil in water, in which oil is difficult to adhere back to the sandstone surface. If surfactant concentration is high enough, some of the oil may also be dissolved in micelles of the surfactant and be taken away. This is what happened when oil stained clothes were being cleaned by washing powder (the main ingredient is sodium dodecylbenzene sulfonate). It can be seen that washing is the combined effect of some functions of surfactant.

In addition, surfactants also possess functions of emulsification (As shown in section 1.4), dispersion, bactericidal, etc.

1.4 Emulsion

Emulsion is a thermodynamically unstable multiphase dispersion system with certain dynamic stability. It is very similar to colloid dispersion system in terms of interface properties and coalescence instability, and thus is included in the field of colloid and interface chemistry research. Since emulsions are also with huge phase interfaces, interface phenomena play an important role in their formation and application.

In oil and gas field development, emulsions mainly involves in applications of emulsified drilling and completion fluids, emulsified acid, emulsified fracturing fluid, viscosity reduction by heavy oil emulsification and microemulsion in tertiary oil recovery.

1.4.1 Basic concepts of emulsion

Emulsion is a mixture of two (or more) immiscible substances (a multiphase dispersion system), which is composed by one liquid dispersed in another liquid in form of fine droplets. Generally, diameter of the fine droplets is larger than 0.1μm.

The process of mixing water with "oil" (generalized oil) to form an emulsion is called emulsification. But sometimes it also needs demulsification, which means to destroy the emulsion and separate oil and water. For example, process of emulsified crude oil dehydration, degreasing milk to make cream, etc. are all demulsification.

In fact, it is impossible to prepare a stable emulsion with only oil and water. Since interface area of droplets is increased after dispersion, surface energy is correspondingly increased. However, surface energy tends to decrease spontaneously when droplets touch each other, aggregation will start and surface energy will be reduced. Thus the formation of two layers (oil and water) in emulsion is a spontaneous process. In general, to prepare a relatively stable emulsion, some stabilizing agents must be added to reduce oil-water interfacial tension and enhance protection of the droplets.

A substance that could emulsify two incompatible phases like oil and water and form a stable emulsion is called emulsifier. Most of emulsifiers are surfactants. However, some macromolecule substances (gelatin, protein, etc.) and some certain solid powders (clay, carbon black, calcium carbonate, etc.) can also stabilize the emulsion.

Normally, there are three necessary conditions for preparing of a stable emulsion (Figure 1-19):

1.4.1 乳状液的基本概念

乳状液是一种多相分散体系，它是一种液体以细小液滴形式分散于与其不相混溶的液体中所构成的，液滴直径一般大于0.1μm。

凡由水和"油"（广义的油）混合生成乳状液的过程，称为乳化。但有时也需要破乳，即将乳状液破坏，使油水分离。如乳化原油脱水，牛奶脱脂制奶油等均需破乳。

实际上，单用油和水不能制得稳定的乳状液。因为液滴分散后界面积随之增大，表面能相应增大。而表面能要自发趋向减小，当液滴彼此接触时就会聚并，从而降低表面能，因此乳状液分层是自发过程。要制得比较稳定的乳状液，必须加入一些稳定物质以降低油-水界面张力和增强对液滴的保护作用。

能使不相溶的油水两相发生乳化而形成稳定乳状液的物质称为乳化剂，乳化剂大多是表面活性剂。但也有一些高分子物质（明胶、蛋白质等）和某些固体粉末（黏土、炭黑、碳酸钙等）亦能稳定乳状液。

通常，稳定乳状液的形成必须满足下述三个条件（图1-19）：

(1) There are two immiscible phases, usually these two phases belong to aqueous phase and oil phase.

(2) There is an emulsifier.

(3) An intense shaking condition.

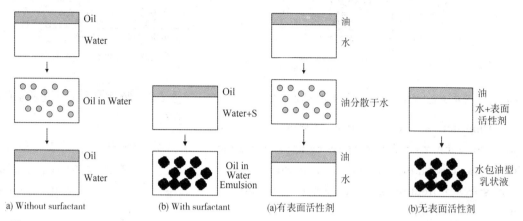

Figure 1-19 Mixing of oil and water by shaking

图1-19 乳状液的形成示意图

1.4.2 Emulsion type and their influencing factors

1.4.2.1 Types of emulsion

In emulsions, all insoluble organic liquids to water (such as benzene, carbon tetrachloride, crude oil, etc.) are collectively referred to as "oil". Emulsions can be divided into two categories:

(1) Water-in-oil (O/W).

The dispersion phase (also known as the internal phase) is oil, and the dispersion medium (also known as the external phase) is water, such as crude oil, invert-emulsion drilling fluids, etc.

(2) Oil-in-water (W/O).

The internal phase is water and the external phase is oil, such as milk, oil emulsion mud, etc.

In addition, there is multiple emulsion type (W/O/W or O/W/O). Although this type could be used in special applications, they are quite rare. The emulsion type is shown in Figure 1-20.

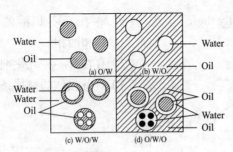

Figure 1-20 Types of emulsion

1.4.2.2 Influencing factors of emulsion type

Emulsion is a complex multi-dispersion system, hence there are many factors affecting its classification. These factors mainly include phase volume ratio, emulsifier molecular structure, hydrophilicity of emulsifier and so on.

(1) Phase volume ratio.

According to viewpoint of spherical droplets close packing, the maximum volume of the droplets can only account for 74% of the total volume, and the remaining 26% is dispersion medium. If volume of dispersion phase is greater than 74.02%, the emulsion will be destroyed or changed into a variant. In case that aqueous phase volume accounts for 26%~74% of the total volume, both emulsion types can be formed; If aqueous phase volume is less than 26%, only W/O type could be formed; if aqueous phase volume is greater than 74%, only O/W type could be formed.

(2) Emulsifier molecular structure.

According to "oriented wedge" theory, the small end of emulsifier molecule cross section always points to the dispersion phase, and the large end of the molecule cross section always directs the dispersion medium (Figure 1-21). Therefore, when a fatty acid salt of monovalent metal ion such as Cs^+, Na^+ or K^+ is used as emulsifier, O/W emulsion will be easily formed. High-valent metal soaps such as Ca^{2+}, Mg^{2+}, Al^{3+} or Zn^{2+} are tend to help forming W/O emulsions. But there are exceptions. For example, Ag soap should be with O/W emulsion, but in fact it is W/O emulsion.

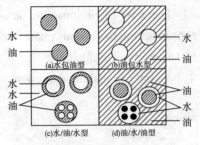

图1-20 乳状液类型示意图

1.4.2.2 影响乳状液类型的主要因素

乳状液是一个复杂的多分散体系，影响其类型的因素很多，主要有相体积比、乳化剂分子构型及亲水性等因素。

（1）相体积比。

根据球形液滴密集堆积观点，在最密集堆积时，液滴的最大体积只能占总体积的74%，其余26%为分散介质。若分散相体积大于74.02%，乳状液就发生破坏或变型。如果水相体积占总体积的26%~74%时，两种乳状液均可形成；若水相体积<26%，则只形成W/O型，若水相体积>74%，则只能形成O/W型。

（2）乳化剂分子构型。

根据"定向楔"理论，乳化剂截面小的一头总是指向分散相，截面大的一头总是伸向分散介质（图1-21）。因此，Cs^+、Na^+、K^+等一价金属离子的脂肪酸盐作为乳化剂时，容易形成O/W型乳状液；Ca^{2+}、Mg^{2+}、Al^{3+}、Zn^{2+}等高价金属皂则易生成W/O型乳状液。但也有例外，如Ag皂应为O/W型，实际上却得到的是W/O型。

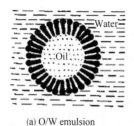
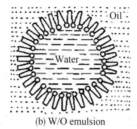

(a) O/W emulsion (b) W/O emulsion

Figure 1-21 Schematic of "oriented wedge" theory

(3) Hydrophilicity of emulsifier.

According to Bancroft rule, if emulsifier solubility in a certain phase is large, the phase will easily become an external phase. Generally speaking, emulsifier with strong hydrophilicity has an *HLB* value between 8 and 18, and is easy to form O/W emulsion; however, emulsifier with strong hydrophobicity has an *HLB* value between 3 and 6, and is easy to form W/O emulsion.

1.4.3 Properties of emulsion

Some physical properties of emulsion are taken as important basis for classifying emulsion type, measuring droplet size and studying its stability.

(1) Size and appearance of droplets.

The average droplet diameter of emulsion is in the range from 0.1 to 10μm. In fact, few droplets are with diameter less than 0.25μm. Different sizes of droplets will absorb and scatter differently from incident light, thereby showing different appearances. Droplet size of emulsion can be roughly estimated from appearances listed in Tables 1-6.

Tablet 1-6 Size and appearance of emulsion droplets

Droplet size /μm	Appearance	Droplet size /μm	Appearance
≫1	Could see two phases	0.05~0.1	Translucent grey
>1	Milky	<0.05	Transparent
0.1~1	Light blue		

(2) Viscosity.

From emulsion composition, it could be known that viscosity of the emulsion will be affected by external phase viscosity, internal phase viscosity, volume fraction of internal phase, property of emulsifier, droplet size, etc. Among them, viscosity of external phase plays a leading role, especially when volume fraction of internal phase is not large.

Emulsifiers are tending to increase emulsion viscosity greatly, this is mainly because emulsifier will enter oil phase leading to a formation of gel, or oil is solubilized in aqueous phase by micelles of emulsifier, etc.

(3) Conductivity.

Since conductivity of emulsion is determined by external phase, conductivity of O/W emulsion is much larger than that of W/O emulsion, which can be used as a basis for emulsion type identification. The conductivity method could also be used to measure water content in crude oil with low water content.

1.4.4　Identification of emulsion types

(1) Dilution method.

Dribble a few drops of emulsion into distilled water. If the drops disperse immediately, the emulsion shall be O/W; otherwise it shall be W/O.

(2) Staining method.

Add a few drops of water-soluble dye (such as methyl blue solution) into emulsion. If the emulsion is dyed uniformly, it shall be O/W. If only some droplets were dyed, the emulsion shall be W/O.

(3) Conductive method.

This is because O/W emulsions are with good conductivity but W/O emulsions not. However, some W/O emulsions may show good conductivity due to large aqueous phase volume fraction, or high content of ionic emulsifier in oil phase. Therefore, application of this method is with certain limitations.

（2）黏度。

从乳状液的组成可知，外相黏度、内相黏度、内相的体积分数、乳化剂性质、液滴大小等都会影响乳状液的黏度。其中，外相黏度起主导作用，特别是当内相体积分数不大时。

乳化剂往往会大大增加乳状液的黏度，这主要是因为乳化剂进入油相形成凝胶，或是水相中的乳化剂胶束增溶了油等。

（3）电导。

乳状液的导电性能决定于外相，故 O/W 型乳状液的电导率远大于 W/O 型乳状液，这可作为鉴别乳状液类型的依据。利用电导率也可测定含水率较低原油中的含水量。

1.4.4　乳状液类型的鉴别

（1）稀释法。

将数滴乳状液滴入蒸馏水中，若在水中立即散开则为 O/W 型乳状液，否则为 W/O 型乳状液。

（2）染色法。

往乳状液中加数滴水溶性染料（如甲基蓝溶液），若被染成均匀的蓝色，则为 O/W 型乳状液，若只是液珠带色，则为 W/O 型乳状液。

（3）导电法。

O/W 型乳状液的导电性好，W/O 型乳状液差。但有的 W/O 型乳状液水相体积分数很大，或油相中离子型乳化剂含量较多时也会有好的导电性。因此，使用该法有一定的局限性。

1.4.5 Emulsion stability theory

Stability here refers that the formulated emulsion will not be destroyed or changed into a variant under certain conditions. According to function of emulsifier, formation and stability of emulsion is determined by the following aspects: Reduction of interfacial tension, formation of interfacial film, establishment of diffused electric double layer, wetting and adsorption of solids and so on.

(1) Formation of interfacial film.

After surfactant is added into oil-water system, interfacial tension is reduced while adsorption will inevitably occurred on the interface, then an interfacial film is formed. Strength and tightness of the film are the determinants of emulsion stability. If adsorbed molecules arrangement of the interface film is tight and not easily desorbed, the interface film will have certain strength and viscoelasticity. Such interfacial film could protect dispersion phase droplets so that they will not easily aggregate and merge when colliding with each other.

When surfactant concentration is low, molecules adsorbed on interface are less and loosely arranged in the film. In this case strength of the film is poor, and emulsion formed is unstable. If surfactant concentration increased, strength of the film is enhanced, and stability of emulsion will also improve. In practice, it has been found that the composite film formed by mixed emulsifier has particularly high strength and will not break easily, which could form stable emulsion.

(2) Establishment of EDL.

Similar to sols, the emulsion droplets are with charges by ionization, adsorption and frictional contact. For emulsions, ionization and adsorption occur simultaneously. For example, when an anionic emulsifier is adsorbed at the interface, the polar group deep into water will make the droplets negatively charged due to

1.4.5 乳状液的稳定性理论

所谓稳定,是指所配制的乳状液在一定条件下,不破坏、不改变类型。根据乳化剂的作用,乳状液的形成、稳定原因可归纳为以下几个方面:界面张力的降低;界面膜的形成;扩散双电层的建立;固体的润湿吸附作用等。

(1)界面膜的形成。

在油-水体系中加入表面活性剂后,在降低界面张力的同时必然在界面发生吸附,形成界面膜。膜的强度和紧密程度是乳状液稳定的决定因素。若界面膜中吸附分子排列紧密,不易脱附,则膜具有一定的强度和黏弹性,对分散相液珠起保护作用,使其在相互碰撞时不易聚结合并。

当表面活性剂浓度较低时,界面上吸附的分子较少,膜中分子排列松散,膜的强度差,形成的乳状液不稳定。当表面活性剂的浓度增高时,膜的强度较好,乳状液的稳定性较好。实践中人们发现,混合乳化剂形成的复合膜具有相当高的强度,不易破裂,所形成的乳状液很稳定。

(2)扩散双电层的产生。

与溶胶相似,乳状液液滴通过电离、吸附和摩擦接触可带有电荷。对乳状液来说,电离和吸附是同时发生的。例如,阴离子乳化剂在界面上吸附时,深入水中的极性基团因电离而使液滴带负电,正电离子(Na^+、Ca^{2+}等)部分在其周围,形成双电层。液

ionization, and positive ion (Na$^+$, Ca^{2+}, etc.) is partially around it, forming an electric double layer. Repulsive effect of the electric double layer will prevent droplets to coalesce, which is helpful for stability of the emulsion.

(3) Wetting and adsorption of solids.

Many solid powders are also good emulsifiers. Like surfactant emulsifier, solid powder emulsifier only works as emulsifier when it is at the oil-water interface.

Whether solid powder stays in oil phase, water phase or two-phase interface is depending on wettability of the solid powder. Only when it could be wetted by both water and oil, it can stay on oil-water interface. The main reason why solid powder has stabilizing effect is that it can form a stable and firm interface film at the interface (Figure 1-22) and has certain Zeta potential. For oil-water systems, water-wet solids (such as Cu, Zn, Al and so on.) are emulsifiers for forming O/W emulsions, while oil-wet solids (such as carbon black, soot powder and rosin, etc.) are emulsifiers for forming W/O emulsions.

（3）固体的润湿吸附作用。

许多固体粉末也是良好的乳化剂。固体粉末乳化剂和表面活性剂乳化剂一样，只有当它处于油–水界面上时才能起到乳化剂的作用。

固体粉末处在油相、水相还是两相界面上，取决于固体粉末的润湿性。只有当它既能被水也能被油润湿时才能停留在油–水界面上。固体粉末的稳定作用主要在于它在界面上形成了稳定坚固的界面膜（图1-22）和具有一定的Zeta电位。对于油水体系，Cu、Zn、Al等水湿固体是形成O/W型乳状液的乳化剂，而炭黑、煤烟粉、松香等油湿固体是形成W/O型乳状液的乳化剂。

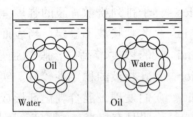

Figure 1-22 Schematic of solid powder emulsifiers at the oil-water interface

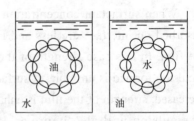

图1-22 固体粉末乳化剂作用示意图

Polymer chemistry 高分子化学 第2章

Polymer chemistry is a science that studies chemical synthesis, chemical reactions and properties of polymers and polymer materials. In the narrow sense, polymer chemistry refers to the synthesis and chemical reaction of polymeric compounds (also known as polymers). At present, polymers have been widely applied in petroleum production because they have the function of viscosity increasing, resistance reduction, suspending, fluidity control, water loss control and flocculation. Polymers have been widely used in acidification, fracturing, oil recovery and drilling; It is also very effective in stabilizing water quality and cementing; The polymer materials are widely used in the industry, such as phenolic resin, epoxy resin, furan resin and other resins used for sand control; Partially hydrolyzed polyacrylamide and its derivatives used for water plugging; Tackifier for viscosity increasing of the injected water; Stabilizer for improving stability of emulsions and foams; Coagulant used with coagulant for increasing water treatment speed; And engineering plastics used in place of metals.

2.1 Polymeric compounds

2.1.1 Basic concepts and characteristics of polymeric compounds

(1) Basic concepts.

Generally, if a compound is with its relative

高分子化学是研究高分子以及高分子材料的化学合成、化学反应与性能的一门科学。狭义的高分子化学，则是指高分子化合物（也称聚合物）的合成和化学反应。目前，聚合物在石油生产中已经得到广泛的应用，这是由于这些聚合物具有增黏、降阻、悬浮作用，控制流动性，控制失水和絮凝作用。聚合物已经广泛用于酸化、压裂、采油、钻井；在稳定水质和注水泥方面也很见成效。如防砂中所使用的酚醛树脂、环氧树脂、呋喃树脂等各种树脂；堵水用的部分水解的聚丙烯酰胺及其衍生物等；提高注入水黏度用的增黏剂；提高乳状液和泡沫稳定性用的稳定剂；配合混凝剂使用从而增加水处理速度的助凝剂；以及代替金属使用的工程塑料等都是高分子材料。

2.1 高分子化合物

2.1.1 高分子化合物的基本概念和特征

（1）基本概念。

通常将那些由众多原子或原

molecular weight more than 10,000 and is mainly formed by combining many atoms or atomic groups with covalent bonds, the compound could be called a polymer (also called macromolecular compound or high polymer). While the compound with relative molecular weight that less than 1000 is called low molecular compound. The compound with a relative molecular weight ranging between polymer and low molecular compound is called oligomer. Normally relative molecular mass of polymer is in the range from 10^4 to 10^6. The compound that relative molecular mass is larger than 10^6 is called ultra high molecular weight polymer.

(2) Characteristics of polymeric compounds.

Polymeric compounds have some characteristics different from organic low molecular compounds due to their high relative molecular mass and long chain structure:

① Polymeric compounds are almost no volatility, mainly in solid and liquid state at room temperature;

② Molecular chains are long and intertwined, and intermolecular forces are large, thus showing strong toughness and wear resistance;

③ Polymeric compounds are with many conformations (Figure 2-1) and often showing a certain degree of elasticity;

④ Dissolution process is very slow, sometimes only show swelling process; Solution viscosity is much larger than that of normal low molecular compound solution.

子团主要以共价键结合而成的相对分子质量高于10000的化合物称为高分子（又称聚合物或高聚物），相对分子质量低于1000的称为低分子。相对分子质量介于高分子和低分子之间的称为低聚物。一般高聚物的相对分子质量为 $10^4 \sim 10^6$，相对分子质量大于这个范围的称为超高相对分子质量聚合物。

（2）高分子化合物的特征。

高分子化合物由于其很高的相对分子质量与长链结构，具有某些不同于低分子有机化合物的特征：

①几乎无挥发性，常温下主要以固态和液态存在；

②分子链长且相互缠绕，分子间作用力大，从而表现出较强的韧性和耐磨性；

③构象多（图2-1），常显示出一定程度的弹性；

④溶解过程很慢，有时只发生溶胀，且溶液的黏度比一般低分子化合物溶液的大得多。

Figure 2-1　Schematic of a curled conformation for polyethylene

图2-1　聚乙烯的一种蜷曲构象

2.1.2 Synthesis and reaction of polymers

2.1.2.1 Synthesis of polymers

Polymers could be prepared by polymerizing of one or several low molecular weight molecules (i.e. monomers). Reaction of preparing polymers from low molecular weight molecules is called polymerization.

Polymerization reaction could be further divided into polyaddition and polycondensation.

(1) Polyaddition.

Polyaddition is a process in which many of the same (or different) low molecular weight molecules are synthesized into polymer without producing any new low molecules. The low molecular weight molecules reacted in polyaddition reaction is mostly unsaturated compounds, such as polyaddition of ethylene to polyethylene:

$$n\,CH_2=CH_2 \longrightarrow +CH_2-CH_2+_n$$

Since there is no precipitation of new low molecular weight molecules in polyaddition, chemical composition of polymer is same as the corresponding monomer. Polystyrene, polyvinyl chloride, polyacrylamide and other polymers are produced by polyaddition.

(2) Polycondensation.

Polycondensation is a reaction in which many of the same (or different) low molecular weight molecules are synthesized into polymer. In this reaction, new low molecular weight molecules (such as water, ammonia or hydrogen chloride) will be produced. For example, the reaction of synthesizing phenolic resin by phenol and formaldehyde is polycondensation:

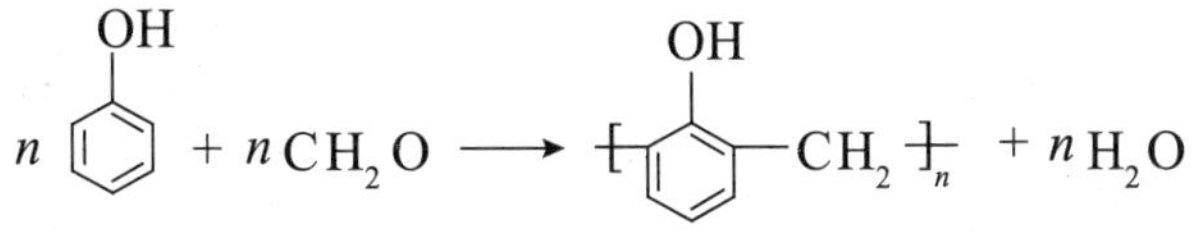

During the process of polycondensation, the polymer will lose a certain amount of low molecular

2.1.2 高分子的合成和反应

2.1.2.1 高分子化合物的合成

高分子可由一种或几种低分子（即单体）通过聚合反应生成。由低分子生成高分子的反应叫作聚合反应。

聚合反应又可分为加聚反应和缩聚反应。

（1）加聚反应。

加聚反应是由许多相同或不同的低分子化合为高分子但无任何新的低分子产生的过程。进行加聚反应的低分子多是不饱和化合物，例如乙烯聚合成聚乙烯的反应：

由于在加聚反应中没有新的低分子析出，所以高分子的化学组成与相应的单体相同。聚苯乙烯、聚氯乙烯、聚丙烯酰胺等高分子都是通过加聚反应生成的。

（2）缩聚反应。

缩聚反应是由许多相同或不相同的低分子化合为高分子但同时有新的低分子（例如水、氨或氯化氢）析出的过程。例如，苯酚和甲醛合成酚醛树脂的反应：

在进行缩聚反应时，高分子每引入一个重复结构单位就失去

weight molecules whenever it combines a repeating structural unit. Thus polymer and monomer molecules are with different chemical compositions. Polymers like polyethylene glycol, phenolic resin etc. are produced by polycondensation.

2.1.2.2 Chemical reactions of polymer

Chemical reactions of polymer include the chemical reaction process that functional groups on or between the molecular chains are converting into each other. According to degree of polymerization and change of groups, chemical reactions of polymer can be divided into reactions of polymer functional groups, reactions in which relative molecular weight of a polymer increase and reactions in which relative molecular weight of a polymer decrease. Researches on chemical reactions of polymer could modify natural polymers, synthesize new polymers and help to obtain polymers with special functions.

(1) Reactions of polymer functional groups.

A reaction in which side or end groups of polymer are changed but degree of polymerization is basically not is called functional group reaction of polymer, also known as similar transformation. New polymers could be prepared by combining some functional groups or elements into polymer molecular chains through group reaction. This method could also be used to change properties or applications of polymers. For example, when synthesizing sulfonated polyacrylamide, the hydroxymethylation reaction shall be carried out first for combining hydroxymethyl into polyacrylamide molecular chain, that is:

一定数量的低分子，因此高分子和单体分子具有不同的化学组成。聚乙二醇、酚醛树脂等高分子等都是缩聚反应产生的。

2.1.2.2 聚合物的化学反应

聚合物的化学反应包括聚合物分子链上或分子链间官能团相互转化的化学反应过程。聚合物的化学反应根据聚合物的聚合度和基团的变化可分为聚合物官能团反应、聚合物变大的反应及聚合物变小的反应。研究聚合物化学反应可以对天然高聚物进行改性，合成新的高聚物，有助于获得具有特殊功能的高聚物。

（1）聚合物的官能团反应。

聚合物的侧基或端基发生改变而聚合度基本不变的反应称为聚合物的官能团反应，又称为相似转变。通过基团反应，在高聚物分子链引入某些官能团或元素，制备新的高聚物或改变高聚物的性能与用途。例如，制备磺化聚丙烯酰胺，先进行羟甲基化反应，在聚丙烯酰胺分子链上引入羟甲基，即

$$\text{\textlbrackdbl} CH_2 - CH \text{\textrbrackdbl}_n + n\,HCHO \longrightarrow \text{\textlbrackdbl} CH_2 - CH \text{\textrbrackdbl}_n$$
$$\quad\quad\quad | \quad\quad\quad\quad\quad\quad\quad\quad\quad\quad\quad | $$
$$\quad\quad CONH_2 \quad\quad\quad\quad\quad\quad\quad\quad CONHCH_2OH$$

Then the sulfonation reaction shall be carried out to combine sulfonic group.

然后再进行磺化反应，引入磺酸基。

$$\{CH_2-CH\}_n + nNa_2SO_3 \longrightarrow \{CH_2-CH\}_n + n\,NaOH$$
$$\quad\quad\quad\;\; |\quad\quad\quad\quad\quad\quad\quad\quad\quad\quad\quad\quad |$$
$$\quad\quad CONHCH_2OH \quad\quad\quad\quad\quad\quad\quad CONHCH_2SO_3Na$$

Sulfonated polyacrylamide (SPAM) has three groups in its molecular chain, including —$CONH_2$, —$CONHCH_2OH$ and —CH_2SO_3Na. It is widely used in oil fields, textile printing and dyeing.

(2) Reactions in which relative molecular mass of a polymer increase.

Chemical reactions in which relative molecular weight of a polymer will increase includes crosslinking, grafting, etc.

Crosslinking of polymer refers to the reaction that under condition of heat, light, radiation or crosslinking agent, the molecular chains of polymer were connected into three dimensional network or spatial structure by chemical bonds. Crosslinking can greatly increase relative molecular weight, and is often used for improving material properties including strength, elasticity, hardness, deformation stability, and chemical resistance, etc. For example, in vulcanization reaction of rubber, reacting process between double bond of rubber molecule and vulcanizing agent is crosslinking. i.e.:

磺化聚丙烯酰胺（代号 SPAM）分子链上含有 —$CONH_2$、—$CONHCH_2OH$、—CH_2SO_3Na 三种基团，在油田、纺织印染等应用很广。

（2）聚合物变大的反应。

聚合度变大的化学转变包括交联、接枝等。

聚合物的交联是指大分子在热、光、辐射或交联剂作用下，分子链间通过化学键连接起来构成三维网状或体形结构的反应。交联可以使相对分子质量大大增加，常常可以提高材料的强度、弹性、硬度、变形稳定性、耐化学物质等性能。例如，橡胶的硫化反应就是利用橡胶分子中的双键与硫化剂作用而产生交联。即

$$\begin{array}{c}\sim\sim CH_2=CH_2\sim\sim\\ \sim\sim CH_2=CH_2\sim\sim\end{array} + 3S \longrightarrow \begin{array}{c}\vdots\\ S\\ \sim\sim CH-CH\sim\sim\\ |\\ S\\ \sim\sim CH-CH\sim\sim\\ |\\ S\\ \vdots\end{array}$$

Grafting of polymer refers to a reaction in which specific structures are attached to main chain of the polymer by chemical reaction and forming different branches. Result of this reaction is called a graft copolymer. Grafting reaction is also an important method for polymer modification. This is because attaching a side chain with special properties to the main

聚合物的接枝是指通过化学反应，在某种聚合物主链上接上特定的结构，组成不同的支链的反应，所形成的产物叫接枝共聚物。接枝反应也是聚合物改性的重要方法，如在主链上增加有特殊性质的侧链可改善聚合物的表

chain can improve surface, dyeing, flame retardancy and other properties of the polymer. For example, amino polystyrene (PS) is reacted with polymethyl methacrylate (PMMA) polymer containing pendant isocyanate groups to give a graft copolymer. i.e.:

面性质、染色性能、阻燃性能等。例如，氨基聚苯乙烯（PS）与含异氰酸酯侧基的聚甲基丙烯酸甲酯（PMMA）聚合物反应，得到接枝共聚物，即

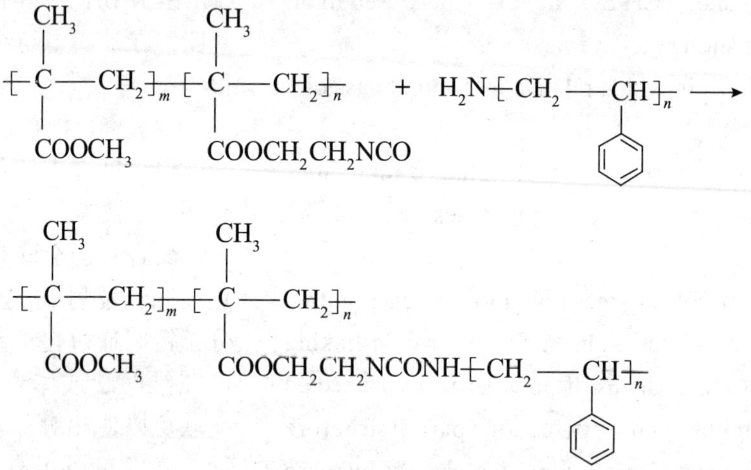

$$\begin{array}{c} CH_3 \\ {+}C{-}CH_2{+}_m \\ COOCH_3 \end{array} \begin{array}{c} CH_3 \\ {+}C{-}CH_2{+}_n \\ COOCH_2CH_2NCO \end{array} + H_2N{+}CH_2{-}CH{+}_n\text{-Ph} \longrightarrow$$

$$\begin{array}{c} CH_3 \\ {+}C{-}CH_2{+}_m \\ COOCH_3 \end{array} \begin{array}{c} CH_3 \\ {+}C{-}CH_2{+}_n \\ COOCH_2CH_2NCONH{+}CH_2{-}CH{+}_n\text{-Ph} \end{array}$$

(3) Degradation of polymer.

Degradation of polymer refers to a process in which the bond of molecular chains or between main chain and functional groups broke under physical, chemical or biological factors. Degradation is the general term for chemical reactions in which relative molecular weight of a polymer decrease, including depolymerization, random chain breaking, removal of side groups and low molecular weight substances, and so on.

Degradation conducted under physical factors such as light, heat, vibration and agitation is called physical degradation; Conducted under air, acid, alkali, oxidant or other similar factors is called chemical degradation; Conducted under microbial enzymes is known as biodegradation. For example, the conversion of starch to glucose, i.e.:

（3）聚合物的降解。

聚合物的降解是指在物理、化学或生物因素作用下。聚合物分子链或官能团与主链连接键发生断裂的过程。降解是聚合物相对分子质量变小的化学反应的总称，包括解聚、无规则断链、侧基和低分子物质脱除等反应。

在光、热、振动、搅拌等物理因素作用下的降解叫物理降解；在空气、酸、碱、氧化剂等因素作用下的降解叫化学降解；在微生物酶作用下的降解称为生物降解。例如，淀粉转化为葡萄糖的反应，即

$$[C_6H_{10}O_5]_n \xrightarrow{H_2O} [C_6H_{10}O_5] \xrightarrow{H_2O} C_{12}H_{22}O_{11} \xrightarrow{H_2O} C_6H_{12}O_6$$
　　Starch　　　　　Dextrin　　　　Malt dust　　　　Glucose

$$[C_6H_{10}O_5]_n \xrightarrow{H_2O} [C_6H_{10}O_5] \xrightarrow{H_2O} C_{12}H_{22}O_{11} \xrightarrow{H_2O} C_6H_{12}O_6$$
　　淀粉　　　　　糊精　　　　　麦芽糖　　　　葡萄糖

Degradation will greatly reduce relative molecular weight of the polymer, resulting in the polymer property missing. Sometimes degradation will be used for certain purpose, such as recycling waste polymer monomers, or treating "three wastes" (waste gas, waste water and waste residues) by biodegradation.

2.1.3　Classification of polymeric compounds

For easy selection, polymeric compounds can be classified by structure, property or source.

(1) Classified by structure.

According to structure, polymeric compounds can be classified into linear polymer and three-dimensional polymer (Figure 2-2). Since some linear polymers are with branches but others are not, linear polymer can be further divided into straight chain linear polymer and branched chain linear polymer.

聚合物降解反应会使高聚物相对分子质量大大降低，失去高分子的性质。有时是有目的地利用聚合物的降解，例如，利用聚合物降解反应回收废聚合物单体，利用生物降解进行三废处理等。

2.1.3　高分子化合物的分类

为了便于选用高分子，可按结构、性质或来源对高分子进行分类。

（1）按高分子结构分类。

若按结构，高分子可分为线型高分子和体型高分子（图2-2）。由于线型高分子中有些有支链，有些无支链，故线型高分子可进一步分为直链线型高分子和支链线型高分子两类。

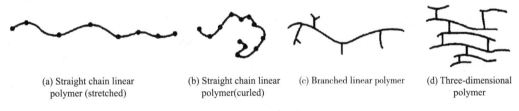

(a) Straight chain linear polymer (stretched)　　(b) Straight chain linear polymer(curled)　　(c) Branched linear polymer　　(d) Three-dimensional polymer

Figure 2-2　Schematic of polymer structure

（a）直链线型（伸直）　　（b）直链线型（蜷曲）　　（c）支链线型　　（d）交联体型

图2-2　高分子结构示意图

(2) Classified by property.

According to different properties, polymeric compounds could be put in different classifications. For example, according to solubility, polymeric compounds could be classified into water soluble type, oil soluble type and oil and water insoluble type; According to property of being heated, they could be divided into

（2）按高分子性质分类。

按不同的性质，高分子有不同的分类。例如，按溶解性分，高分子可分为水溶性高分子、油溶性高分子和油水都不溶的高分子；按对热的性质分，高分子可分为热塑性高分子和热固性高分

thermoplastic type and thermosetting type. Molecular chains of the thermoplastic type are linear structure or structure with branches and there is no chemical bond between the chains. This type will flow after being heated and solidify after cooling. The flow-solidify process could be repeated. Most of plastics used in our daily life belong to this category. Thermosetting type is generally three-dimensional polymer. After heated and solidified, they will not melt or dissolve in solvents in case of being reheated, such as phenolic resins, vulcanized rubber, etc.

(3) Classified by source.

According to the source, polymeric compounds could be divided into biochemical polymer, natural polymer and synthetic polymer. Biochemical polymer is obtained by biochemical methods (e.g. xanthan gum is obtained by bacteria fermentation and can be used as oil displacement agent). Natural polymer is from the nature (such as sesbania gum and guar gum that commonly used in oilfield fracturing) and this category also includes related modified compounds. The most widely used synthetic polymeric compounds in oil fields are polyacrylamide and its derivatives, phenolic resins and so on.

2.2　Solution of polymer

Macromolecule material is generally used as solution, which is mainly aqueous. Water soluble macromolecule materials are widely used in oil recovery due to functions including viscosity increasing, resistance reduction, suspension, fluidity control and so on. In oilfield chemistry, the macromolecule materials used are often called polymers, and this book is no exception.

Polymer solution is a binary or ternary dispersion system formed by dissolving macromolecule substance in low molecular weight solvent. Historically, polymer solutions have been mistakenly regarded as colloidal

子。热塑性高分子分子链都是线型或带支链的结构，分子链之间无化学键产生，加热后可以流动，冷后固化，并可反复进行，日常生活中使用的大部分塑料属于这个范畴。热固性高分子一般是体型聚合物，加热固化后，再加热时不熔化也不溶于溶剂，如酚醛树脂、硫化橡胶等。

（3）按高分子的来源分类。

若按来源，高分子可分为生化高分子、天然高分子和合成高分子三类。生化高分子是由生物化学方法（如黄胞胶是由细菌发酵而得，可用作驱油剂）得到。天然高分子来自自然界（如油田压裂改造中常用的田菁胶和胍胶），此外还应包括它的改性产物。油田上用得最多的合成高分子是聚丙烯酰胺及其衍生物、酚醛树脂等。

2.2　聚合物溶液

高分子一般配成溶液使用，主要是配成水溶液使用。水溶性高分子在采油中得到广泛的应用，这是由于这些高分子具有增黏、降阻、悬浮、控制流度等作用。油田化学中将所用的高分子常称为聚合物，本书也采用此习惯称谓。

聚合物溶液是由高分子物质溶解于低分子溶剂所形成的二元或三元分散体系。历史上人们曾经将高分子溶液当成胶体溶液。

solutions. This misunderstanding is caused by the similar behaviors between polymer solution and colloidal solution due to similarity of polymer molecular volume and colloidal particles. After repeated research, it was finally known that the polymer solution is a true solution in which single solute molecules are dispersed in the solvent.

Properties of low concentration polymer solution are quite different from high concentration polymer solution. It is generally considered that solution with less than 5% polymer mass fraction is low concentration solution; above 5% will be high concentration solution. According to this judgment, most of polymer solutions used in oilfield drilling, completion, fracturing, and oil recovery shall be classified as dilute solutions.

2.2.1 Dissolution process of polymers

Dissolution process of low molecular weight substances is relatively simple. When low molecular weight solid is put into solvent as solute, due to thermal motion and solvent solvation, molecules or ions on its surface will overcome attraction of molecules or ions inside the solute and gradually leave surface of the solid solute and diffuse. These molecules or ions are then dispersed into the solvent by diffusion to form a homogeneous solution. However, since molecular volume accumulated inside polymer is much larger than that of low molecular weight substance, and there is more interaction between molecules, which made dissolution process much more complicated.

(1) Dissolution and swelling.

Compared with low molecular weight substances, the most remarkable feature of polymers is dissolution process with slow speed. This slow process includes two stages: swelling and dissolution. This is because when polymer is in contact with the solvent, molecular chain segments on its surface will be firstly solvated.

经过反复研究，终于认识到高分子溶液是以单个溶质分子分散于溶剂中的一种真溶液。只是由于高分子物质的分子体积和胶体粒子接近，使得高分子溶液和胶体溶液的某些行为有相似之处。

聚合物低浓度溶液的性质与高浓度溶液截然不同。一般认为聚合物质量分数低于5%的溶液为稀溶液，在此以上者为浓溶液。以此为标准，在油田钻井、完井、压裂以及采油中使用的聚合物溶液大多应归属于稀溶液范围。

2.2.1 聚合物的溶解过程

当把低分子固态溶质投入溶剂中时，溶质表面的分子或离子由于本身的热运动和溶剂的溶剂化作用，克服了溶质内部分子或离子的吸引力而逐渐离开固态溶质表面，并通过扩散作用分散到溶剂中，成为均匀的溶液。因而，低分子物质的溶解过程较为简单。但对于高分子物质来说，由于在其内部聚集的分子体积比低相对分子质量物质大得多，分子之间的作用力也不止一种，因而其溶解过程要复杂得多。

（1）溶解和溶胀。

与低分子物质相比，高分子物质的溶解过程最显著的特点是其溶解速度很慢，包括溶胀和溶解两个阶段。这是因为当高分子物质与溶剂接触时，它与溶剂接触的外表面上的分子链段最先被溶剂化。但因整个分子链很长，

However, since the entire molecular chain is quite long, the inner chain segment could not be solvated and dissolved quickly. In this process, on the one hand, the segments on surface of polymer are solvated by solvent molecules, on the other hand, thermal movement of polymer chain segments are causing solvent diffusion in interior of the polymer, which will result in a gradual solvation of internal chain segments. This volume expansion phenomenon of polymer caused by solvation of molecular chain segments is called "swelling". As solvent molecules continue to diffuse into solute interior, more polymer chain segments will become loose. After polymer chain segments on surface are completely solvated and left, a new surface will appear. The solvent will then work on the new surface until all molecular chains are dissolved, which means the whole dissolve process is over. Therefore swelling is the first step of dissolution and is a peculiar phenomenon in polymer dissolution process.

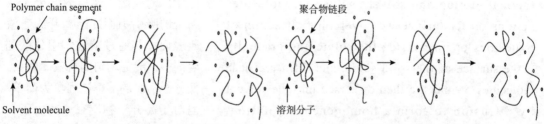

Figure 2-3　Schematic of polymer dissolution process

(2) Principle of solvent selection.

It has been proved that polymer solubility depends not only on molecular structure but also on type of solvent selected to a large extent. Although there is no mature theory for polymer dissolution, some dissolution theories of low molecular weight substance are also applicable to polymers.

Principle of polarity similarity: Polymers with polar molecular structure will be dissolved in polar solvents, polymers with non-polar molecular structure will be dissolved in non-polar solvents, and polymers with strong polar molecular structure will be dissolved in strong polar solvents. This is the principle of polarity similarity. For example, polymers with polar group on the molecular chain such as polyacrylamide (PAM), polyacrylic acid (PAA), and xanthan gum (XC) could be dissolved in polar water but could not be dissolved in ester which is non-polar. Non-polar natural rubber could be dissolved in non-polar hydrocarbons, but could not be dissolved in water which is polar.

Principle of solubility parameter similarity: There is interaction energy between molecules of all substances. And substance molecules including solutes and solvents could maintain a certain aggregation state with this energy, which is called cohesive energy. Obviously, in order to disperse solute molecules in the solvent, it is necessary to make interaction energy between solute and the solvent larger than cohesive energy between molecules inside the solute or in solvent itself. Solubility parameter is a measurement of interaction energy between molecules within substance. Table 2-1 shows solubility parameters of several polymers, and Table 2-2 shows solubility parameters of several solvents.

解于非极性溶剂中，强极性分子结构的高分子物质溶解于强极性溶剂中，这就是极性相似原则。例如，分子链上带有极性基团的聚丙烯酰胺（PAM）、聚丙烯酸（PAA）以及黄胞胶（XC）均可溶解于极性的水而不溶解于非极性的酯。非极性的天然橡胶可溶解于非极性的烃而不溶解于极性的水。

溶解度参数相近原则：任何物质内部分子之间都存在着作用能，物质分子正是通过这种作用能而保持一定的聚集状态。溶质是这样，溶剂也是这样，这种能被称为内聚能。显然，要使溶质分子分散于溶剂中，就必须使溶质与溶剂之间的作用能大到足以克服溶质或溶剂内部分子之间的内聚能。物质的溶解度参数就是物质内部分子间作用能的一种量度。表2-1中列出了几种聚合物的溶解度参数，表2-2中列出了几种溶剂的溶解度参数。

Table 2-1 Solubility parameters of several polymers

Polymer	$\delta/(J/m^3)^{1/2}$	Polymer	$\delta/(J/m^3)^{1/2}$
Polyacrylamide	23.1	Polyvinyl chloride	19.2-22.1
Polyacrylonitrile	25.6-31.5	Polyacrylamide	16.8-18.8
Polyacrylic acid	29.1	Polystyrene	17.4-19.0
Methylene polyacrylamide	30.8	Styrene-butadiene rubber	16.6-17.8
Hydroxymethyl polyacrylamide	34.9	Chloroprene rubber	16.8-18.9
Cellulose diacetate	23.3	Epoxy resin	19.8
Polyvinyl alcohol	25.8-29.1	Polyethylene	15.8-17.1

表2-1 几种聚合物的溶解度参数

聚合物	$\delta/(J/m^3)^{1/2}$	聚合物	$\delta/(J/m^3)^{1/2}$
聚丙烯酰胺	23.1	聚氯乙烯	19.2~22.1
聚丙烯腈	25.6~31.5	聚丙烯	16.8~18.8
聚丙烯酸	29.1	聚苯乙烯	17.4~19.0
甲叉基聚丙烯酰胺	30.8	丁苯橡胶	16.6~17.8
羟甲基聚丙烯酰胺	34.9	氯丁橡胶	16.8~18.9
二醋酸纤维素	23.3	环氧树脂	19.8
聚乙烯醇	25.8~29.1	聚乙烯	15.8~17.1

Table 2-2 Solubility parameters of several solvents

Solvent	$\delta/(J/m^3)^{1/2}$	Solvent	$\delta/(J/m^3)^{1/2}$
Benzene	18.7	Water	47.4
Toluene	18.2	Ethylene glycol	32.1
O-xylene	18.4	Methanol	29.6
M-xylene	18	Ethanol	26
P-xylene	17.9	Acetic acid	25.7
Chloroform	19	Acetone	29.4
Carbon tetrachloride	17.6	Chloroethanol	27.6

表2-2 几种溶剂的溶解度参数

溶剂	$\delta/(J/m^3)^{1/2}$	溶剂	$\delta/(J/m^3)^{1/2}$
苯	18.7	水	47.4
甲苯	18.2	乙二醇	32.1
邻二甲苯	18.4	甲醇	29.6
间二甲苯	18	乙醇	26
对二甲苯	17.9	乙酸	25.7
氯仿	19	丙酮	29.4
四氯化碳	17.6	氯乙醇	27.6

According to principle of solubility parameter similarity, when solubility parameter of polymer solute and that of solvent is the same, molecule chain of polymer could fully expand and diffuse in the solvent, and viscosity of the solution reaches to the largest. In this case, the solvent is a good solvent for the polymer.

Principle of solvation: When solute is in contact with solvent, solvent molecules will exert a force on molecules of the solute. If this force is greater than cohesion between solute molecules, the solute molecules will be separated from each other and dissolved in the solvent. Polar solvent molecules and polar groups of polymer solute are able to attract each other and resulting in solvation. This is mainly relying on combination between acidic groups or alkaline groups on polymer molecular chains of solute and alkaline groups or acidic groups of the solvent. Acidic and alkaline mentioned here are concepts in broadly sense. Acidic refers to electron acceptor (i.e., the electrophile), and alkaline is electron donor (i.e., the nucleophile).

The above three principles of solvent selection

根据溶解度参数相近原则，当聚合物溶质的溶解度参数和溶剂的溶解度参数相等时，高分子链在溶剂中能够充分伸展、扩散，溶液的黏度最大，该溶剂就是该聚合物的良溶剂。

溶剂化原则：当溶质与溶剂接触时，溶剂分子对溶质分子产生作用力，当此作用力大于溶质内部分子的内聚力时，溶质分子便彼此分离而溶解于溶剂中。极性溶剂分子和聚合物溶质的极性基团相互吸引能够产生溶剂化作用，主要是聚合物溶质分子链上的酸性基团或碱性基团能够与溶剂中的碱性基团或酸性基团起作用的结果。这里所指的酸和碱是广义的，酸是指电子接受体（即亲电子体），碱是电子给予体（即亲核体）。

上述溶剂的选择三原则是通

are empirical rules that summarized from different perspectives by dissolution experiments. They are interrelated with each other, but with many exceptions. Therefore, in order to select the best solvent, the three principles should be applied comprehensively in combination with necessary dissolution experiments.

2.2.2　Viscosity of polymer solution

(1) Definition of fluid viscosity.

The fluids (liquids, gases) are with different degrees of viscosity. When adjacent two layers are moving at different speeds, there will be friction between the layers. The fast moving layer will accelerate the slow moving layer and the slow moving layer has a retarding effect on the fast moving layer. This property of fluid is called viscosity. Viscosity of fluid is the evidence of internal frictional resistance caused by movement of the fluid molecules. According to Newton's law of flow, viscosity of liquid can be defined as the ratio of shear stress to shear rate. Its unit is $(N/m^2)\cdot s$ (poise) or $Pa \cdot s$.

(2) Measurement of fluid viscosity.

There are many methods for fluid viscosity measurement. The common methods are capillary viscometer method (Figure 2-4) and rotational viscometer method (Figure 2-5). Rotational viscometer is the essential equipment for polymer solution viscosity measurement.

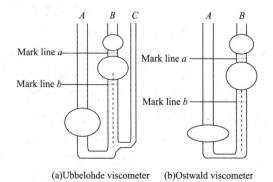

(a) Ubbelohde viscometer　　(b) Ostwald viscometer

Figure 2-4　Schematic of capillary viscometer structure

2.2.2　聚合物溶液的黏度

（1）流体黏度的定义。

各种流体（液体、气体）都具有不同程度的黏性，当其相邻两流层各以不同速度运动时，层间就有摩擦力产生，运动快的流层对运动慢的流层有加速作用，运动慢的流层对运动快的流层有阻滞作用，流体的这种性质称为黏性。流体的黏度是由于流体分子之间受到运动的影响而产生内摩擦阻力的表现。根据牛顿流动定律，液体的黏度可定义为剪切应力与剪切速率的比值，其单位为$(N/m^2)\cdot s$（泊）或$Pa \cdot s$（帕·秒）。

（2）流体黏度的测量。

有许多测定流体黏度方法，实验室测定流体黏度的方法有毛细管黏度计法（图2-4）和旋转黏度计法（图2-5）。旋转黏度计是测定聚合物溶液黏度的必备仪器。

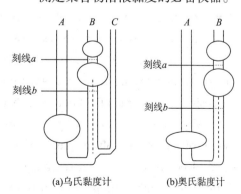

(a) 乌氏黏度计　　(b) 奥氏黏度计

图2-4　毛细管黏度计结构示意图

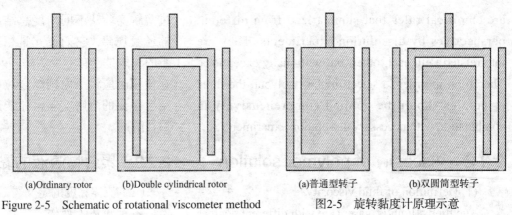

(a)Ordinary rotor (b)Double cylindrical rotor (a)普通型转子 (b)双圆筒型转子

Figure 2-5 Schematic of rotational viscometer method   图2-5 旋转黏度计原理示意

2.2.3 Rheological properties of polymer solutions

It is well known that elastic deformation (strain) will occur when solids are subjected to force (stress). If relationship between the stress applied to a solid and the strain produced is in accordance with Hooke's law, such a solid is called an elastomer. Shear deformation will occur when fluids were subjected to forces, and degree of the deformation will vary according to viscosity of the fluid. Therefore, fluids are also called viscous substances. According to different fluid viscosity, quantitative relationship between the shear stress applied to the fluid and the shear deformation rate (shear rate) is also different. Rheology is the science that studying the relationship between shear stress and shear rate during flow of fluids. Relationship between the shear stress and the shear rate is called rheological characteristic of the fluid. Rheology is the core of polymer flooding theory.

In case that the ratio of shear stress to shear rate is always a constant, that is, the shear stress is proportional to the shear rate, the fluid is a Newtonian fluid. The viscosity of Newtonian fluid does not change with shear rate. Most low molecular weight liquids such as pure water, oil, inorganic solvents, and so on are Newtonian fluids. Rheological characteristic diagram of Newtonian fluid is a straight line passing through origin of the coordinates (Figure 2-6).

2.2.3 聚合物溶液的流变学特性

众所周知,当固体受到作用力(应力)后会产生弹性形变(应变)。如果一种固体所受应力与所产生应变的关系符合虎克定律,这样的固体称为弹性体。而流体受到作用力后会产生剪切形变,变形程度随流体黏度的大小而不同,因此流体又称黏性体。流体黏性不同,施加于流体上的剪切应力与剪切变形率(剪切速率)之间的定量关系也不同。流变学就是研究流体流动过程中剪切应力与剪切速率变化关系的科学。流体所表现出的这种剪切应力与剪切速率的变化关系称为流体的流变学特征。流变学是聚合物驱油理论的核心内容。

当剪切应力与剪切速率的比值始终为一常数时,即剪切应力与剪切速率成正比关系,或者说黏度不随剪切速率而改变的流体称为牛顿流体。大多数低分子液体如纯水、油、无机溶剂等都属于牛顿流体。牛顿流体的流变特征曲线是一条通过坐标原点的直线(图2-6)。

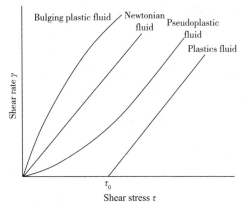

Figure 2-6 Four basic flow patterns rotational viscometer method

However, for some other liquids it is a different situation. Viscosity of these liquids will change with the shear rate. When the shear rate is changed, relationship between shear stress and shear rate deviates greatly from Newton's law of flow. These fluids are called non-Newtonian fluids. In order to distinguish from Newtonian fluid viscosity, the viscosity which changes with shear rate is called apparent viscosity. Drilling fluids, completion fluids, fracturing fluids, and polymer oil displacement agents applied in petroleum industry are all non-Newtonian fluids. Non-Newtonian fluids can be divided into the following three categories:

（1）Fluids that the relationship between shear stress and shear rate is independent with shear time.

These fluids include Bingham plastics fluids, pseudoplastic fluids and bulging plastic fluids (Fig. 2-6).

（2）Fluids that the relationship between shear stress and shear rate is related with shear time.

These fluids include thixotropic fluids (Figure 2-7) and rheopectic fluids (Figure 2-8).

（3）Viscoelastic fluid.

Viscoelastic fluids are fluids that have rheological properties of both viscous substance and elastomer under certain conditions. This fluid has the ability to restore original elastic characteristics from deformation caused by flow process. If the deformation is severe, recovery of this elasticity may be only partial. The fluid is usually used in drag reduction of liquid in pipes, because the polymer in this fluid can cause the inhibition of turbulence, which indicates obvious drag reduction effect.

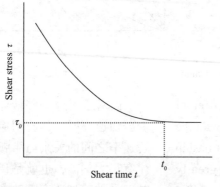

Figure 2-7   Rheological curve of thixotropic fluid

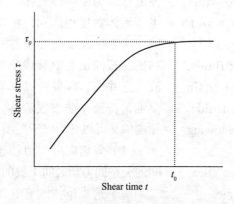

Figure 2-8   Rheological curve of rheopectic fluid

Since non-Newtonian fluid viscosity values strongly depend on shear rate, whenever referring to viscosity value of a non-Newtonian fluid, the corresponding shear rate must be indicated, otherwise the viscosity value is meaningless. Because of this, during apparent viscosity measurement of non-Newtonian fluid, it is necessary to use a rotational viscometer since it can optionally set shear rates as needed.

Rheological curve of typical polymer HPAM (partially hydrolyzed polyacrylamide) solution in porous media (Figure 2-9) can be divided into

有弹性特征的能力，如果形变很剧烈，这种弹性的恢复也可能只是局部的。这种流体中的高分子聚合物对紊流有抑制作用，具有明显的减阻效应，常被应用于管道输送的减阻上。

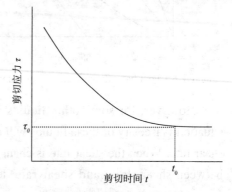

图2-7   触变性流体流变曲线

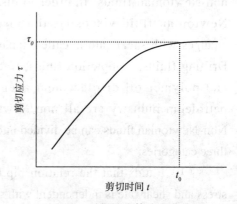

图2-8   震凝性流体流变曲线

由于非牛顿流体黏度值都强烈地依赖于剪切速率，因而当说到非牛顿流体的黏度时，必须同时指明是在何种剪切速率下的黏度，否则黏度就没有意义。也正是由于这种剪切依赖性，在测定非牛顿流体的表观黏度时，必须使用可以根据需要随意设定剪切速率的旋转黏度计。

典型聚合物 HPAM（部分水解聚丙烯酰胺）溶液在孔隙介质

five zones, namely zero shear zone (first Newton zone), pseudoplastic zone, limit shear zone (second Newton zone), bulging plastic zone and degradation zone.

中的流变曲线（图2-9）可分为5个区，即零剪切区（第一牛顿区）、假塑性区、极限剪切区（第二牛顿区）、胀流区和降解区。

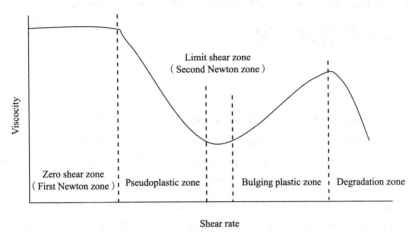

Figure 2-9　Rheological curve of HPAM solution in porous media

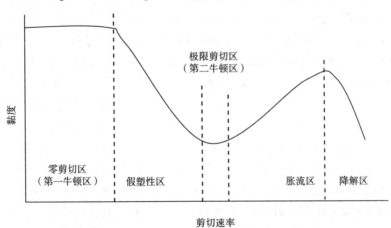

图2-9　HPAM溶液在孔隙介质中的流变曲线

第3章 钻井液化学
Chemistry and chemicals of drilling fluids

According to American Petroleum Institute (API), a drilling fluid is defined as a circulating fluid used in rotary drilling to perform any or all of the various functions required in drilling operations. Drilling fluids are mixtures of natural and synthetic chemical compounds used to cool and lubricate the drill bit, clean the hole bottom, carry cuttings to the surface, control formation pressure and improve the function of the drill string and tools in the hole. The type of fluid base used depends on needs of drilling and formation, as well as the requirements for disposition of the fluid after it is no longer needed.

Chemistry of drilling fluid is the study of compositions and properties of drilling fluids, and chemical methods of controlling and adjusting them, in order to obtain high quality, fast, safe and economic drilling operations. Therefore, it is necessary to understand the type, structure, performance and mechanism of various drilling fluid additives, which are the main contents of drilling fluid chemistry.

3.1 Basic knowledge of clay minerals

Anyone concerned with drilling fluids technology should have a good basic knowledge of clay mineralogy, as clay provides the colloidal base of nearly all aqueous muds, and is also modified for use in oil-based drilling fluids. Drill cuttings from argillaceous formations become incorporated in any drilling fluid, and can

按照美国石油学会(API)的定义，钻井液是一种在钻井过程中能满足多种功能需求的可循环流体。钻井液通常为天然和人工合成化学品的混合物，具有冷却和润滑钻头、清理井底、将岩屑携至地表、平衡地层压力以及改善井下钻杆和钻具的功能等作用。在生产过程中，可以根据钻具、储层以及钻井完成后的后处理需要而确定所需的钻井液体系。

钻井液化学主要是研究钻井液的组成、性能及其控制与调整，达到优质、快速、安全、经济钻井的目的。因此，必须了解各种钻井液处理剂的类型、结构、性能及其作用机理，这是钻井液化学的主要内容。

3.1 黏土矿物基础

黏土矿物结构和性质是钻井液技术的重要基础。黏土矿物不仅为几乎所有的水基钻井液提供了所需要的胶体性质，而且还可以通过表面改性分散于油基钻井液中。来自泥质地层的钻屑混入钻井液，并显著地改变其性质。井眼的稳定性在很大程度上取决

profoundly change its properties. The stability of the borehole depends to a large extent on interactions between the drilling fluid and exposed shale formations. The development of "inhibitive" aqueous drilling fluids was initiated to control clay hydration and swelling. In nonaqueous fluids, the introduction of water-wet solids can upset the properties of the fluid. The interactions between the mud filtrate, whether water or oil, and the clays present in producing horizons may restrict productivity of the well.

3.1.1 The basic structure of clay minerals

Most clay types have a mica-like structure. Their flakes are composed of tiny crystal platelets, normally stacked together face-to-face. A single platelet is called a unit layer, and consists of octahedral sheet and silica tetrahedral sheet.

3.1.1.1 Silica tetrahedra and its sheet

Silica tetrahedra consists of one silicon atom being coordinated with four oxygen atoms, as shown in Figure 3-1(a). The sheet of tetrahedron on the base forms a hexagonal network of oxygen atoms of indefinite areal extent with ordered arrangement, as shown in Figure 3-1(b) and Figure 3-1(c).

3.1.1 黏土矿物的基本构造

大多数黏土具有云母型结构。它们由微小的晶片组成，通常面对面地重叠在一起。单个血小板称为单位层，由硅氧四面体片和铝氧八面体片组成。

3.1.1.1 硅氧四面体与四面体片

硅氧四面体由一个硅原子等距离地配上四个比它大得多的氧原子构成，硅原子在四面体中心，四个氧原子在四面体顶点［图3-1（a）］。硅氧四面体在平面上有序排列可形成六方网格的连续结构，又称为硅氧四面体片［图3-1（b）、图3-1（c）］。

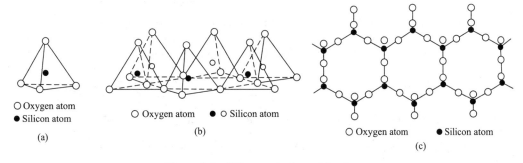

Figure 3-1 Silica tetrahedra and its sheet

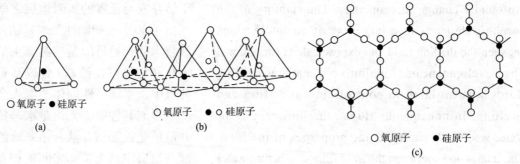

图3-1 硅氧四面体与四面体片

3.1.1.2　Aluminum oxide octahedra and its sheet

An octahedral is made up of either aluminum or magnesium atoms in octahedral coordination with oxygen atoms (or hydroxyls) as shown in Figure 3-2(a). The ordered arrangement of octahedron on the base form a octahedral sheet, as shown in Figure 3-2(b). (If the metal atoms are aluminum, the structure is the same as the mineral gibbsite, $Al_2(OH)_6$. In this case, only two out of three possible sites in the structure can be filled with the metal atom, so the sheet is termed dioctahedra. If, on the other hand, the metal atoms are magnesium, the structure is that of brucite, $Mg_3(OH)_6$. In this case all three sites are filled with the metal atom, and the structure is termed as trioctahedral.)

3.1.1.2　铝氧八面体与八面体片

铝氧八面体由一个铝原子与六个氧原子(或羟基)配位而成[图3-2（a）]。多个铝氧八面体在平面上有序排列可形成铝氧八面体片[图3-2（b）]。(如果金属原子是铝，则是叶蜡石结构$Al_2(OH)_6$，这时，这种结构中三个可能的位置中仅有两个被金属原子填充，被称为二八面体片；如果，金属原子是镁原子，则是水镁石结构$Mg_3(OH)_6$，这时，三个位置都被金属原子填充，被称为三八面体片。)

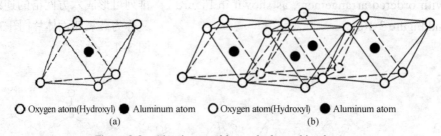

Figure 3-2　Aluminum oxide octahedra and its sheet

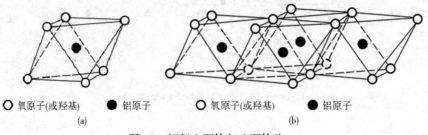

图3-2　铝氧八面体与八面体片

3.1.1.3 The crystalline layer of clay minerals

The sheets of silica tetrahedra and alumina octahedral, which are the basic structure of clay minerals, are tied together by sharing common oxygen atoms, which is called unit layer, and the phyllosilicates are further formed. Clay minerals can be divided into two categories according to the proportion of silica tetrahedral and aluminum octahedral sheets:

(1) 1:1 layer structure consists of the repetition of one tetrahedral and one octahedral sheet, and kaolinite is composed of such unit layers.

(2) 2:1 layer structure that one octahedral sheet is sandwiched between two tetrahedral sheets, and smectite and illite are composed of such unit layers.

3.1.2 Common clay minerals

Clay minerals include smectite, kaolinite, illite, chlorite, palygorskite, sepiolite, etc., among which the former three are the most common.

Smectite is one of 2:1 layer clay minerals, which contains water and some exchangeable cations between the crystal structure layers (Figure 3-3), and has a high capacity of water absorption and expansion. In nature, the main component of bentonite is montmorillonite, which is by far the best known member of the smectite group. The First Grade bentonite is mainly sodium montmorillonite, and Secondary Grade bentonite is mainly calcium montmorillonite. They can be used as a thickener or viscosifier for water-based drilling fluids.

Illite is also one of 2:1 layer clay minerals, similar to smectite. The difference between illite and smectite is that the crystalline layer is more tightly bound and the water sensitivity is weaker. However, due to microfractures, in the process of drilling in the shale formation with high illite content, wellbore instability often occurs, such as sloughing, hard and brittle collapse. The inhibitive drilling fluid with enhanced sealing performance should be chosen in order to keep wellbore stability.

Kaolinite is one of 1:1 layer clay minerals (Figure 3-4). One tetrahedral sheet is tied to one octahedral in the usual manner, so that the octahedral hydroxyls on the face of one layer are juxtaposed to tetrahedral oxygens on the face of the next layer. In consequence, there is strong hydrogen bonding between the layers, which prevent lattice expansion. There is little substitutions, and very few cations are adsorbed on the basal surfaces.

塌等井壁失稳现象，需要采用强封堵的抑制性防塌钻井液。

高岭石属于1:1层型晶体结构(图3-4)。一个四面体片以常见的方式连接到一个八面体上，这样一层表面上的八面体的羟基与下一层表面上的四面体氧并置。因此，层与层之间有很强的氢键，阻止了晶格的膨胀。晶格取代很少，吸附在表面的阳离子也很少。

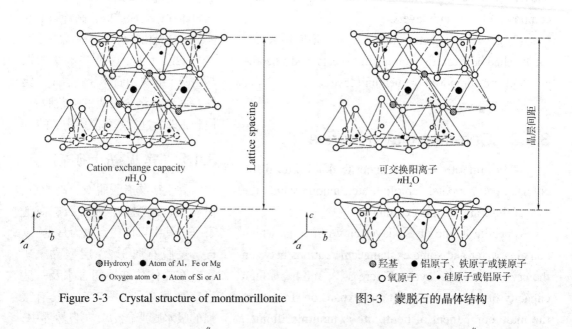

Figure 3-3　Crystal structure of montmorillonite　　图3-3　蒙脱石的晶体结构

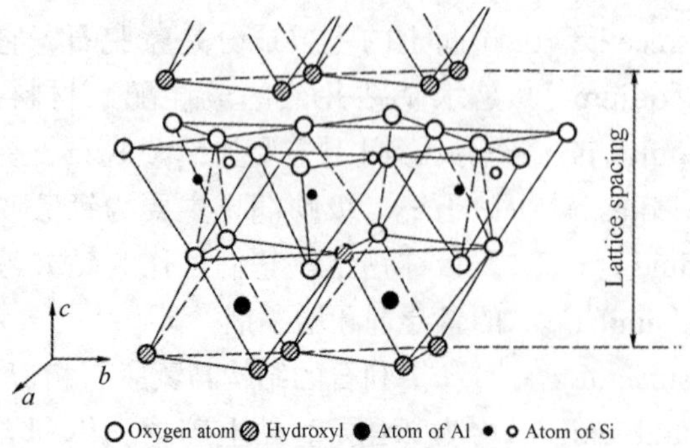

Figure 3-4　Crystal structure of kaolinite 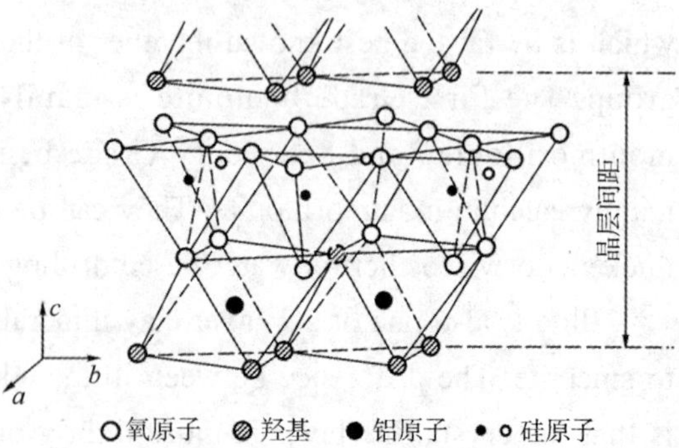　　图3-4　高岭石的晶体结构

The properties of three clay minerals are shown in table 3-1.

上述三种黏土矿物的特点见表3-1。

Table 3–1 Characteristics of crystal structure and physicochemical properties of three clay minerals

Mineral name	Layer structure	lattice spacing /nm	Layer attraction	Cation Exchange Capacity / (mmol/100g)	Swelling Capacity /%
Kaolinite	1:1	0.72	Hydrogen bonding, strong	3–5	<5
Montmorillonite	2:1	0.96–4.0	intermolecular forces, weak	70–130	90–100
Illite	2:1	1.0	attraction, strong	20–40	2.5

表3–1 三种黏土矿物的晶体构造和物理化学性质的特点

矿物名称	晶 型	晶层间距/nm	晶层间引力	阳离子交换容量/(mmol/100g)	膨胀量/%
高岭石	1:1	0.72	氢键力，引力强	3~5	<5
蒙脱石	2:1	0.96~4.0	分子间力，引力弱	70~130	90~100
伊利石	2:1	1.0	引力较强	20~40	2.5

3.1.3 Properties of Clay Minerals

There are many physical and chemical properties of clay minerals, and the main ones closely related to the properties of drilling fluid are surface electrical properties, adsorption properties, expansibility and aggregation behavior. The first two properties have been described in the theory of colloidal chemistry.

3.1.3.1 Expansibility of clay minerals

Expansibility is the property of clay minerals that increase in volume after contact with water.

The expansibility of clay minerals varies widely. According to the crystal structure, clay minerals can be divided into expansive and non-expansive. In addition, the type of exchangeable cation has great influence on the swelling of clay minerals.

Smectite belongs to expansive clay minerals, and its expansibility is caused by large amount of exchangeable cations. When it comes into contact with water, water can enter the interlamellar between the crystal layers, so that the exchangeable cation can be dissociated, and a diffusion double layer is established on the surface of the crystal layer, resulting in electronegativity. The interlamellar electronegativity is repulsive to each other, which results in the increase of interlamellar spacing and the expansion of smectite.

3.1.3 黏土矿物的性质

黏土矿物的物理化学性质特点较多，与钻井液性能密切相关的主要有表面电性、吸附性、膨胀性和凝聚性。前两个性质在胶体化学理论部分已有叙述，本节主要介绍后两个性质。

3.1.3.1 膨胀性

膨胀性是指黏土矿物与水接触后体积增大的特性。

黏土矿物的膨胀性有很大的不同。根据晶体结构，黏土矿物可分为膨胀型黏土矿物和非膨胀型黏土矿物。同时，黏土矿物中的可交换阳离子类型对黏土矿物的膨胀有很大的影响。

蒙脱石属于膨胀型黏土矿物，它的膨胀性是由于其有大量的可交换阳离子所产生的。当它与水接触时，水可进入晶层间，使可交换阳离子解离，在晶层表面建立了扩散双电层，从而产生负电性。晶层间负电性互相排斥，引起晶层间距加大，使蒙脱石表现出膨胀性。一般而言，高价阳离

In general, high-valent cations are more easily absorbed than low-valent cations. Therefore, smectite containing low-valent exchangeable cations is more likely to expand than those containing high-valent exchangeable cations, as shown in Figure 3-5.

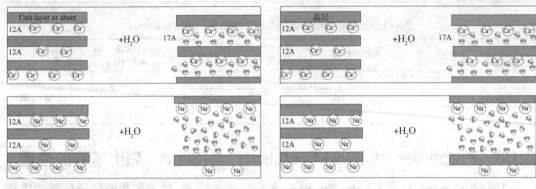

Figure 3-5 Hydration Expansion Diagram of Calcium Montmorillonite (Upper) and Sodium Montmorillonite (Lower)

Kaolinite, illite and chlorite belong to non-expansive clay minerals. For Kaolinite, it only has a small amount of lattice substitution, and there are hydrogen bonds between layers; For illite, its lattice substitution occurs in siloxtetrahedral sheets, and the intercrystalline layers exchange cations are potassium ions, which can also make the crystalline layers closely connected; Chlorite is because the presence of hydrogen bonds in its crystal layers and the unbalanced electricity caused by lattice substitution by compensation for exchangeable cation of the substitution of brucite.

3.1.3.2 The Aggregation Behavior of Clay Particles in suspension

The aggregation behavior refers to the associative state and property of clay mineral particles in aqueous suspension under certain conditions. The certain condition here mainly refers to a certain concentration of electrolytes (such as sodium chloride, calcium chloride, etc.).The associative state of clay mineral particles in aqueous suspension mainly depends on the resultant of repulsion force (electronegative repulsion force, etc.) and gravitation force (van der Waals gravitation, etc.) between clay particles. The associations of clay particles have significant influences on the viscosity, yield point, filtration loss and other properties of drilling fluid.

  Clay particle powder (solid powder aggregation state) is hydrated and dispersed in water. Due to the repulsive force between the particles, a stable suspension is formed, showing the dispersion state (or deflocculation state).With the increase of electrolyte concentration, the diffusion double-layer on the surface of clay mineral particles is compressed, the electrical properties on the edges and faces are reduced, and the gravity is increased, resulting in edge-edge and edge-face association state. When the electrolyte concentration exceeds a certain level, the face-face aggregation of clay mineral particles (complete flocculation and aggregation) can be caused （Figure 3-6）.

力（范德华引力等）的总和，黏土颗粒的凝聚状态对钻井液的黏度、屈服值（动切力）和滤失量等性能有显著影响。

  黏土颗粒粉末（固体粉末聚集态）在水中水化分散，由于颗粒间的斥力占主要作用，形成稳定的悬浮液，呈分散状态（或解絮凝态）；由于随着电解质浓度增加，黏土矿物颗粒表面的扩散双电层被压缩，边、面上的电性减小，引力增加，黏土颗粒发生边－边和边－面的联结（絮凝态）。当电解质超过一定浓度时，就可引起黏土矿物颗粒发生面－面的联结（完全絮凝和聚集态）（图3-6）。

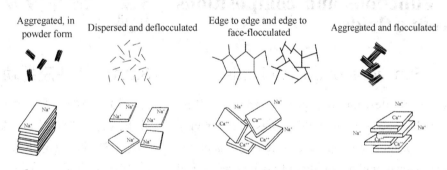

Figure 3-6　Models of particle association of clay

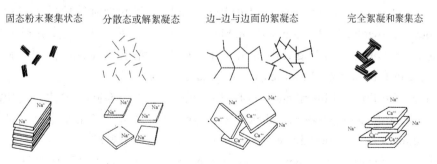

图3-6　水中黏土颗粒凝聚特性示意图

Formation mineralogy can affect the performances of drilling fluids through the aggregation of clay mineral particles, and the chemical additives can adjust the properties of drilling fluids by improving aggregation of clay mineral particles. For example, calcium contamination (drilling fluids are contaminated by Ca^{2+} in the formation fluids) can cause small flaky particles of clay mineral to be connected by edge-edge and edge-face, resulting in spatial networking and increasing viscosity. Further contamination of calcium can lead to face-face association of clay mineral particles, resulting in reduced viscosity. The addition of viscosity reducer (also known as dispersant) can improve the surface electronegativity of clay minerals and increase the thickness of hydration layer through its adsorption, thus causing the re-dispersion of clay minerals and restoring the performance of drilling fluid.

地层因素可通过黏土矿物的凝聚性影响钻井液性能，处理剂又可通过黏土矿物的凝聚性调整钻井液的性能。例如，地层中的Ca^{2+}侵入钻井液（称为钙侵）可引起黏土矿物小片边－边联结、边－面联结，产生空间结构，导致黏度增加。进一步钙侵，又可引起黏土矿物颗粒的面－面联结，导致黏度减小。降黏剂（又称分散剂）的加入，可通过它的吸附提高黏土矿物表面的负电性并增加水化层的厚度，进而引起黏土矿物小片的重新分散，恢复钻井液的使用性能。

3.2 Functions and compositions of drilling fluids

3.2 钻井液功能与组成

3.2.1 Functions of drilling fluids

3.2.1 钻井液的功能

The principle functions performed by drilling fluid include controlling downhole pressure, carrying cuttings, suspending cuttings and weighting materials, which are maintaining the stability of open borehole and transmitting hydraulic force. Such functions of drilling fluid are closely related to properties of drilling fluid such as density, rheology, filtration and inhibition. The associated functions of drilling fluid are to cool and lubricate the drill bit and drill tool respectively, and to assist in the collection and interpretation of information available from drill cuttings, cores, and electrical logs.

In conjunction with the above functions, certain limitations—or negative requirements—are placed on the drill fluid. The fluid should:

(1) Not injure drilling personnel nor be

钻井液的主要功能包括控制井下压力、携带岩屑、悬浮岩屑和加重材料、稳定井壁、传递水动力五个方面，钻井液的这类功能与钻井液的密度、流变、滤失和抑制等性质密切相关。钻井液两个附带功能分别是冷却和润滑钻头和钻具、保障获得井眼信息。

连同上述钻井液功能，也同样带来了某些限制，对钻井液有以下要求：

（1）既不伤害钻井作业人员也不破坏和影响环境。

（2）不需特殊或昂贵的方法

Chemistry and chemicals of drilling fluids
钻井液化学

damaging or offensive to the environment.

（2）Not require unusual or expensive methods to complete and produce the drilled hole.

（3）Not interfere with the normal productivity of the fluid-bearing formation.

（4）Not corrode or cause excessive wear of drilling equipment.

3.2.2 compositions of drilling fluids

All drilling fluids systems are composed of:

（1）Base fluids—Water, Nonaqueous, Pneumatic.

（2）Solids—Active and Inactive (inert).

（3）Additives to maintain the properties of the system.

Additives in a drilling fluid are used to control one or more of the properties measured by the drilling fluids specialist. These properties can be classified as controlling:

①Mud weight—Specific Gravity (Density).

②Viscosity—Thickening, Thinning, Rheology Modifiction.

③Fluid loss—API Filtrate, Seepage, Lost Circulation, Wellbore Strengthening.

④Chemical reactivity—Alkalinity, pH, Lubrication, Shale Stability, Clay Inhibition, Flocculation, Contamination Control, Interfacial/Surface Activity, Emulsification.

The chemistry of the additive is entirely determined by the base fluid used to make the drilling fluid. Figure 3-7—Figure 3-9 show most of the additives used to control each of the above functions for water, brine, and nonaqueous base fluid.

来完成钻孔。

（3）不干扰流体承载地层的正常产能。

（4）不会腐蚀或导致钻井设备过度磨损。

3.2.2 钻井液的组成

所有钻井液体系都由以下几部分组成：

（1）基础流体——水、油（包括柴油、矿物油、非石油烃类等）、气体。

（2）固相——活性固相和惰性固相。

（3）保持钻井液体系性能的添加剂。

钻井液添加剂主要用于控制一种或多种钻井液性能。这些性能包括：

①钻井液密度——比重或密度。

②黏度——增稠，稀释，流型调节。

③滤失——API滤失，渗漏，漏失，井眼强化。

④化学反应性——碱度，pH，润滑，页岩稳定，黏土抑制，絮凝，污染控制，表/界面活性，乳化等。

应根据配制钻井液所选择不同基液，选择相应的化学处理剂。图3-7~图3-9显示了用于淡水、盐水和油基/合成基等基液的上述各项功能的常用处理剂。

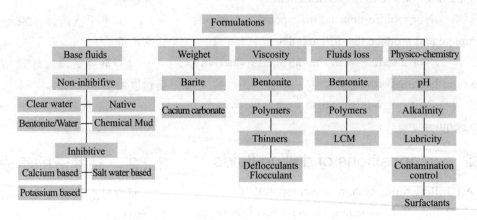

Figure 3-7　Additives for water-based drilling fluids

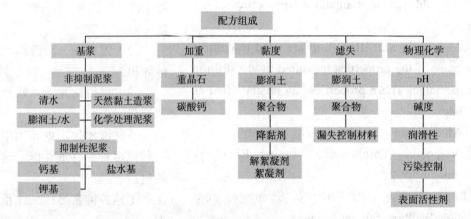

图3-7　水基钻井液配方组成

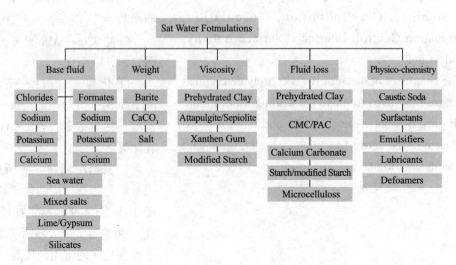

Figure 3-8　Additives for brine-based drilling fluids

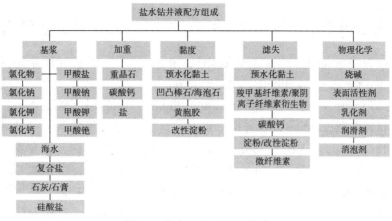

图3-8 盐水钻井液配方组成

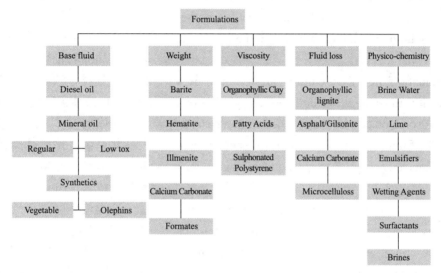

Figure 3-9 Additives for non-aqueous drilling fluids

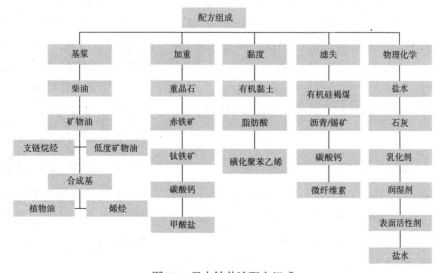

图3-9 无水钻井液配方组成

3.3 Drilling fluids systems

A drilling fluid or mud is a specially designed fluid that is circulated through a wellbore as the wellbore is being drilled to facilitate the drilling operation. Fluid systems are classified according to their base fluid.

Water-Based Muds (WBM). Solid particles are suspended in water or brine. Oil may be emulsified in the water, in which case water is termed as the continuous phase.

Nonaqueous-Based Drilling Fluids (NADF). Solid particles are suspended in oil. Brine water or another low-activity liquid is emulsified in the oil, i.e., oil is the continuous phase.

Pneumatic fluids. Drill cuttings are removed by a high-velocity stream of air, natural gas, nitrogen, carbon dioxide, or some other fluid that is injected in a gaseous phase. When minor inflows of water are encountered, these systems are converted to gas/liquid systems—mist or foam.

Over the years a considerable number of drilling fluid formulations have been developed to suit various subsurface conditions. The best fluid that can meet anticipated conditions will minimize well costs and reduce the risk of catastrophes, such as wellbore instability, stuck drill pipe, loss of circulation, and gas kicks. Consideration must also be given to obtaining adequate formation evaluation and maximum productivity.

These drilling fluids systems are explained below.

3.3.1 Water–based muds

In general, water is taken as the continuous phase in WBMs. Such mud typically comprise an aqueous base, either of fresh water or brine, and agents or additives for suspension, weight or density, oil-wetting,

3.3 常见钻井液体系

钻井液体系是指一般地层和特殊地层（如岩盐层、石膏层、页岩层、高温层、煤层、地热等）钻井用的各类钻井液。钻井液体系通常按分散介质（连续相）分成水基、油基和气基三大类。

水基钻井液：淡水或盐水为连续相，固相颗粒和添加剂分散和溶解在淡水和盐水中，以及油被乳化分散在水中。

油基/合成基钻井液：柴油、矿物油和非石油烃为连续相，固相颗粒悬浮在油相中，盐水或其他低活度液体被乳化在油相中。

气体型流体：以气态形式注入钻井系统，钻屑被高速气流携带，气体可以是空气、氮气、二氧化碳或其他气体流体。当有水侵入这类流体时，形成气-液混合系统即雾化或泡沫流体。

多年来，为适应不同的地下条件，开发了许多钻井液配方。选择满足钻遇条件的最佳钻井液可以降低钻井成本和灾难性风险（包括井壁垮塌、卡钻、漏失和气侵等）。必须从足够的地层评价信息和产能最大化等方面考虑。

以下将分别介绍上述钻井液体系。

3.3.1 水基钻井液

水基钻井液是以水作分散介质的钻井液，由水、膨润土和处理剂配成。

Chemistry and chemicals of drilling fluids

fluid loss or filtration control, and rheology control.

(1) Freshwater drilling fluid.

Freshwater drilling fluids range from clear water having no additives to high-density drilling fluids containing clays, barite, and various organic additives. The composition of the drilling fluid is determined by the type of formation to be drilled. When a viscous fluid is required, clays or water soluble polymers shall be added. Fresh water is ideal for formulating stable drilling fluid compositions, as many mud additives are most effective in a system of low ionic strength. Inorganic and/or organic additives could control the rheological behavior of the clays, particularly at elevated temperature. Water-swellable and water-soluble polymers and/or clays may be used for filtration control. The pH of the mud is generally alkaline and, in fact, viscosity control agents like montmorillonite clays are more efficient at a pH more than 9. Sodium hydroxide and sodium carbonate are by far the most widely used alkalinity control agent.

(2) Low-solids nondispersed polymer drilling fluids.

The so called nondispersed polymer drilling fluids (LSND systems) is composed by fresh water, clay, and polymers for viscosity enhancement and filtration control. Low-solids muds are maintained using minimal amounts of clay and require removal of all but modest quantities of drill solids. Low-solids muds can be weighted to high densities but are used primarily in the unweighted state. The main advantage of these systems is the high drilling rate that can be achieved because of the lower colloidal solids content. High- and low-molecular-weight long-chain polymers are used in these systems to provide the desired rheology and fluid-loss control, especially xanthan has proven to be an effective solids suspending agent. Low-colloidal solids are encapsulated and flocculated for more efficient removal at the surface, which in turn decreases dilution

（1）淡水钻井液。

淡水钻井液包括从无添加剂的清水到含有膨润土、重晶石和各种有机添加剂的高密度钻井泥浆。钻井液的组成由所钻地层的类型决定。当需要增加钻井液黏度时，加入黏土或水溶性聚合物。由于许多钻井液添加剂在低离子强度体系中最有效，淡水是配制稳定钻井液成分的理想选择。特别是在高温下，无机和/或有机添加剂控制黏土的流变性。亲水膨胀和水溶性聚合物和/或黏土可用于滤失控制。钻井液pH值通常是碱性的，事实上，像蒙脱土这样的黏度控制剂在pH值大于9时更有效。氢氧化钠和碳酸钠是目前应用最广泛的碱度控制剂。在淡水泥浆中加入不水溶的加重材料，使其达到控制地层压力所需的密度。

（2）低固相不分散聚合物钻井液。

用于提高黏度和滤失控制的淡水、黏土和聚合物组成了低固相和所谓的非分散聚合物钻井液。低固相钻井液用最小量的黏土维护并且需要清除劣质固相。低固相钻井液通常在未加重的状态下使用，其密度可增加到高密度。这类钻井液的主要优点是由于膨润土含量较低，可以达到较高的机械钻速。高分子量和低分子量的天然或合成聚合物被用来提供所需的流变性和滤失性，特别是黄原胶已被证明是一种有效的固体悬浮剂。劣质固相通常被包被絮凝以便于在地面清除，从而降

requirements. Specially developed high-temperature polymers are available to help overcome gelation issues that might occur on high-pressure, high-temperature (HP/HT) wells. With proper treatment, some LSND systems can be weighted to 17.0 to 18.0 ppg and run at 350°F and higher. These low-solids muds are normally applied in hard formations where the penetration rate increase can reduce drilling costs significantly and the tendency for solids buildup is minimal.

(3) Salt water drilling fluid.

In many drilling areas both onshore and offshore, salt beds or salt domes are penetrated. Saturated salt drilling fluids are used to reduce the hole enlargement that would result from formation-salt dissolution through contact with an undersaturated liquid. Salt formations are primarily made up of sodium chloride, and also may be composed of mixed salts, predominantly magnesium and potassium chlorides. It has become quite common to use high (20%~23% NaCl) salt drilling fluids in wells being drilled. The reasons are twofold: stabilization of water-sensitive shales and inhibition of the formation of gas hydrates. The high salinity of salt water muds may require different clays and organic additives from those used in fresh or seawater drilling fluids. Salt water clays and organic polymers contribute to viscosity. The filtration properties are adjusted using starch or cellulosic polymers. The pH ranges from that of the makeup brine, which may be somewhat acidic, to 9-11 by the use of sodium hydroxide or lime.

(4) Potassium-treated drilling fluids.

Generally potassium-treated systems combine one or more polymers and a potassium ion source, primarily potassium chloride, in order to prevent problems associated with drilling certain water-sensitive shales. The flow and filtration properties may be quite different from those of the other water-base fluids. Potassium muds have been applied in most active drilling regions around the world. Environmental regulations in the

低了稀释比例。特殊研制的抗高温聚合物可用于克服高温高压条件下的钻井液胶凝稠化。通过恰当的处理，低固相钻井液可以在176℃（350°F）或更高的温度下使用。低固相钻井液通常应用于硬地层，在硬地层中，增加钻速可以显著降低钻井成本，且固相堆积趋势最小。

（3）盐水钻井液。

在陆地和海洋的许多钻井区域，经常钻遇盐层或盐穹窿。饱和盐泥浆是用来减少由于地层盐溶解与不饱和液体接触而引起的孔洞扩大。盐层主要由氯化钠组成，也可能由混合盐组成，主要是氯化镁和氯化钾。在此类地层钻井中使用高盐钻井液（氯化钠含量20%~23%）已经变得相当普遍。其原因有两方面：稳定水敏页岩和抑制天然气水合物的形成。高盐度的盐水钻井液可能需要不同于淡水或海水钻井液的黏土和有机添加剂。抗盐的黏土和有机聚合物有助于提高黏度。滤失性能的调整使用淀粉或纤维素类聚合物。使用氢氧化钠或石灰调整pH值，使偏酸性的浓盐水pH值到9~11。

（4）钾盐钻井液。

通常钾盐钻井液含有一种或多种聚合物和钾离子，主要是氯化钾，以防止与钻遇水敏感页岩相关的井下复杂。其流动和滤失性能可能与其他水基钻井液有很大不同。钾盐钻井液已得到非常广泛的应用。但美国的环境法规限制了在近海钻井中使用钾质泥

US have limited the use of potassium muds in offshore drilling owing to the apparent toxicity of high potassium levels in the bioassay test required by discharge permits.

(5) High performance or enhanced performance water-based muds (HPWBM).

High performance or enhanced performance water-based muds are formulated to solve drilling problems such as wellbore stability, solids inhibition, wellbore strengthening, drilling rate enhancement, HTHP stability, and reservoir drill-in fluids (RDIF). Most high performance fluids are formulated with a brine base fluid and polymers for rheology and fluid loss control. Depending on the end use of purpose, the fluid additional surfactants, wellbore strengthening materials, antiaccretion agents, film formers, etc. could be added. Some high-performance water-based drilling fluids can reduce the transmission of formation pore pressure as well as oil-based drilling fluids. High-performance water-based drilling fluids can be used in areas where the environment is sensitive, waste disposal costs are high, and the application of oil-based and synthetic based drilling fluids is limited.

3.3.2 Non-aqueous drilling fluids (NADF)

(1) Oil-based drilling fluids.

Oil-based drilling fluids (OBFs) were developed and introduced in the 1960s to help to solve several drilling problems: formation clays that react, swell, or slough after exposure to WBFs; increasing downhole temperatures; contaminants; and stuck pipe and torque and drag.

Oil-based fluids used today are formulated by diesel, mineral oil, or low-toxicity linear paraffins (that are refined from crude oil). The emulsifier is used to emulsify water phase into oil phase to form an oil-in-water emulsion. The electrical stability of the internal brine or water phase is monitored to ensure strength of the emulsion is maintained at or near a predetermined value. The emulsion should be stable enough to incorporate additional water volume if a downhole water flow is encountered.

Barite is used to increase system density, and specially-treated organophilic bentonite is the primary viscosifier in most oil-based systems. The emulsified water phase also contributes to fluid viscosity. Organophilic lignitic materials are added to help control low-pressure/low-temperature (LP/LT) and HP/HT fluid loss. Oil-wetting is essential for ensuring that particulate materials remain in suspension; the surfactants used for oil-wetting also can work as thinners. Oil-based systems usually contain lime to maintain an elevated pH, resist adverse effects of hydrogen sulfide (H_2S) and carbon dioxide (CO_2) gases, and enhance emulsion stability.

Shale inhibition is one of the key benefits of using an oil-based system. The high-salinity-water phase helps to prevent shales from hydrating, swelling, and sloughing into the wellbore. Most conventional oil-based mud (OBM) systems are formulated with calcium-chloride brine, which appears to offer the best inhibition properties for most shales.

The ratio of the oil percentage to the water percentage in an oil-based system is called its oil/water ratio. Oil-based systems generally function well with an oil/water ratio in range from 65/35 to 95/5, but the most commonly observed range is from 80/20 to 90/10.

(2)Synthetic-based drilling fluids.

Synthetic-based fluids were developed out of an increasing desire to reduce the environmental impact of offshore drilling operations, but without sacrificing the cost-effectiveness of oil-based systems. Like traditional OBFs, SBFs help maximize ROPs, increase lubricity in directional and horizontal wells, and minimize wellbore-stability problems such as those caused by reactive shales. Field data gathered since the early 1990s confirm that SBFs could provide exceptional drilling

地层水的侵入和污染，油包水乳状液应足够的稳定，以乳化额外的水量。

重晶石用于增加钻井液的密度，经过特殊处理的有机土是油基钻井液的主要增黏剂。乳化水相液滴也有助于提高钻井液黏度。添加亲油的氧化沥青或褐煤，帮助控制低压/低温（LP/LT）和高温高压（HP/HT）滤失量。亲油润湿性对于确保颗粒材料保持悬浮状态是必不可少的；用于油润湿的表面活性剂也可以作为稀释剂。油基钻井液通常含有石灰以保持较高的pH值，抵抗硫化氢（H_2S）和二氧化碳（CO_2）气体的不利影响，并提高乳化液的稳定性。

泥页岩抑制性是油基钻井液突出的优点。高浓度盐水有助于防止泥页岩水化、膨胀或坍塌。大部分的常规油基钻井液都是用氯化钙盐水配制的。

油基钻井液中油相体积与水相体积的比例被称为油水比。油基钻井液的油水比一般在65/35~95/5，但最常见的范围是80/20~90/10。

（2）合成基钻井液。

合成基钻井液的开发是为了减少钻井作业对海洋环境的影响，并且不降低油基钻井液的成本效益。与常规油基钻井液一样，合成基钻井液有助于最大限度地提高机械钻速、提高定向井和水平井的润滑性，并最大限度地减少由活性页岩造的井眼稳定性问题。自20世纪90年代初收集的现场

performance, similar with that of diesel- and mineral-oil-based fluids.

In many offshore areas, regulations that prohibit the discharge of cuttings drilled with OBFs do not apply to some of the synthetic-based systems. SBFs' cost per barrel can be higher, but they have proved economical in many offshore applications for the same reasons that traditional OBFs have: fast penetration rates and less mud-related nonproductive time (NPT). SBFs that are formulated with linear alphaolefins (LAO) and isomerized olefins (IO) exhibit the lower kinematic viscosities that are required in response to the increasing importance of viscosity issues as operators move into deeper waters. Early ester-based systems exhibited high kinematic viscosity, a condition that is magnified in the cold temperatures encountered in deepwater risers. However, a shorter-chain-length (C_8), low-viscosity ester that was developed in 2000 exhibits viscosity similar to or lower than that of the other base fluids, specifically the heavily used IO systems. Because of their high biodegradability and low toxicity, esters are universally recognized as the best base fluid for environmental performance.

Until operators began drilling in deepwater locations, where the pore pressure/fracture gradient (PP/FG) margin is very narrow and mile-long risers are not uncommon, the standard synthetic formulations provided satisfactory performance. However, the issues that arose because of deepwater drilling and changing environmental regulations prompted a closer examination of several seemingly essential additives.

When cold temperatures are encountered, conventional SBFs might develop undesirably high viscosities as a result of the organophilic clay and lignitic additives in the system. The introduction of SBFs formulated with zero or minimal additions of organophilic clay and lignitic products allowed rheological and fluid-loss properties to be controlled

数据证实，合成基钻井液具有优异的钻井性能，与柴油和矿物油基钻井液相媲美。

在许多地区尤其是海上作业区，禁止油基钻井液及其岩屑排放的规定不适用于合成基钻井液。合成基钻井液的每桶成本更高，但事实证明，在许多海上应用中，其经济性与常规油基钻井液的经济性是相同的，原因是：机械钻速快，与钻井液相关的非生产时间少。由线性α-烯烃（LAO）和异构化烯烃（IO）配制的合成基钻井液具有较低的运动黏度，这与深水区域作业时较低的温度相适应。酯类钻井液表现出高运动黏度，在深水立管遇到的低温中被放大。然而，2000年研制了较短链长（C_8）低黏度酯基基础油的黏度与其他基液（特别是大量使用的IO系统）相似或更低，由于其高生物降解性和低毒性，酯被公认为环境性能最佳的基液。

随着作业者在深水领域开展钻井作业，由于孔隙压力/破裂压力梯度越来越窄并且英里级长度的立管也不常见，标准化的合成基钻井液配方提供了所需的性能。然而，由于深水钻井和不断变化的环境法规引起的问题促使人们对一些重要的处理剂进行更为细致的研究。

低温条件下，常规合成基钻井液中有机土、亲油氧化沥青和褐煤类产品造成了不被期望的高黏度。此类产品低加量甚至零添加的合成基钻井液被研发出来，这类钻井液的性能特点包括具有

through the fluid-emulsion characteristics. The performance advantages of these systems include high, flat gel strengths that break with minimal initiation pressure; significantly lower equivalent circulating densities (ECDs); and reduced mud losses while drilling, running casing, and cementing.

(3) All-oil fluids.

Normally, the high-salinity water phase of an invert-emulsion fluid helps to stabilize reactive shale and prevent swelling; however, drilling fluids that are formulated with diesel or syntheticbased oil and without water phase are used to drill long shale intervals where the salinity of the formation water is highly variable. By eliminating the water phase, the all-oil drilling fluid can preserve shale stability throughout the interval.

3.3.3 Pneumatic fluids

Also called air or gas drilling, in which one or more compressors are used to clean cuttings from the hole. Pneumatic fluids can be divided into three categories: air or gas only, aerated fluid, or foam. In its simplest form, a pneumatic drilling fluid is a dry gas, either air, natural gas, nitrogen, or carbon dioxide, etc. The dry gas drilling is best suited for formations that have a minimal amount of water. Once water is encountered while drilling, the dry gas system is converted by the addition of a surfactant and called a "mist" system. Increasing amounts of water in the formation require additional amounts of foaming surfactants and polymer foam stabilizers. Pneumatic-drilling operations require specialized equipment to help ensure safe management of the cuttings and formation fluids that return to surface, as well as tanks, compressors, lines, and valves associated with the gas used for drilling or aerating the drilling fluid or foam.

Except when drilling through high-pressure hydrocarbon or fluid-laden formations that demand a high-density fluid to prevent well-control issues,

高且平缓的凝胶强度，并且用较低的启动压力便可打破，可以显著地降低当量循环密度，在钻井、下套管和固井作业中降低钻井液损失。

（3）全油基钻井液。

通常情况下，油包水钻井液中的高矿化度水相有助于稳定反应性泥页岩并防止膨胀；然而，无水的柴油或合成基钻井液被用于钻穿地层水矿化度变化大的长段泥页岩地层。由于没有水相，全油基钻井液可以保持整个井段的泥页岩稳定。

3.3.3 气体钻井流体

也被称为空气或气体钻井，一个或多个压缩机被用来清理井筒的岩屑。气体钻井流体可分为三种类型：纯空气或气体、充气流体或泡沫流体。最简单的形式是，气动钻井液是一种干气，可以是空气、天然气、氮气、二氧化碳等。干气钻井最适合于含水量最少的地层。一旦在钻井过程中遇到水，干气系统就会通过添加表面活性剂来转换，这被称为"雾化"系统。在地层中增加的水量需要额外量的发泡表面活性剂和聚合物泡沫稳定剂。气动钻井作业需要专门的设备来帮助确保岩屑和返回地面的地层流体的安全管理，以及用于钻井或给钻井液或泡沫充气的气体的储罐、压缩机、管道和阀门。

除了钻穿需要高密度流体以防止井控问题的高压含烃或含流

pneumatic fluids offers several advantages: little or no formation damage, rapid evaluation of cuttings for the presence of hydrocarbons, prevention of lost circulation, and significantly higher penetration rates in hard-rock formations.

3.4 Special additives for drilling fluids

An efficient drilling fluid must exhibit numerous characteristics, such as desired rheological properties (plastic viscosity, yield value, low-end rheology, and gel strengths), fluid loss prevention, stability under various temperature and pressure operating conditions, and stability against contaminating fluids such as salt water, calcium sulfate, cement, and potassium contaminated fluids. The performances of drilling fluids depend on the structure, chemical properties of the additives and their compatibility and synergistic effects. The chemistry and mechanisms of the major additives are described and explained. The additives are discussed in detail with regard to their chemical nature and functionality.

3.4.1 Weighting materials

Density is defined as weight per unit volume. To prevent the inflow of formation fluids, the pressure of the mud column must exceed the pore pressure, which could be adjusted according to the need of balancing formation pore pressure and structural stress. A reasonable drilling fluid density can prevent serious wellbore kick, blowout, or lost circulation, as well as borehole collapse.

Adjusting drilling fluid density includes reducing and increasing. The density of drilling fluid can be diluted with water or oil, and aerated with gas. Mechanical or (and) chemical flocculation can be used to remove drilling solids to reduce drilling fluid density; Drilling fluid density can also be increased by adding high-density materials.

Weighting materials are two types. One is insoluble high density minerals which can be suspended in the drilling fluid to increase density.(Figure 3-2, Figure 3-3) Barite has become the most widely used high-density material because of its wide source and low cost. The other category is high-density water-soluble salts which can increase the base fluid density.

高密度的材料有两类。一类是高密度的不溶性矿物或矿石的粉末(表3-2、表3-3)。这些粉末可悬浮在黏土矿物颗粒形成的空间结构中提高钻井液密度。由于重晶石来源广、成本低，所以它成为目前使用最多的高密度材料。另一类是高密度的水溶性盐。这些盐可溶于钻井液中提高钻井液密度。

Table 3-2 High density insoluble minerals

Minerals	Main components	Density/(g/cm^3)
Limestone	$CaCO_3$	2.7-2.9
Barite	$BaSO_4$	4.2-4.6
Siderite	$FeCO_3$	3.6-4.0
Ilmenite	$TiO_2 \cdot Fe_3O_4$	4.7-5.0
Magnetite	Fe_3O_4	4.9-5.2
Pyrite	FeS_2	4.9-5.2

表3-2 高密度不溶性矿物

名称	主要成分	密度/(g/cm^3)
石灰石	$CaCO_3$	2.7~2.9
重晶石	$BaSO_4$	4.2~4.6
菱铁矿	$FeCO_3$	3.6~4.0
钛铁矿	$TiO_2 \cdot Fe_3O_4$	4.7~5.0
磁铁矿	Fe_3O_4	4.9~5.2
黄铁矿	FeS_2	4.9~5.2

Table 3-3 High density water soluble salts and their saturated solution density

Water-soluble salt	Density of salt/(g/cm^3)	Density of saturated aqueous solution/(g/cm^3)
KCl	1.398	1.16(20℃)
NaCl	2.17	1.20(20℃)
$CaCl_2$	2.15	1.40(60℃)
$CaBr_2$	2.29	1.80(10℃)
$ZnBr_2$	4.22	2.30(40℃)

表3-3 高密度水溶性盐及其饱和溶液密度

水溶性盐	盐的密度/(g/cm^3)	饱和水溶液密度/(g/cm^3)
KCl	1.398	1.16(20℃)
NaCl	2.17	1.20(20℃)
$CaCl_2$	2.15	1.40(60℃)
$CaBr_2$	2.29	1.80(10℃)
$ZnBr_2$	4.22	2.30(40℃)

Corrosion inhibitor should be added to prevent corrosion of drilling tools in high density brines drilling fluids. Crystallization temperature of the salts shall be concerned in this case. In low-density brine, when the temperature of brine drops, ice will be precipitated, the temperature is called the freezing point of brine. With the increase of brine density, the freezing point of brine decreases. In high-density brine, when the temperature of brine drops, the solid precipitated is not ice but salt, the temperature is called salts crystallizing temperature. As the brine density of increases, the salts crystallizing temperature rises abruptly. Therefore, the temperature of drilling fluid should be higher than salts crystallizing temperature when employed high density brines drilling fluids. In order to prevent the influence of salt extraction on drilling fluid performance, salt crystallization inhibitors can be added into drilling fluid. The available salt crystallization inhibitors are amino polycarboxylate, such as DTPA, which can selectively adsorb on the surface of newly precipitated grain, distort it, and inhibit grain growth, to controlling salt crystallization.

在使用水溶性盐提高钻井液密度时要加入缓蚀剂，防止盐对钻具的腐蚀。同时，要注意盐从钻井液中的析出温度。对于低密度盐水，盐水温度降至一定程度所析出的是冰，该温度称为盐水的冰点，随着盐水密度的增加，盐水的冰点降低；对于高密度盐水，盐水温度降至一定程度所析出的固体不是冰而是盐，该温度称为析盐温度，随着盐水密度的增加，盐水的析盐温度陡然上升。因此，在使用水溶性盐作高密度材料时，应要求钻井液的使用温度高于该钻井液密度下的析盐温度。为防止析盐对钻井液性能的影响，可在钻井液中加入盐结晶抑制剂，可用的盐结晶抑制剂是氨基多羧酸盐，如二乙烯三胺五乙酸盐（DTPA）：

Diethylenetriamine pentaacetic acid, DTPA

二乙烯三胺五乙酸盐，DTPA

分子式中的 M 可为 K^+, Na^+ 或 NH_4^+。氨基多羧酸盐溶于钻井液后，即通过离子交换转变为相应的盐（如高密度材料为钙盐时即转变为钙盐），可选择性地吸附在刚析出的晶粒表面，使它发生畸变，抑制晶粒生长，起控制析盐作用。

3.4.2 Alkaline control chemicals of drilling fluid

The acidity and alkalinity of drilling fluids are closely related to the dispersion of clay the existence of Ca^{2+}, Mg^{2+} and the performance of additives, the fluid rheological behavior and the corrosion on drilling

3.4.2 钻井液的酸碱调节剂

钻井液的酸碱性与钻井液中黏土的分散程度、Ca^{2+} 及 Mg^{2+} 和钻井液处理剂的存在状态、钻井液的流变性和钻井液对钻具的腐

tools in drilling fluids. The pH value of drilling fluid is generally controlled in the weakly alkaline (8 ~ 11) range. The pH value control agents used in drilling fluids include sodium hydroxide, potassium hydroxide, sodium carbonate, etc.

(1) Sodium/potassium hydroxide.

Sodium hydroxide can be dissociated in water to provide OH^-:

$$NaOH \longrightarrow Na^+ + OH^-$$

So sodium hydroxide has a very strong pH control capability. Na^+ produced by the dissociation of sodium hydroxide can change the calcium ion in Ca-bentonite clay minerals in drilling fluid into Na-bentonite, which is conducive to improve drilling fluid stability. However, it can also make the shale swell disperse around the wellbore, which is not conducive to the wellbore stability.

Potassium hydroxide can also be dissociated in water to provide OH^-:

$$KOH \longrightarrow K^+ + OH^-$$

And potassium hydroxide has similar capability to control pH as sodium hydroxide. Compared with Na^+, K^+ produced by potassium hydroxide dissociation has the effect of inhibiting expansion and dispersion of shale around the wellbore, to enhance the wellbore stability.

(2) Sodium carbonate.

Sodium carbonate indirectly produces OH^- to adjust the pH value of drilling fluid through dissociation in water and hydrolysis of carbonate in water:

$$Na_2CO_3 \longrightarrow 2Na^+ + CO_3^{2-}$$
$$CO_3^{2-} + H_2O \longrightarrow HCO_3^- + OH^-$$
$$HCO_3^- + H_2O \longrightarrow H_2CO_3 + OH^-$$

Besides controlling the pH value, sodium carbonate can also reduce the concentration of Ca^{2+} and Mg^{2+} in drilling fluid:

$$Ca^{2+} + CO_3^{2-} \longrightarrow CaCO_3 \downarrow$$
$$Mg^{2+} + 2OH^- \longrightarrow Mg(OH)_2 \downarrow$$

蚀密切相关。钻井液的 pH 值一般控制在弱碱性（8~11）范围，因为在此范围内，钻井液中的黏土有适当的分散性，钻井液处理剂有足够的溶解性，对 Ca^{2+} 和 Mg^{2+} 在钻井液中的浓度有一定的抑制性，钻井液对钻具有低的腐蚀性。钻井液使用的 pH 值控制剂有氢氧化钠、氢氧化钾、碳酸钠等。

（1）氢氧化钠/钾。

氢氧化钠可在水中解离，提供 OH^-：

$$NaOH \longrightarrow Na^+ + OH^-$$

因此氢氧化钠有很强的 pH 值控制能力。氢氧化钠解离产生的 Na^+ 可使钻井液中的钙土转变为钠土，有利于提高钻井液的稳定性，但它也可使井壁的页岩膨胀、分散，因而不利于井壁稳定。

氢氧化钾也可在水中解离，提供 OH^-：

$$KOH \longrightarrow K^+ + OH^-$$

因此氢氧化钾控制 pH 值的能力与氢氧化钠相同。与氢氧化钠不同的是，氢氧化钾解离产生的 K^+ 对井壁的页岩有抑制膨胀、分散作用，有利于提高井壁的稳定性。

（2）碳酸钠。

碳酸钠是通过在水中解离和碳酸根在水中水解，间接产生 OH^- 起调整钻井液 pH 值的作用的：

$$Na_2CO_3 \longrightarrow 2Na^+ + CO_3^{2-}$$
$$CO_3^{2-} + H_2O \longrightarrow HCO_3^- + OH^-$$
$$HCO_3^- + H_2O \longrightarrow H_2CO_3 + OH^-$$

碳酸钠在控制钻井液 pH 值的同时，还可起降低钻井液中 Ca^{2+} 和 Mg^{2+} 浓度的作用：

$$Ca^{2+} + CO_3^{2-} \longrightarrow CaCO_3 \downarrow$$

$Mg^{2+} + CO_3^{2-} \longrightarrow MgCO_3 \downarrow$

Therefore, sodium carbonate can be used as a calcium or magnesium remover. Sodium carbonate has the same effect on the Ca-bentonite in the drilling fluid and the shale around wellbore as sodium hydroxide because of interaction of Na^+ and clay minerals.

3.4.3 Fluid loss additives

Filtration property of a drilling fluid refers to the property whether drilling fluid is easy to filter into the formation. It can be measured and evaluated by API drilling fluid filtration test. In general, lower fluid loss of drilling fluid can form thin and tight filter cake with compact structure, erosion resistance and low friction coefficient.

The filtration property of drilling fluid has important influence on reservoir protection, wellbore stability and filter cake forming on highly permeable formations. The fluid loss additives can be used to control the fluid loss.

3.4.3.1 Fluid loss additive and their properties

Conventional drilling fluid loss additives include modified starch, modified cellulose, modified resin, synthesized polymer modified and modified humate.

(1) Starch.

Most oilfield grade starches are from processed corn or potato feedstocks. Regular starch is not dispersible in cold water. It must be pregelatenized to make it dispersible. Also, starches are highly susceptible to bacterial degradation. Oilfield grade starches are pregelatenized and treated with a bactericide before supplied to the rig site. Modified starches, carboxymethyl and hydroxypropyl, have been processed during manufacture. Also, the modified starches do not need a bactericide while the drilling fluid is being actively circulated. They do need a preservative if stored.

Temperature stability—natural starch has a relatively low temperature stability, around 200–212°F (93–100°C). Modified starches have a higher temperature stability of up to 250-300°F (121–149°C). Higher temperature stability is obtained in saturated brines, especially in formate brines. The molecular weights of regular and modified starches are 100,000 to 500,000.

Their primary use is for fluid loss control, either to augment bentonite in a freshwater system, or to replace bentonite in a brine system. The modified starches can enhance the low shear rate viscosity (LSRV) of most water-based systems.

(2) Modified celluloses.

Carboxy methyl cellulose (CMC) can be used as a fluid loss additive and bulk fluid viscosifier. Oilfield cellulosic products are manufactured from plant natural cellulose modified with added short side chains—carboxymethyl (CM), hydroxypropyl (HP), and hydroxyethyl (HE). The modification is necessary because natural cellulose is insoluble in water.

Cellulosics can be made in various molecular weight ranges and degrees of substitution (number of side chains). The maximum DS is three side chains per saccharide unit. CMCs that have a higher DS, above approximately 2.0, are called PACs, polyanionic cellulose.

Higher cellulose MWs result in a higher bulk viscosity in the fluid. CMCs and PACs are available as low viscosity (LV), medium viscosity (MV), and high viscosity (HV) products. The PACs, in general, perform better in sauturated brines, including saturated calcium chloride. Most cellulosics have a moderate temperature stability ~250°F (120°C).

CMCs are not as shear thinning as the fermentation

添加防腐剂。

抗温稳定性：淀粉的温度稳定性相对较低，约为200~212°F（93~100℃）。改性淀粉具有更高的温度稳定性，高达250~300°F（121~149℃）。在饱和卤水中，特别是在甲酸盐中具有较高的温度稳定性。普通淀粉和改性淀粉的分子量在（10~50）×10^4。

它们的主要用途是控制钻井液滤失量，或在淡水钻井液中增加膨润土，或在盐水钻井液中取代膨润土。改性淀粉可以提高大多数水基钻井液的低剪切速率黏度（LSRV）。

（2）改性纤维素。

羧甲基纤维素（CMC）可以作为降滤失和增黏剂使用。油田用纤维素产品是以天然纤维素为原料，经短侧链改性而成，短侧链为羧甲基（CM）、羟丙基（HP）、羟乙基（HE）。由于天然纤维素不溶于水，因此这种改性是必要的。

纤维素能以不同的分子量范围和取代度（侧链的数目）制成。最大的取代度为每个糖单元三个侧链。具有较高取代度（大约2.0以上）的CMC称为聚阴离子纤维素，即PAC。

纤维素相对分子质量越高，钻井液黏度越高。CMC和PAC有低黏度（LV）、中等黏度（MV）和高黏度（HV）三类产品。总体来说，PAC在盐水中表现更好，包括饱和氯化钙。大多数纤维素产品有一个中等的温度稳定性，最高250°F（120℃）。

CMC不像发酵生物聚合物那

biopolymers and show little to no suspending ability. They are excellent fluid loss control materials.

(3) Synthetic polymers.

Synthetics polymer can be made in a wide range of molecular weights and charge densities, with both anionic and cationic surface charges. The synthetic polymer with medium molecular weight, 100,000 can be used as fluid loss reducer. Some AM-AMPS-NVP terpolymers can be made to withstand very high temperatures, up to 400-600 °F (200-300 °C).

(4) Ammonium humates.

Ammonium humates is a kind of filtration-reducing additives for oil muds. This material, described as consisting of n-alkyl ammonium humates, was prepared by reactions between quaternary amines with 12 to 22 carbon atoms in one of the alkyl chains, and the alkalisoluble fraction of lignite (leonardite, brown coal). This reaction product was readily dispersed in refined or crude oils, and markedly reduced filtration without substantially increasing viscosity.

3.4.3.2 Mechanisms of fluid loss prevention

(1) Viscosification.

Water-soluble macromolecule dissolved in drilling fluid can improve the viscosity of drilling fluid, which is conducive to the decrease of drilling fluid filtration.

(2) Stability Enhancement of colloidal and suspension.

Polymeric fluid loss additives can be absorbed on the surface of clay particle, increase on the of the surface electronegative increase and thicken hydration layer of the clay particles, then result in enhancement of coalescence stability and reasonable particle size distribution of the clay particles.

(3) Bridging.

The mud contains particles of the size required to bridge the pores of the rock, and thus establish a base on which the filter cake can form. Only particles of a certain size relative to the pore's size can bridge. Particles of a certain critical size will stick at bottlenecks in the flow channels, and form a bridge just inside the surface pores.

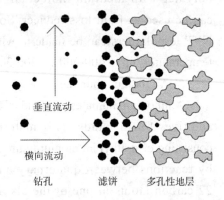

Figure 3-10 The process of drilling fluid to form a filter cake in a porous formation

图3-10 钻井液在多孔性地层形成滤饼的过程

3.4.4 Rheology modification additives

The rheological control is mainly to adjust the viscosity and gel strength.

In the process of drilling, higher or lower viscosity and gel strength of drilling fluid will result in adverse effects. Excessive viscosity and gel strength will cause excessive flow friction and high energy consumption, which will seriously reduce rate of penetration. In addition, it will cause problems such as bit balling, sticking, removing cuttings, and difficult degassing of drilling fluid. If the viscosity and gel strength of drilling fluid are too low, it will affect the cuttings carrying capacity and wellbore stability.

The viscosity and gel strength of drilling fluid can be adjusted by adjusting the solid content of drilling fluid. Although this is the first method to consider, its usage has a certain limit, because the change of solid content will also affect other properties of drilling fluid.

3.4.4 钻井液流变性调控

钻井液流变性的调整主要是调整钻井液的黏度（指表观黏度）和切力（即剪切应力）。

在钻井过程中，钻井液黏度和切力过大或过小都会产生不利的影响。钻井液黏度和切力过大会使钻井液流动阻力过大、能耗过高，严重影响钻速。此外，还会引起钻头泥包、卡钻、钻屑在地面不易除去和钻井液脱气困难等问题。钻井液黏度和切力过小，则会影响钻井液携岩和井壁稳定。

可用调整钻井液中固相含量的方法调整钻井液的黏度和切力。虽然这是首先考虑使用的方法，但它的使用有一定的限度，因为固相含量的变化还会影响钻井液

On the basis of solid control, rheological modifiers can also be used to adjust the viscosity and gel strength of drilling fluid. Rheological modifiers can be divided into viscosity reducers and viscosifiers (or thickeners).

3.4.4.1 Viscosity reducer

Viscosity reducer refers to the rheological modifier that can reduce the viscosity and shear force of drilling fluid, also known as thinner or deflocculant. The commonly used viscosity reducers are lignosulfonic acid and anionic oligomers. Such low molecular weight, heavily sulfonated polymers are believed to aid in coating clay edges in the subterranean formation with a lasting or effectively permanent negative charge.

Phosphates are effective only in small concentrations. The mud temperature must be less than 55℃. The salt contamination must be less than 500 mg/L sodium chloride. The concentration of calcium ions should be kept as low as possible. The pH should be between 8 and 9.5. Some phosphates may decrease the pH, so adding more NaOH is required.

Lignite muds are high-temperature resistant up to 450℉ (230℃). Lignite can control viscosity, gel strength, and fluid loss. The total hardness must be lower than 20 ppm.

Lignosulfonate freshwater muds contain ferrochrome lignosulfonate for viscosity and gel strength control. These muds are resistant to most types of drilling contamination because of the thinning efficiency of the lignosulfonate in the presence of large amounts of salt and extreme hardness. Because chromium is potentially toxic, its release into the natural environment and the thereof is continuously being reviewed by various government agencies around the world.

Anionic oligomers are polymers with a relatively small molecular weight ($1.0 \times 10^3 \sim 6.0 \times 10^3$ D).

Oligomers such as sodium acrylate (AA) - sodium (2-acrylamido-2-methyl)propyl sulfonate (AMPS) copolymer, styrene sulfonate (SS) - maleic anhydride (MA) copolymer, styrene sulfonate - itaconic acid (IA) copolymer and sodium acrylate–sodium allyl sulfonate (SAS) copolymer can be used as thinner or deflocculant. Maleic anhydride and styrene copolymers were developed for use in geothermal drilling operations. It is stable to temperatures above 600℉ (315℃).

D）的聚合物。丙烯酸钠–AMPS共聚物、苯乙烯磺酸钠–顺丁烯二酸酐共聚物、苯乙烯磺酸钠–衣康酸共聚物和丙烯酸钠–丙烯磺酸钠共聚物等低聚物可用作降黏剂。顺丁烯二酸酐–苯乙烯的共聚物可用于地热钻井作业，抗温能力达到600℉（315℃）以上。

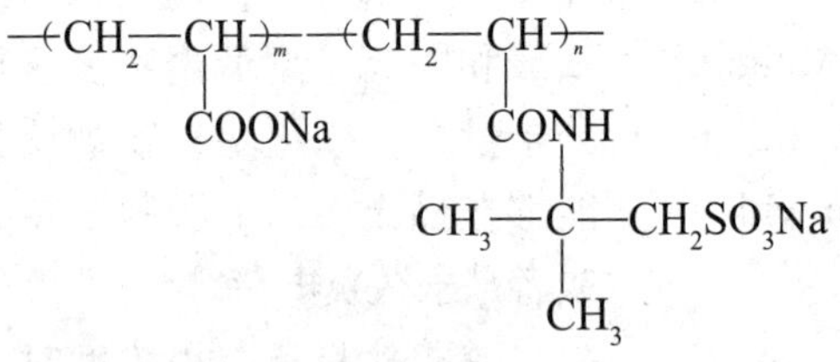

AA/AMPS copolymer

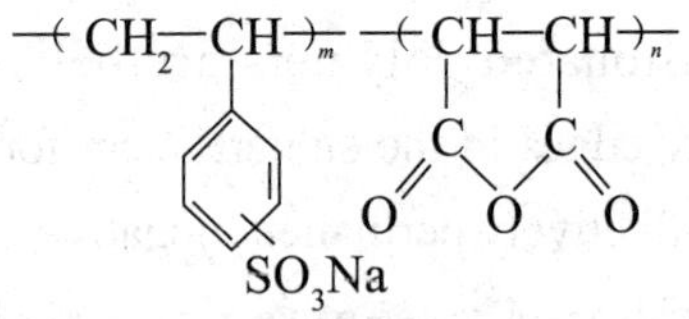

SS/MA copolymer

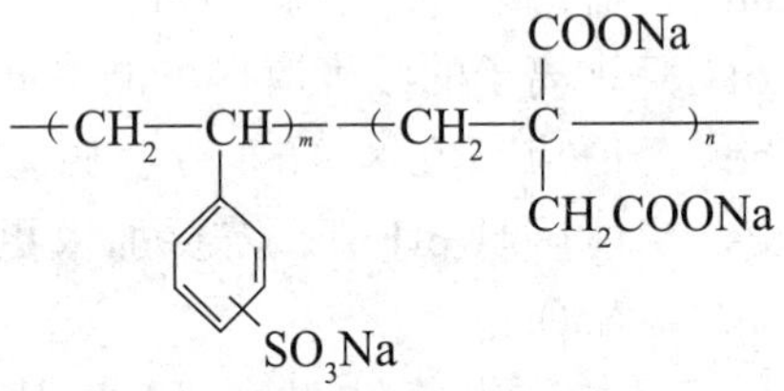

SS/IA copolymer

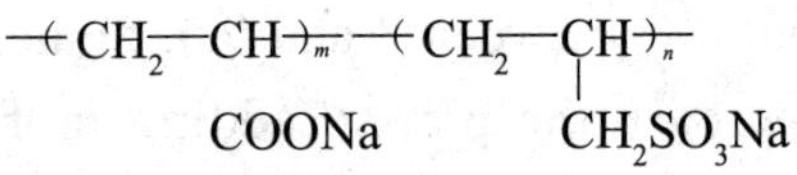

AA/SAS copolymer

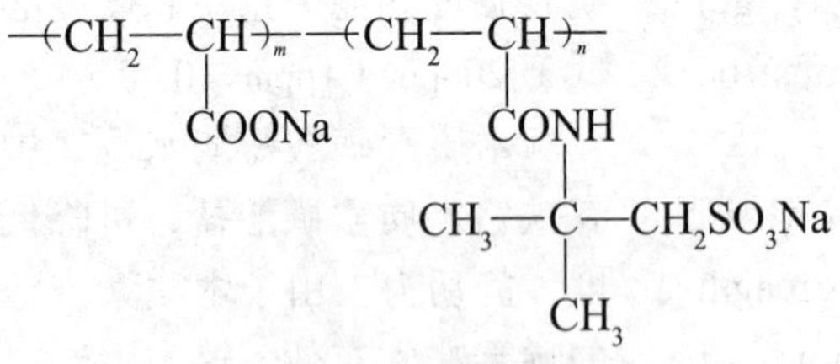

丙烯酸钠与(2-丙烯酰胺基-2-甲基)丙基磺酸钠共聚物

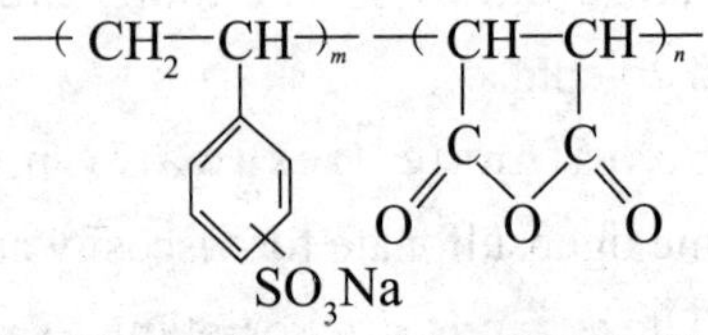

苯乙烯磺酸钠与顺丁烯二酸酐共聚物

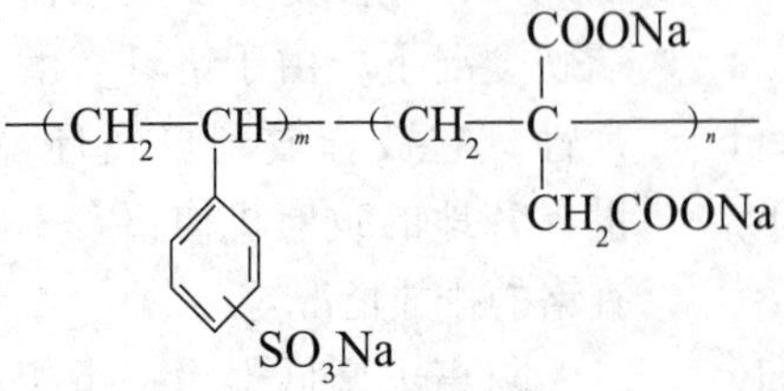

苯乙烯磺酸钠与衣康酸共聚物

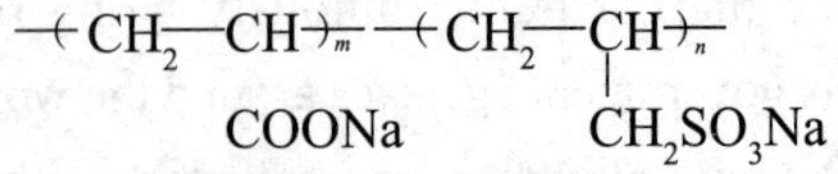

丙烯酸钠与丙烯磺酸钠共聚物

3.4.4.2 Thickener and viscosifiers

Thickeners and viscosifiers are rheological modifiers that increase the viscosity and gel strength of drilling fluids, including Xanthan gum, modified cellulose and synthetic polymers.

(1)Biopolymer xanthan gum（XC）.

Xanthan gum （XC）have extremely complex structures and high molecular weights (500 to over 2 million). Their side chains are slightly anionic. When dispersed, XC forms a double helix molecular arrangement. The unique property is as a rheology modifier, imparting significantly increased LSRV for hole cleaning and suspension properties. The increased LSRV characteristic of these biopolymers is due to the complex nature of the polymer molecules, both from complex side chains and high molecular weights. In addition, when hydrated in either fresh or brine water base fluids, they tend to associate and the long chain backbones wrap around each other, thereby greatly slow down any particle settling that tends to occur.

XC exhibits excellent shear and temperature resistance. XC was used to replace bentonite to enhance the rate of penetration while drilling. At very high salt concentrations complete hydration of xanthan is not achieved, but it does hydrate rapidly at temperatures above 250℉ (120℃) and high shear.

XC has moderate temperature stability—250-300℉ (120-150℃). A preservative is not normally needed while circulating.

(2)Modified cellulose.

Modified cellulose with higher MWs result in a higher bulk viscosity in the fluid, such as HV-CMC and HV-PAC. HECs are exclusively as a bulk viscosifer of saturated brines, such as chlorides and bromates, for completion fluids. They have no suspending ability and low particle carrying capacity in annular flow regimes.

(3)Synthetic polymeric viscosifiers.

Synthetics polymer can be made in a wide range of molecular weights and charge densities, with both anionic and cationic surface charges. The polymers with high molecular weight, 1,000,000 can be used as bentonite extender for low solids, unweighted drilling fluids, such as acrylates and acrylate copolymers. Some AM-AMPS-NVP terpolymers can be made to withstand very high temperatures, up to 400-600 ℉ (200-300℃).

(4)Mixed metal hydroxides/silicates (MMH/MMS).

By addition of mixed metal hydroxides, typical bentonite muds are transformed to an extremely shear-thinning fluid. At rest, these fluids exhibit a very high viscosity but are thinned to an almost water-like consistency when shear stress is applied. In theory, the shear-thinning rheology of mixed metal hydroxides and bentonite fluids is explained by the formation of a three-dimensional, fragile network of mixed metal hydroxides and bentonite. The positively charged mixed metal hydroxide particles attach themselves to the surface of negatively charged bentonite platelets. Typically, magnesium aluminum hydroxide salts are used as mixed metal hydroxides. Mixed metal hydroxides demonstrate the following advantages in drilling:High cuttings removal, Suspension of solids during shutdown, Lower pump resistance, Stabilization of the borehole, High drilling rates, Protection of the producing formation.Mixed metal hydroxide drilling muds have been successfully used in horizontal wells, for drilling in fluids, for drilling large-diameter holes, with coiled tubing.

(5)Organophilic clays.

Organophilic Clays are viscosifiers or thickeners for nonaqueous drilling fluids (NADFs). The organophilic Clays were formed by replacing the exchangeable cations of the bentonite with the cationic groups of the amines and by further adsorption of the

（3）合成聚合物增黏剂。

合成聚合物具有较宽的分子量和电荷密度（包括阴离子和阳离子），只有大等分子量的聚合物可以用作低固相未加重钻井液的膨润土增效剂，诸如聚丙烯酸盐和聚丙烯酰胺。一些丙烯酰胺、AMPS 和 N-乙烯基吡咯烷酮共聚物产品可以抗更高温度，达到 400~600 ℉（200~300℃）。

（4）混合金属氢氧化物/混合金属硅酸盐。

通过加入混合金属氢氧化物，典型的膨润土钻井液被转化为一种剪切稀释性的流体。在静止状态下，这些流体呈现出很高的黏度，但在施加剪切应力时，它们被稀释到几乎与水一样。从理论上讲，混合金属氢氧化物和膨润土钻井液的剪切稀释流变性是由混合金属氢氧化物和膨润土形成的一个三维脆性弱链接网架结构所解释的。带正电荷的混合金属氢氧根颗粒附着在带负电荷的膨润土片状颗粒表面。通常，氢氧化镁铝盐被用作混合金属氢氧化物。混合金属氢氧化物钻井具有以下优点：高效携岩、关停悬浮固体、稳定井壁、高钻速和保护产层。混合金属氢氧化物钻井液已成功地应用于水平井、储层钻井、大井眼和连续油管钻井。

（5）有机土。

有机土是油基/合成基钻井液的增黏剂或增稠剂。有机土是将膨润土的交换阳离子置换为胺的阳离子基团，并将烃类链进一步吸附在黏土层面上，形成亲油

hydrocarbon chain on the clay lamina surface. The oil-dispersible clays would suspend solids in oil, without requiring additional soaps and emulsifiers.

性的有机黏土。这种可在油相中分散的有机土可以悬浮固体，不需要额外的肥皂和乳化剂。

3.4.5 Flocculants

The flocculants of water-based drilling fluids are the chemicals to cause colloidal particles in suspension to group into bunches, or “flocks”, causing solids to settle out.，such as salts, hydrated lime, gypsum and anion polymers, and partially hydrolyzed polyacrylamide (HPAM) is by far the most effective polymer.

The degree of hydrolysis of the HPAM is important, presumably because it spaces the negative sites along the chain so that they match the positive sites on the clay platelet edges. Another important benefit of HPAM is that it coats the drill cuttings (usually referred to as encapsulation) thereby inhibiting their dispersion and incorporation into the mud.

3.4.5 钻井液用絮凝剂

钻井液絮凝剂是一类能够使悬浮在钻井液中胶体颗粒束缚在一起，形成絮体，以便于从钻井液中分离出去的化学物质，比如盐、石灰乳、石膏和聚合物，其中目前最有效的是部分水解聚丙烯酰胺。

HPAM 的水解程度很重要，这可能是因为它将链上的阴性位点隔开，以便与黏土血小板边缘的阳性位点相匹配。HPAM 的另一个重要优点是它覆盖了钻屑(通常称为包被)，从而抑制了钻屑的分散到钻井液中去。

3.4.6 Shale inhibitors

It is important to maintain the wellbore stability during drilling, especially in water-sensitive shale and clay formations. The rocks within these types of formations absorb the fluid used in drilling. This absorption causes the rock to swell and may lead to a wellbore collapse. Clay stabilizers are shown below.

(1) Salts.

The salts herein mainly refer to inorganic salt (such as sodium chloride, ammonium chloride, potassium chloride, calcium chloride, etc.) and organic salt (such as sodium formate, potassium formate, sodium acetate, potassium acetate, etc.), and potassium and ammonium salts are the most effective of stabilizing shale, which is because their cations can enter the crystal layer of clay and are not easy to be released, so that the clay pieces can be connected together .

3.4.6 黏土和页岩抑制剂

在钻井过程中，尤其在水敏页岩和黏土地层中，保持井壁稳定是非常重要的。这些类型地层中的岩石吸收钻井中使用的流体。这种吸收导致岩石膨胀，并可能导致井筒坍塌。常用的页岩抑制剂由以下几类。

(1)盐。

这里的盐主要指无机盐(如氯化钠、氯化铵、氯化钾、氯化钙等)和有机盐(如甲酸钠、甲酸钾、乙酸钠、乙酸钾等)。在水溶性盐中，稳定页岩效果最好的是钾盐和铵盐，这是因为它们的离子可进入黏土的晶层而不易释出，使黏土片能联结在一起，有效地抑制页岩的膨胀。

(2) Quaternary ammonium salts and polymers.

Quaternary ammonium salts are effective antiswelling drilling fluid additives for underbalanced drilling operations, such as choline. Choline is addressed as a quaternary ammonium salt containing the N,N,N-trimethylethanolammonium cation. An example of choline halide counterion salts is choline chloride. Polyquaternary ammonium salts also have a good shale inhibition effect. There are several kinds of such polymers as the following.

Quaternary and polyquaternary ammonium salts stabilize shale mainly by neutralizing the negative electrical charge on shale surface and bridging and adsorption between clay sheets. In addition, quaternary ammonium salt cationic surfactant can reverse the wettability of clay surface.

（2）季铵盐和聚合物。

季铵盐是有效的抑制黏土膨胀的钻井液添加剂，例如胆碱，适用于欠平衡钻井作业。胆碱是一种含N，N，N－三甲基乙醇铵阳离子的季铵盐。卤化胆碱反离子盐的一个例子是氯化胆碱。聚季铵盐也具有良好的页岩抑制效果，这类聚合物有以下几种。

季铵盐和聚季铵盐主要通过中和页岩表面的负电性和在黏土片间通过桥接吸附起稳定页岩的作用。另外，季铵盐型阳离子表面活性剂还具有使黏土表面润湿性反转的作用。

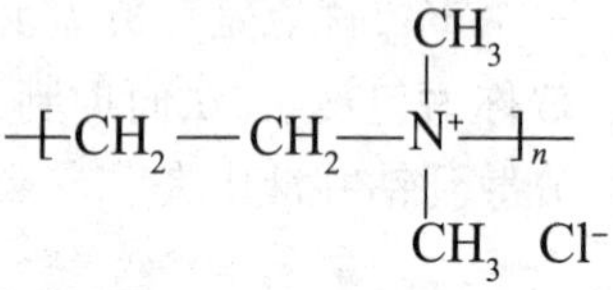

Poly 1,2-ethylene dimethyl ammonium chloride

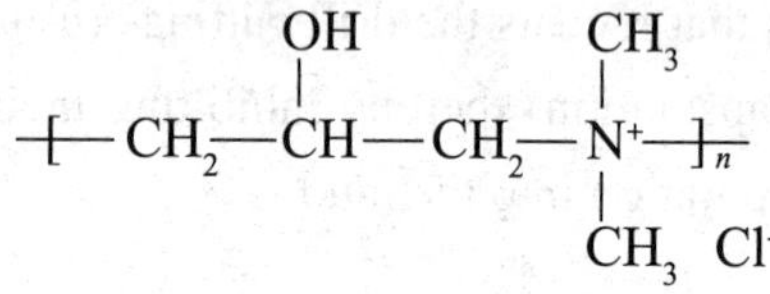

Poly 2-hydroxy-1,3-propylene dimethyl ammonium chloride

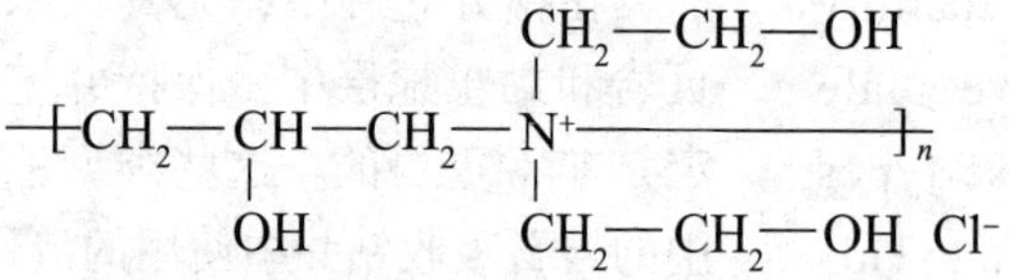

Poly 2-hydroxy-1,3-propylene dihydroxyethyl ammonium chloride

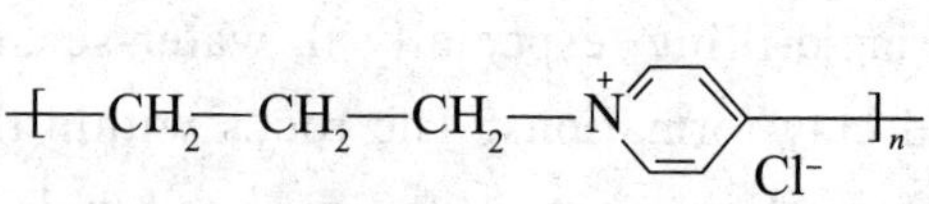

Poly 1,3-propylene pyridine chloride

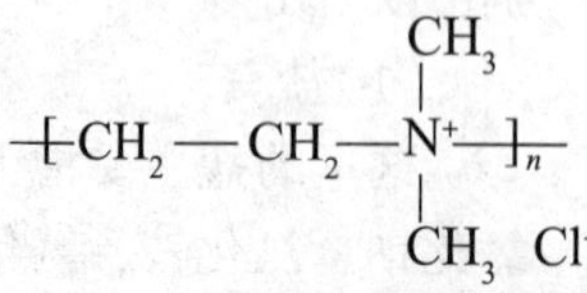

聚1,2-亚乙基二甲基氯化铵

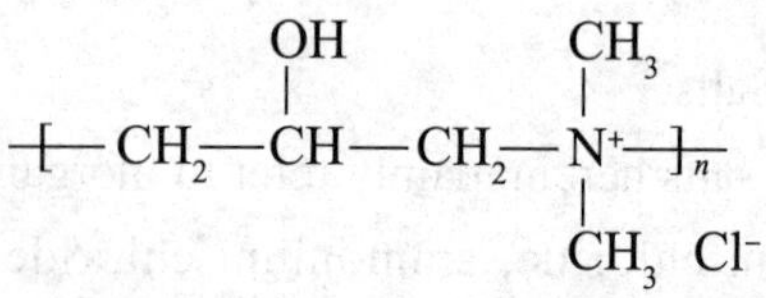

聚2-羟基-1,3-亚丙基二甲基氯化铵

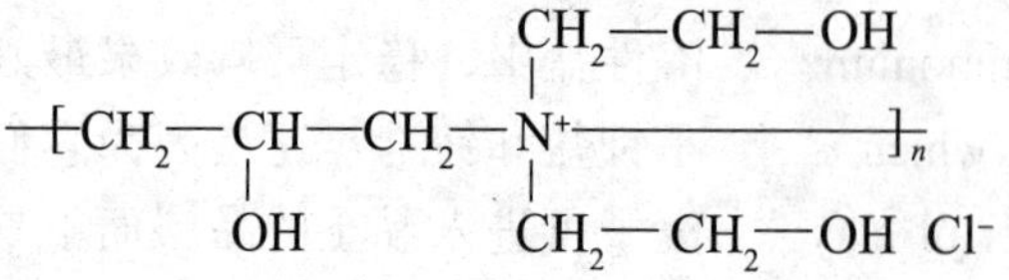

聚2-羟基-1,3-亚丙基二羟乙基氯化铵

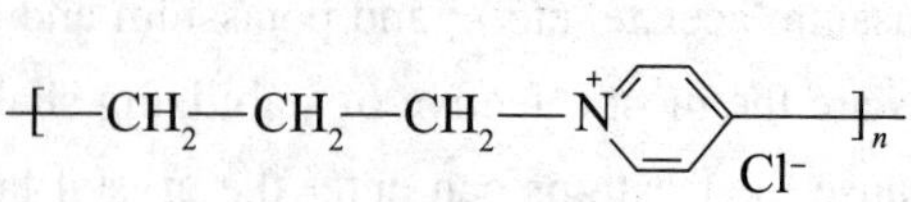

聚1,3-亚丙基氯化吡啶

(3) Nonionic polyether PEG.

Nonionic polyether, such as polyethylene glycol (PEG), polypropylene glycol (PPG) and polyethylene-propylene glycol (PEG/PG), are commonly used as shale inhibitors. These ether polymers are dissolved in water by hydrogen bonding to H_2O molecules. In addition, A drilling fluid additive, which is the reaction product of methyl glucoside and alkylene oxides, such as EO, PO, or 1,2-butylene oxide, acts as a clay stabilizer.

$$+CH_2-CH_2-O+_n \qquad +CH_2-CH-O+_n$$
$$|$$
$$CH_3$$

polyethylene glycol Polypropylene glycol

$$CH_3-CH-O+(C_3H_6O)_m+(C_2H_4O)_n H$$
$$|$$
$$CH_2-O+(C_3H_6O)_m+(C_2H_4O)_n H$$

Polyoxyethylene polyoxypropylene glycol ether

$$+CH_2-CH_2-O+_n \qquad +CH_2-CH-O+_n$$
$$|$$
$$CH_3$$

聚乙二醇 聚丙二醇

$$CH_3-CH-O+(C_3H_6O)_m+(C_2H_4O)_n H$$
$$|$$
$$CH_2-O+(C_3H_6O)_m+(C_2H_4O)_n H$$

聚氧乙烯聚氧丙烯丙二醇醚

Such an additive is soluble in water at ambient conditions, but becomes insoluble at elevated temperatures. Because of their insolubility at elevated temperatures, these compounds concentrate at important surfaces such as the drill bit cutting surface, the borehole surface, and the surfaces of the drilled cuttings. The insoluable compounds can seal the pores of shale and reduce the contact between shale and water, thus stabilizing shale.

(4) Poly (oxyalkylene Amine)s.

Poly(oxyalkylene amine)s are a general class of compounds that contain primary amino groups attached to a poly(ether) backbone. They are also addressed as

（3）非离子型聚合物。

在非离子型聚合物中，主要用醚型聚合物，如：聚乙二醇（PEG）、聚丙二醇（PPG）和聚乙二醇丙二醇（PEG/PG）。另外，一种由甲基葡萄糖苷和环氧乙烷或环氧丙烷反应的钻井液添加剂也具有页岩抑制作用。

这类添加剂在环境条件下可溶于水，但在升高至一定温度后不溶于水。由于这些化合物在高温下不溶于水，它们吸附在钻头切削表面、钻孔表面和钻屑表面。这些析出的醚型聚合物可黏附在页岩表面，封堵页岩的孔隙，减小页岩与水的接触而起稳定页岩的作用。

（4）聚醚胺化合物。

聚醚胺是一类化合物，它含有伯胺基在聚氧乙烯醚的主链上，也被称为聚氧化乙基胺。它们

poly(ether amine)s. They are available in a variety of molecular weights, ranging up to 5 kDa.

Poly(oxyalkylenediamine)s have been proposed as shale inhibitors. These are synthesized from the ring-opening polymerization of oxirane compounds in the presence of amino compounds. A typical poly(ether amine) is shown. Such products belong to the Jeffamine product family. A related shale hydration inhibition agent is based on an N-alkylated 2,20-diaminoethylether.

的分子量多种多样，最高可达 5kDa。

聚醚二胺认为是页岩抑制剂。这些化合物是在氨基化合物存在的情况下，由环氧烷化合物的开环聚合而成的。一个典型的聚醚胺如下所示。这些产品属于聚醚胺产品系列。一种页岩水化抑制剂是基于 N 烷基化 2,20- 二乙醚。

$$H_2N-(CH_2CH_2O)_2-CH_2-CH_2-N-CH_2-CH_2-(OCH_2CH_2)_2-NH_2$$
$$|$$
$$CH_2$$
$$|$$
$$CH_2$$
$$|$$
$$(OCH_2CH_2)_2-NH_2$$

poly ether amine

$$H_2N-(CH_2CH_2O)_2-CH_2-CH_2-N-CH_2-CH_2-(OCH_2CH_2)_2-NH_2$$
$$|$$
$$CH_2$$
$$|$$
$$CH_2$$
$$|$$
$$(OCH_2CH_2)_2-NH_2$$

聚醚胺

It is believed that the organic shale inhibitor molecules are adsorbed on the surfaces of clays, with the added organic shale inhibitor competing with water molecules for clay reactive sites and thus serving to reduce clay swelling.

(5) Modified asphalt.

Air blown asphalt and sulfonated asphalt are traditional modified asphalt products. The former is mainly used in non-aqueous drilling fluid, while the latter can be dispersed in water-base drilling fluid.

Sulfonated asphalt can be obtained by reacting asphalt with sulfuric acid and sulfur trioxide. By neutralization with alkali hydroxides, such as NaOH, KOH or NH_3, sulfonate salts are formed. Sulfonated asphalt is predominantly utilized for water-based drilling fluids but also for those based on oil. Apart from reduced filtrate loss and improved filter cake properties,

这类有机页岩抑制剂分子吸附在黏土表面，添加的有机页岩抑制剂与水分子竞争黏土活性位点，从而减少黏土膨胀。

（5）改性沥青。

常用沥青改性产品有氧化沥青和磺化沥青两种。前者主要用在油基钻井液中，后者可分散在水基钻井液中使用。

磺化沥青是用沥青与浓硫酸和三氧化硫反应而成，用 NaOH、KOH 或 NH_3 等碱中和即成磺化沥青产品。磺化沥青主要用于水基钻井液，也用于油基钻井液。磺化沥青作为一种钻井液添加剂，除了降低滤失和改善滤饼性能外，还具有良好的钻头润滑性能和降

good lubrication of the drill bit and decreased formation damage are important features assigned to sulfonated asphalt as a drilling fluid additive. In particular, clay inhibition is enhanced by sulfonated asphalt in the case of water-based drilling fluids.

The mechanism of action of sulfonated asphalt as a clay inhibitor in a drilling fluid is explained by the fact that the electronegative sulfonated macromolecules attach to the electropositive ends of the clay platelets. Thereby, a neutralization barrier is created, which suppresses the absorption of water into the clay. In addition, because the sulfonated asphalt is partially lipophilic, and therefore water repellent, the water influx into the clay is restricted by purely physical principles.

3.4.7 Shale encapsulators

A shale encapsulator is added to a WBM in order to reduce the swelling of the subterranean formation. In order to be effective, a shale encapsulator should be at least partially soluble in the aqueous continuous phase. The conventional encapsulators are HPAM, quaternary PAM and quaternized PVA. Suitable examples of anions that are useful include halogen, sulfate, nitrate, formate, etc. By varying the molecular weight and the degree of amination, a wide variety of products can be tailored. It is possible to create shale encapsulators for the use in low salinity, including freshwater.

低地层损害等重要特点。特别是在水基钻井液中，磺化沥青对黏土的抑制作用更强。

磺化沥青作为黏土抑制剂在钻井液中的作用机理可以用电负性磺化大分子附着在黏土片状颗粒的正电荷部位来解释。这种电性中和产生了一种屏障，抑制了黏土对水的吸收。此外，由于磺化沥青具有部分亲油性，因此具有疏水性，因此流入黏土中的水受到物理的限制。

3.4.7 页岩包被抑制剂

页岩包被抑制剂是在水基钻井液中主要用来降低地层的膨胀。页岩包被抑制剂应至少部分溶于水相，这样才有效。常见的包被抑制剂是部分水解聚丙烯酰胺、季铵化的聚丙烯酰胺和季铵化聚乙烯醇。季铵盐配对阴离子可以是卤根、硫酸根、硝酸根和甲酸根等。由于分子量和季铵化度的不同，有较多的产品可被设计和生产。都可以在低盐度和淡水中用作页岩包被剂。

Quatemized etherified poly(vinyl alcohol) and quaternized poly(acrylamide)

季铵化醚化聚乙烯醇和季铵化聚丙烯酰胺

3.4.8 Lubricants

During drilling, the drill string may develop an unacceptable rotational torque or, in the worst case, become stuck. When this happens, the drill string cannot be raised, lowered, or rotated. Common factors leading to these situations include:cuttings or slough buildup in the borehole，an undergauge borehole，irregular borehole development embedding a section of the drill pipe into，the drilling mud wall cake，unexpected differential formation pressure.

The lubrication of drilling fluid can be improved by adding the lubricant into drilling fluid.

(1) Liquid lubricants.

Liquid lubricants are mainly oils, which include vegetable oils (such as soybean oil, cottonseed oil, and castor oil), animal oils (such as lard), and mineral oils (such as kerosene, diesel, and mechanical lubricants). Surfactants can be added to drilling fluids to form a uniform oil film on the friction surface. Adding surfactant only to drilling fluid can also improve the lubrication of drilling fluid, but its effect is far less than that of using oil and surfactant at the same time.

(2) Solid lubricants.

Solid lubricants are mainly solid beads (such as plastic, glass beads) and graphite. Among them, graphite is suitable for lubricant of WBMs and NADFs.

3.4.9 Emulsifiers

Emulsion is a very important type fluid in oilfield drilling fluid, including water-in-oil drilling fluid and oil-in-water drilling fluid. Whether an oil-in-water (O/W) or a water-in-oil (W/O) emulsion is formed depends on the relative solubility of the emulsifier in the two phases. Thus, a preferentially water-soluble surfactant, such as sodium oleate, will form an O/W

3.4.8 钻井液润滑剂

在钻井过程中，钻柱不可避免地产生旋转扭矩，在最坏的情况下，可能卡住。当这种情况发生时,钻柱不能上升、下降或旋转。导致这种情况的共同因素包括：钻屑或泥浆在钻孔中堆积；井径过小的井眼；嵌入部分钻杆不规则井眼；井壁虚厚的钻井液滤饼；意外的地层压力差。

可在钻井液中加入润滑剂来改善钻井液的润滑性。

（1）液体润滑剂。

液体润滑剂主要是油，其中包括植物油（如豆油、棉籽油、蓖麻油）、动物油（如猪油）和矿物油（如煤油、柴油和机械润滑油）。为了使油在摩擦面上形成均匀的油膜，可在钻井液中加入表面活性剂。在钻井液中只加入表面活性剂也有改善钻井液润滑性的作用，但其效果远比不上油与表面活性剂同时使用的效果。

（2）固体润滑剂。

固体润滑剂主要有固体小球（如塑料、玻璃）和石墨两种。其中，石墨适用于作各类钻井液的润滑剂。

3.4.9 乳化剂

乳状液在油田钻井液中是一种非常重要的流体，包括水包油钻井液和油包水钻井液。水包油（O/W）和油包水（W/O）乳液的形成取决于乳化剂在两相中的相对溶解度。因此，油酸钠等水溶性强表面活性剂，由于降低了油

emulsion because it lowers the surface tension on the water side of the oil-water interface, and the interface curves towards the side with the greater surface tension, thereby forming an oil droplet enclosed by water. On the other hand, calcium and magnesium oleates are soluble in oil, but not in water, and thus form W/O emulsions. Similarly, a nonionic surfactant with a large hydrophilic group (HLB number 10 to 12) will be mostly soluble in the water phase, and thus form an O/W emulsion, whereas a nonionic surfactant with a large lipophilic group (HLB number about 4) will form a W/O emulsion.

Typical O/W emulsifiers historically used in freshwater muds are alkyl aryl sulfonates and sulfates, polyoxyethylene fatty acids, esters, and ethers. A polyoxyethylene sorbitan tall-oil ester, sold under various trade names, is used in saline O/W emulsions, and an ethylene oxide derivative of nonylphenol, $C_9H_{19}C_6H_4O(CH_2CH_2O)_{30}H$, known as DME, is used in calcium-treated muds. Fatty acid soaps, polyamines, amides, or mixtures of these are used for making W/O emulsions.

3.4.10 Lost circulation materials

Lost circulation is one of the most troublesome drilling problems. Its economic cost is due to the loss of expensive drilling fluid into the formation and due to nonproductive time spent on regaining circulation. If untreated, losses may lead to well control issues, poor hole cleaning, and stuck pipe. When lost circulation occurs, sealing the zone is necessary. The common lost circulation materials (for short LCMs) that generally are mixed with the mud to seal loss zones.

A variety of LCMs have been proposed and used in the industry over the past 100 years. Examples of commercially available LCMs are calcium carbonate (ground marble), nut hulls, graphite, petroleum coke, fiber etc. LCMs can be added to the drilling fluid and, during well cementing, to the spacer or cement. They can also be applied as part of a lost circulation pill squeezed into the formation. They may be divided into four categories.

(1) Fibrous materials.

Such as shredded sugar cane stalks (bagasse), cotton fibers, wood fibers, sawdust, and paper pulp. These materials have relatively little rigidity, and tend to be forced into large openings. If large amounts of mud containing a high concentration of the fibrous material are pumped into the formation, sufficient frictional resistance may develop sealing effect.

(2) Flaky materials.

Such as shredded cellophane, mica flakes, plastic laminate and wood chips. These materials are believed to lie flat across the face of the formation and thereby cover the openings. If they are strong enough to withstand the mud pressure, they form a compact external filter cake. If not strong enough, they are forced into the openings, and their sealing action is then similar to that of fibrous materials.

(3) Granular materials.

Such as ground nutshells, or vitrified, expanded shale particles. These materials have strength and rigidity and when the correct size range is used, seal by jamming just inside the openings for the bridging of normal porous formations, i.e., there must be some particles approximately the size of the opening and a gradation of smaller particles. The greater the concentration of particles in the mud, the larger the opening bridged, and that strong granular materials, such as nutshells are with larger bridged openings than fibrous or flaky materials; however, weak granular materials, such as expanded perlite, did not.

(4) Settable materials.

Such as hydraulic cement, diesel oil-bentonite-mud mixes, and high filter loss muds. Settable materials

以在固井过程中添加到间隔物或水泥中。它们也可以作为挤压进入地层的漏失封堵塞。堵漏材料一般可分为四类。

（1）纤维材料。

如甘蔗渣、棉纤维、猪毛、木材纤维、锯末和纸浆。这些材料的刚性相对较低，往往会被挤入开度大的裂缝。大量含有高浓度纤维物质的泥浆被泵入地层，就会产生足够的阻力，从而产生堵漏效果。

（2）片状材料。

如碎玻璃纸、云母片、塑料层压板、木屑等。这些材料被认为平躺在地层井筒的表面，从而覆盖裂缝开口。如果它们足够强大，能够承受泥浆压力，就会形成一个紧凑的外部滤饼。如果没有足够的强度，它们会被压入漏层开口，它们的封堵作用与纤维材料相似。

（3）颗粒状材料。

如磨碎的坚果壳、陶瓷和玻璃材料颗粒。这些材料具有强度和刚性，当使用正确的尺寸范围时，通过在普通多孔地层的架桥孔内堵塞来密封。必须有部分颗粒大小近似于或小于漏层开度的颗粒。泥中颗粒浓度越大，架桥开口就越大，而坚硬的颗粒状材料，如坚果壳，桥接的开口就比纤维状或片状材料大；而膨胀珍珠岩等低强度粒状材料则没有。

（4）胶结性材料。

如水泥浆、柴油－膨润土－泥浆混合料浆、高滤失钻井液等。胶结性或沉淀性材料是在漏层形

work by building up a solid seal in the thief zone; e.g., cement seal. Slurries of settable materials usually are easily pumpable since they often contain little or no solids. Moreover, settable slurries can enter fractures and voids of any width, unlike LCM particles that can only enter fractures or voids of sufficient width for the particles.

3.4.11 Other special additives for drilling fluids

(1) Calcium removers.

Caustic soda, soda ash, bicarbonate of soda and certain polyphosphates make up the majority of chemicals designed to prevent and overcome the contaminating effects of anhydrite, gypsum and calcium sulfate.

(2) Bactericides.

Para formaldehyde, caustic soda, lime and starch preservatives are commonly used to reduce bacteria count. Other bactericides for drilling fluids include dithiocarbamic acid, hydroxamic acid, etc.

(3) Corrosion inhibitors.

Hydrated lime and amine salts are often added to inhibit corrosion. The organic corrosion inhibitors can be classified into the following broad groupings:amides and imidazolines，salts of nitrogenous molecules with carboxylic acids (i.e., fatty acids and naphthenic acids)，nitrogen quaternaries，polyoxylated amines, amides, imidazolines，nitrogen heterocyclics.

(4) Hydrogen sulfide remover.

It is necessary to remove hydrogen sulfide from a drilling mud, in order to inhibit corrosion and protect environment. Zinc compounds (usually Zinc carbonate hydroxide hydrate) have a high reactivity with regard to H_2S and therefore are suitable for the quantitative removal of even small amounts of hydrogen sulfide. However, at high temperatures they may negatively affect the rheology of drilling fluids.

Techniques using iron compounds that form sparingly soluble sulfides have been developed, e.g., with iron (II) oxalate and iron sulfate. The sulfur will be precipitated out as FeS. Ferrous gluconate is an organic iron-chelating agent, stable at pH levels as high as 11.5.

例如，用草酸亚铁和硫酸亚铁。硫以 FeS 的形式析出。葡萄糖酸亚铁是一种有机的铁螯合剂，在 pH 值高达 11.5 的条件下可以保持稳定。

Cement slurry chemistry
第 4 章 水泥浆化学

Cement slurry chemistry is a subject that studying composition, property, control and adjustment of cement slurry, the purpose of which is to seal thief zones, complex formation and protect producing formation and casing.

This chapter mainly introduces composition and hydration mechanism of the oil well cement, the performance of cement slurry, the cement admixtures, as well as the cement slurry system and its current application.

4.1 Basic composition of portland cement

Oil well cement refers to silicate cement (Portland cement) and cements of other types that being used in cementing, well servicing, squeeze cementing or other operations under various drilling conditions, including various modified cement or special oil well cement system that contains external admixtures or additions.

4.1.1 Class and application range of oil well cements

The API divided early basic oil well cements into nine classes: A, B, C, D, E, F, G, H and J. Among them, cement of J class is difficult to produce and the quality stability is low. Since cement of G class with sand and appropriate admixture can meet the requirements of deep well cementing, ISO cancelled the class of J in cement

水泥浆化学是通过研究水泥浆的组成、性能及其控制与调整，达到封隔漏失层、复杂地层和保护产层、套管的目的。

本章主要介绍了油井水泥及组成、水泥的水化机理、水泥浆的性能及水泥外加剂等方面的内容，同时对水泥浆体系及目前的应用情况进行了介绍。

4.1 油井水泥及组成

油井水泥是指应用于各种钻井条件下进行固井、修井、挤注等作业的硅酸盐水泥（波特兰水泥）和非硅酸盐水泥，包括掺有各种外掺料或外加剂的改性水泥或特种水泥的油井水泥体系。

4.1.1 油井水泥级别及使用范围

早期 API 将基本油井水泥分为 A、B、C、D、E、F、G、H 和 J 九个级别，由于 J 级水泥存在生产工艺困难、质量稳定性差等缺点，且采用 G 级水泥加砂再加适当的外加剂能达到深井固井的要求，因此 ISO 取消了 J 级水泥。同时，由于

system. In addition, since function of grade D, E and F can be replaced with grade G and H by adding retarder, the common basic cements are only with five classes, namely A, B, C, G and H. These five basic cements can be divided into three types: ordinary type (O), medium sulfate type (MSR) and high sulfate resistant type (HSR). These classes are summarized in Table 4-1.

D、E、F级水泥可由G、H级水泥加缓凝剂来代替，因此常用的基本水泥仅有A、B、C、G、H共5个级别。5种基本水泥可分为普通型（O）、中抗硫酸盐型（MSR）和高抗硫酸盐型（HSR）三类，其适用的井深及使用条件见表4-1。

Table 4-1　Classes of cements and properties

Clsss	Depth/m	Application conditions	Type
A	0-1900	General purpose	Ordinary (O) type
B	0-1900	Sulfate resistant	Medium (MSR) and high sulfate resistant type (HSR)
C	0-1900	High early strength	Ordinary type (O), medium sulfate type (MSR) and high sulfate resistant type (HSR)
G, H	0-2500	Covering a wide range of depth, temperatures and pressures; no additives besides CaSO4; can be used with accelerators or retarders	Medium sulfate type (MSR) and high sulfate resistant type (HSR)

表4-1　油井水泥适用的井深及使用条件

级　别	适用井深	使用条件	类　型
A	由地面至1900m井深的井	不需要特殊性能要求时使用	只有普通型
B	由地面至1900m井深的井	井下条件对水泥要求中抗或高抗硫酸盐时使用	中抗硫酸盐型和高抗硫酸盐型
C	由地面至1900m井深的井	井下条件对水泥要求具有高的早期强度时使用	普通型、中抗硫酸盐型和高抗硫酸盐型
G、H	由地面至2500m井深的井	可与外加剂混合使用，加促凝剂或缓凝剂可在较宽的井深范围内使用	中抗硫酸盐型和高抗硫酸盐型

In above oil well cements, the class of A, B, C, D, E and F are made by grinding cement clinker and appropriate amount of gypsum and grinding aids. G and H class oil well cement are made by grinding cement clinker and appropriate amount of gypsum (1%-3%) and water. The cement clinker is mainly composed by hydraulic calcium silicate.

在上述油井水泥中，A、B、C、D、E和F级油井水泥是在以水硬性硅酸钙为主要成分的水泥熟料中，加入适量的石膏和助磨剂磨细制成的产品。G、H级油井水泥是在以水硬性硅酸钙为主要成分的水泥熟料中，加入适量的石膏（1%~3%）和水磨细制成的产品。

4.1.2　Composition of oil well cement

(1) Chemical composition of oil well cement.
Uniformly mix the calcareous substance with clays

4.1.2　油井水泥组成

（1）油井水泥的化学组成。
凡将石灰质物质与黏土质物

or other substances containing silicon oxide, aluminum oxide and iron oxide, then calcine the mixture under sintering temperature, the final product will be cement clinker. The Portland cement, i.e. silicate cement will be then produced by grinding the cement clinker and gypsum powder. In China, the most frequently used oil well cement is silicate cement. Its chemical composition consists of the main oxides, free oxides and trace oxides. Among them, the main oxides include calcium oxide, silicon dioxide, aluminum oxide and iron oxide; free oxides include free calcium oxide and free silicon oxide; And trace oxides include magnesium oxide, some alkali, etc. The oxide components of typical silicate cement are shown in Table 4-2.

质或其他含氧化硅、氧化铝及氧化铁的物质均匀混合，在烧结温度下煅烧，并将所得的熟料与石膏粉研磨制得的产品称为波特兰水泥，即硅酸盐水泥。我国使用的油井水泥主要是硅酸盐水泥，其化学成分主要由氧化钙、二氧化硅、三氧化二铝和氧化铁等主要氧化物和游离氧化钙、游离氧化硅等游离氧化物及氧化镁、碱质等微量氧化物组成。典型的硅酸盐水泥氧化物成分见表4-2。

Table 4-2 Composition and proportion of oxides in typical silicate cement

Clsss	Proportion / %
Calcium oxide (CaO)	64.77
Silicon dioxide (SiO_2)	22.43
Aluminum oxide (Al_2O_3)	4.76
Ferric oxide (Fe_2O_3)	4.10
Sulfur trioxide(SO_3)	1.67
Magnesium oxide (MgO)	1.14
Potassium oxide (K_2O)	0.08
Other oxides	0.51
Loss on ignition	0.54

表 4-2 典型的硅酸盐水泥氧化物成分及所占比例

氧化物	百分比/ %
氧化钙（CaO）	64.77
二氧化硅（SiO_2）	22.43
三氧化二铝（Al_2O_3）	4.76
三氧化二铁（Fe_2O_3）	4.10
三氧化硫（SO_3）	1.67
氧化镁（MgO）	1.14
氧化钾（K_2O）	0.08
其他	0.51
烧失量	0.54

(2) Mineral composition of oil well cement.

Oil well cement clinker is an unbalanced multi-component solid solution system. It consists of many different minerals and intermediates, which are mainly

（2）油井水泥的矿物组成。

油井水泥熟料是一种不平衡的多组分固溶体系统，由许多不同的矿物和中间体组成，其主要

composed of four solid solutions: tricalcium silicate, dicalcium silicate, tricalcium aluminate and tetracalcium aluminoferrite.

① Tricalcium silicate (C_3S): With a proportion of 40%~60%, $3CaO \cdot SiO_2$ is with the highest content in cement. Strength of the cement paste is mainly determined by it. Hydration rate of C_3S is quite fast, which will influence form of cement compressive strength a lot.

② Dicalcium silicate (C_2S): $2CaO \cdot SiO_2$ takes 15%~35% of the cement component. Its slow hydration rate will not influence initial setting time of the cement, but the final strength may be affected.

③ Tricalcium aluminate (C_3A): $3CaO \cdot Al_2O_3$ takes 6%~15% of the cement component. It is a compound that hydrates quickly, and is a main component that will influence initial setting and thickening time, which responsible for strength of cement in early stage. The setting time can be controlled by addition of gypsum.

④ Tetracalcium aluminoferrite (C_4AF): $4CaO \cdot Al_2O_3 \cdot Fe_2O_3$ takes 8%~10% of the cement component. Increase of this component will reduce strength of the cement paste, but have little effect on compressive strength of the cement.

4.2 Hydration of the cement

4.2.1 Hydration of silicate

Since the main mineral component of oil well cement is tricalcium silicate, the hydration of silicate is mainly referred to hydration of tricalcium silicate. Hydration reaction of C_3S under normal temperature can be expressed by the following equation:

由硅酸三钙、硅酸二钙、铝酸三钙和铁铝酸四钙4种固溶体组成。

① 硅酸三钙（C_3S）：$3CaO \cdot SiO_2$ 占 40%~60%，其在水泥中含量最高，水泥石强度主要由它形成。硅酸三钙水化速度很快，对水泥抗压强度的形成有很大影响。

② 硅酸二钙（C_2S）：$2CaO \cdot SiO_2$ 占 15%~35%，其水化速度缓慢，不影响水泥初凝时间，但对水泥最终强度有影响。

③ 铝酸三钙（C_3A）：$3CaO \cdot Al_2O_3$ 占 6%~15%，是促进水泥快速水化的化合物，也是决定水泥初凝时间和稠化时间的主要成分，对水泥早期强度有影响。加入石膏可以控制 C_3A 的凝结时间。

④ 铁铝酸四钙（C_4AF）：$4CaO \cdot Al_2O_3 \cdot Fe_2O_3$ 占 8%~10%，其含量增高会使水泥石强度降低，对水泥抗压强度影响较小。

4.2 水泥的水化

4.2.1 硅酸盐水化

油井水泥的主要矿物成分是硅酸三钙，因而硅酸盐水化主要是硅酸三钙的水化。C_3S 在常温下的水化反应，可用下列方程式表示：

$$3CaO \cdot SiO_2 + nH_2O = xCaO \cdot SiO_2 \cdot yH_2O + (3-x)Ca(OH)_2$$

即

i.e.

$$C_3S + nH = xC\text{–}S\text{–}H + (3-x)CH \qquad (4\text{-}1)$$
$$C_3S + nH = xC\text{–}S\text{–}H + (3-x)CH \qquad (4\text{-}1)$$

The hydration rate of tricalcium silicate is very fast at first, and then the rate will be changed greatly according to different time periods. The hydration exotherm curve of C_3S is shown in Figure 4-1.

硅酸三钙的水化速度一开始很快，但在随后的不同时间段差别较大，C_3S 的水化放热曲线如图 4-1 所示。

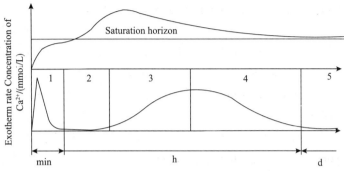

Figure 4-1　Curves of C_3S hydration exotherm rate and Ca^{2+} concentration curve

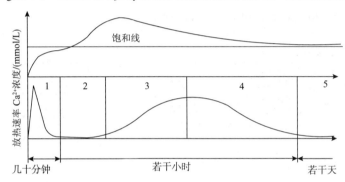

图 4-1　C_3S 的水化放热速度和 Ca^{2+} 浓度变化曲线

Hydration of tricalcium silicate is a process of exotherm, during which the hydration rate can be divided into five stages:

(1) Preinduction period: After being mixed with water, it will react rapidly. At this point, a large amount of heat could be observed and a C—S—H gel hydrated layer will be formed on anhydrous surface of the C_3S.

(2) Induction period: The reaction rate is extremely slow, which helps the silicate cement slurry remain fluidity for several hours. In this period, the excess C—S—H colloid will slowly precipitate, concentration of Ca^{2+} and OH^- will continue to rise, and calcium hydroxide will precipitate when the critical saturation point is reached. Then a new round of hydration reaction starts, marking end of the induction period.

硅酸三钙的水化是一个放热过程，其水化速率可以划分为五个阶段：

（1）诱导前期：与水混合后立即发生急剧反应，此时可观察到大量的热生成，并在无水的 C_3S 表面形成一层 C—S—H 凝胶水化层。

（2）诱导期：反应速率极其缓慢，是硅酸盐水泥浆体能在几小时内保持具有流动性的原因。多出的 C—S—H 胶质缓慢沉淀，Ca^{2+} 和 OH^- 浓度继续上升，当达到临界饱和点时开始析出氢氧化钙沉淀。此时，又开始发生水化

(3) Accelerating period: During this period, the hydration reaction will accelerate again, and the reaction rate will increase with time. The second exothermic peak will appear in the accelerating period, marking ending of this period and begining of hardening macroscopically.

(4) Deceleration period: In this period the reaction rate will decrease with time. Generally the deceleration period will last for 12-24 hours, and the hydration will be gradually controlled by the diffusion rate.

(5) Stable period: In this period, reaction rate is very low and the hydration is completely controlled by diffusion rate.

The dctailed hydration process of C_3S is shown in Figure 4-2.

反应，标志着诱导阶段的结束。

（3）加速期：水化反应重新加快，反应速率随时间而增长，出现第二个放热峰，到达峰顶时加速期结束，宏观上硬化开始。

（4）减速期：反应速率随时间下降的阶段，一般持续12~24h，水化作用逐渐受扩散速率的控制。

（5）稳定期：反应速率很低，水化作用完全受扩散速率的控制。

C_3S的详细水化过程如图4-2所示。

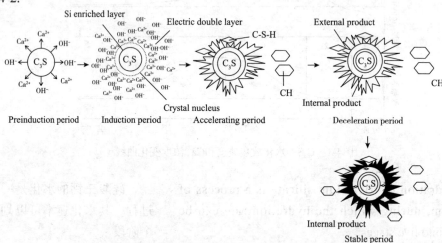

Figure 4-2　Schematic diagram for hydration process of C_3S

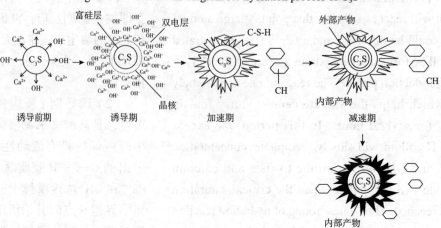

图4-2　C_3S水化机理及水化过程示意图

4.2.2 Hydration of aluminate

Aluminate includes tricalcium aluminate and tetracalcium aluminoferrite. Although hydration of tetracalcium aluminoferrite is similar with that of tricalcium aluminate, the hydration rate of the former is much lower. Because of this, the hydration of tricalcium aluminate could represent hydration characteristics of all aluminates. Hydration activity of tricalcium aluminate is very strong, hence will influence fluidity of cement slurry and early strength of solidified cement greatly. Hydration reaction of C_3A can be expressed by the following reaction formula:

4.2.2 铝酸盐水化

铝酸盐包括铝酸三钙和铁铝酸四钙，由于铁铝酸四钙的水化作用和铝酸三钙很相似，但水化速度却低很多，因此铝酸三钙的水化作用就代表了铝酸盐的水化特征。铝酸三钙水化活性很强，其对水泥浆的流动性和凝固水泥的早期强度有很大影响。C_3A 的水化反应可用如下水化反应式表示：

$$2C_3A+27H \longrightarrow C_2AH_8 + C_4AH_{19} \qquad (4\text{-}2)$$

$$2C_3A+27H \longrightarrow C_2AH_8 + C_4AH_{19} \qquad (4\text{-}2)$$

Hydration products in formula(4-2) are unstable hexagonal crystal which will eventually transform into stable C_3AH_6 crystal cubes. The specific reaction is shown in formula (4-3). Under certain conditions, this reaction will last for several days.

式（4–2）中的水化产物是不稳定的六角形晶体，其最终转变成稳定的正方体晶体 C_3AH_6，具体如式（4–3）所示。在一定条件下，这一反应要持续几天。

$$C_2AH_8 + C_4AH_{19} \longrightarrow 2C_3AH_6 + 15H \qquad (4\text{-}3)$$

$$C_2AH_8 + C_4AH_{19} \longrightarrow 2C_3AH_6 + 15H \qquad (4\text{-}3)$$

Hydrated calcium aluminate is a crystal structure, thus protective layer will not be formed on its surface. Due to this reason, it will reach to full hydration very soon. If not being controlled, it will influence rheology of the cement slurry seriously. Hydration of calcium aluminate can be controlled by adding 3% ~ 5% gypsum to the clinker. After being contact with water, gypsum will react with calcium aluminate quickly and form ettringite. Chemical formula of this reaction is:

水化铝酸钙是晶体结构，它的表面并不形成保护层，很快就达到完全水化。如果不加以控制，则会对水泥浆流变性产生严重影响。在熟料中加入3%~5%的石膏，可控制铝酸钙水化。石膏与水接触后，立即会与铝酸钙反应，生成钙矾石，化学反应式如下：

$$C_3A + 3CS\cdot 2H + 26H \longrightarrow C_3A\cdot 3CS\cdot 32H \qquad (4\text{-}4)$$

$$C_3A + 3CS\cdot 2H + 26H \longrightarrow C_3A\cdot 3CS\cdot 32H \qquad (4\text{-}4)$$

The needle ettringite crystal precipitates on surface of C_3A , which reduce hydration rate and is resulting

钙矾石以针状晶体形式沉淀在 C_3A 表面上，阻止了水化迅速

in the "induction stage". During this stage, amount of gypsum gradually decreases while ettringite continues to precipitate. After all of the gypsum is reacted, the delay period of C_3A hydration ends and the rapid hydration period will begin. Then concentration of SO_4^{2-} drops rapidly, and ettringite becomes unstable. It will convert into flaky calcium sulphoaluminate hydrate:

进行，产生了"诱导阶段"。在这一阶段，石膏逐渐减少，而钙矾石继续沉淀。当石膏耗尽时，C_3A 水化延缓期结束，快速水化期开始。SO_4^{2-} 浓度迅速降低。钙矾石变得不稳定，而转变成片状硫铝酸钙水合物：

$$C_3A \cdot 3CS \cdot 32H + 2\ C_3A + 4H \longrightarrow 3C_3A \cdot CS \cdot 12H \quad (4-5)$$

$$C_3A \cdot 3CS \cdot 32H + 2\ C_3A + 4H \longrightarrow 3C_3A \cdot CS \cdot 12H \quad (4-5)$$

4.2.3 Hydration of the Portland cement

Hydration of Portland cement is actually a composite chemical reaction among mineral component of cement clinker, calcium sulfate and water. By this reaction, the cement will be thicken and harden gradually. The hydration reaction of C_3S is usually used as a model for hydration reaction of Portland cement. Of course, hydration reaction of other components should also be included.

Hydration reaction of Portland cement is a complex dissolution/precipitation process. This process is different from hydration reaction of a single component. This is because the hydration reactions of different components are simultaneously carried out at different rates, and there are interactions or restrictions among these components.

Figure 4-3 is a hydration exotherm curve of the typical Portland cement, which can also be roughly considered as the exotherm curve of C_3S and C_3A according to its concentration.

4.2.3 波特兰水泥的水化

波特兰水泥的水化，实质上是水泥熟料中矿物组分和硫酸钙与水之间发生复合化学反应，而使水泥逐步稠化和硬化。通常，人们用 C_3S 的水化反应作为波特兰水泥水化反应的模型，当然还应包括其他成分的水化反应。

波特兰水泥的水化反应是一个复杂的溶解/沉淀过程。在这一过程中，与单纯一种成分的水化反应是不同的，因各组分是以不同的水化反应速度同时进行的，而且各组分之间还相互影响或制约。

图 4-3 是典型的波特兰水泥水化放热曲线，该曲线也可以粗略地认为是 C_3S 和 C_3A 随其浓度变化的放热曲线。

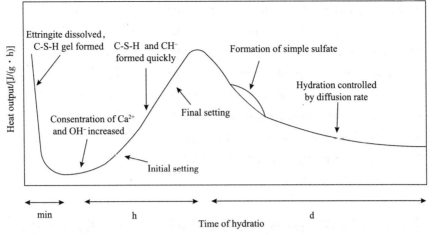

Figure 4-3 Hydration exotherm curve of Portland cement

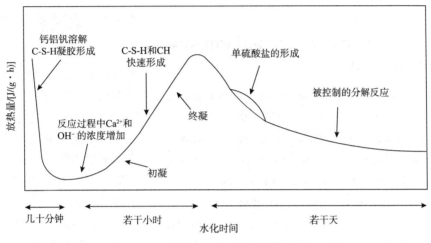

图4-3 波特兰水泥水化放热曲线

4.2.4 Factors affecting hydration of oil well cement

(1) Mineral components of cement: since hydration rate of each mineral component in cement is different, when mineral composition of the oil well cement changes, the hydration rate will also change.

(2) Temperature: hydration rate will increase with temperature. Besides, temperature will also affect properties of the hydrates, as well as stability and morphology of the Portland cement.

(3) Fineness of oil well cement: the finer the cement is, the better the reactivity of the cement particles and the faster the hydration rate will be.

4.2.4 影响油井水泥水化作用的因素

（1）水泥各个矿物组分的影响：水泥各个矿物组分的水化速度不同，当油井水泥矿物组分改变时，其水化速度也会改变。

（2）温度：温度的升高可增加水泥的水化速度，且温度还会影响波特兰水泥的水化物性质、稳定性和形态等。

（3）油井水泥细度：水泥细度值越大，水泥粒子的反应活性越大，其水化速度越快。

(4) Water cement ratio: with the same time duration, the greater the water cement ratio is, the higher the hydration degree of oil well cement will be. i.e., the hydration rate will be accelerated with increase of water cement ratio. However, in order to ensure cementing quality in the actual cementing operations, the water cement ratio is usually controlled in range from 40% to 70%.

(5) Pressure: Increase of pressure will accelerate hydration of the oil well cement.

4.3　Important properties of cement slurry

4.3.1　Requirements on performance of cement slurry

Due to particularity of the cementing operations, the cement slurry needs to meet the safety pumping requirements and has sufficient strength to support casing and seal the formation. Therefore, the cement slurry should be with the following performances:

(1) The required density of cement slurry that will not settle can be prepared, and it is with good fluidity; with proper initial consistency; homogeneous and non-foaming.

(2) Easy for mixing and pumping; with good dispersion and low frictional resistance.

(3) Good rheology and high replacement efficiency;

(4) Could maintain required physical and chemical properties during cement injection, waiting-on, and hardening.

(5) With low water loss, will not permeate after solidification.

(6) With a early strength that fast enough.

(7) With large enough cementation strength that is between cements and between casing and the formation;

(8) With certain stability and able to resist corrosion of formation water.

4.3　水泥浆的重要性能

4.3.1　水泥浆的性能要求

由于固井工程的特殊性，水泥浆需要满足安全泵注，并具备足够的强度以支撑套管，封隔地层。因此，水泥浆应具备以下特性：

（1）能配成需要密度的水泥浆，不发生沉降且具有良好的流动度，初始稠度适宜，且均质和不起泡。

（2）易混合与泵送，分散性要好，摩擦阻力小。

（3）流变性好，顶替效率高；

（4）在注水泥、候凝、硬化期间应保持需要的物理性能及化学性能。

（5）失水量小，固化后不渗透。

（6）具有足够快的早期强度。

（7）具有足够大的套管、水泥和地层间的胶结强度。

（8）具有一定的稳定性，具有抗地层水的腐蚀能力。

（9）满足射孔强度要求。

（10）满足所要求条件下的稠化时间和抗压强度。

(9) Could meet requirements of perforation strength.

(10) Could reach to proper thickening time and compressive strength under all required conditions.

4.3.2　Properties of cement slurry

(1) Density.

In design of cement slurry, density is a basic factor affecting the physical properties of cement slurry. Mass of cement slurry per unit volume is called density of the cement slurry. Density of the dry oil well cement powder is 3.15g/cm^3, and density of normal cement slurry is 1.78-1.98g/cm^3. Since formations are with different pressure bearing capacities, there is a wide range of requirements for density of cement slurry. Thus normal density will be the standard for classification. The ones with density lower than the normal are called low density cement slurries, and if the density is higher than the normal, it will be called high density cement slurry.

The main representative types of ultra-low density cement slurry are foam cement and microbead cement. Density range of foam cement slurry is 0.84-1.32g/cm^3; and density range of microbead cement slurry is 1.08-1.44g/cm^3. A common method for obtaining high density cement slurry is to add weighting agent. For example, by adding sand, barite and hematite, density of cement slurry could reach to 2.16g/cm^3, 2.28 g/cm^3 and 2.4g/cm^3.

(2) Filtrate-loss of cement slurry.

Water loss of cement slurry refers to the amount of free water that can be lost through a certain area of pores within 30 minutes under specified temperature and pressure. In general, filtrate loss of pure cement slurry is about 1000-2000mL/30min at normal temperature and 0.7-6.9MPa. API had summarized the relationship between water loss and cementing control based on extensive cementing practices and studies (Table 4-3).

4.3.2　水泥浆的性能

（1）密度。

水泥浆密度是水泥浆设计中影响水泥浆物理性能的基本因素。单位体积的水泥浆质量称为水泥浆的密度。油井水泥干灰密度为3.15g/cm^3，正常水泥浆的密度为1.78~1.98g/cm^3。由于地层承压能力不同，因此对水泥浆密度有较大范围的要求。低于正常密度的称低密度水泥浆，高于正常密度的称高密度水泥浆。

超低密度水泥浆的主要代表类型为泡沫水泥及微珠水泥。泡沫水泥浆密度范围为0.84~1.32g/cm^3；微珠水泥浆密度范围为1.08~1.44g/cm^3。获得高密度水泥浆常用的方法是掺入加重剂，加砂可获得密度为2.16g/cm^{3}的水泥浆，加重晶石可获得密度为2.28g/cm^{3}的水泥浆，加赤铁矿可获得密度为2.4g/cm^{3}的水泥浆。

（2）水泥浆滤失量。

水泥浆失水是指在指定的温度和压差下，30min内通过一定面积孔隙所能滤失的自由水量。一般说来，纯水泥浆在常温、0.7~6.9MPa压差下的滤失量大约是1000~2000mL/30min。API根据大量的固井实践和研究，总结出失水量和固井控制的关系（表4-3）。

Table 4–3 The relationship between water loss and cementing control

water loss/(mL/30min)	cementing control
0-200	Easy to be controlled
200-500	Can be controlled
500-1000	Hard to be controlled
1000-2000	Cannot be controlled

表4–3 失水量和固井控制关系

失水量/（mL/30min）	固井控制
0～200	很好控制
200～500	中等控制
500～1000	勉强控制
1000～2000	不有控制

(3) Rheology of cement slurry.

The rheological properties of cement slurry are the basic parameters of cementing design, which will directly affect the displacement efficiency. The rheological properties of cement slurries are usually expressed by plastic viscosity, dynamic shear force, consistency coefficient and fluidity index. It will be expressed by fluidity sometimes also since it is relatively intuitive. When fluidity of the cement slurry reaches 20-30cm, it shows good flowability; the flowability will not be good when fluidity of the cement slurry is less than 18cm, and when the fluidity is greater than 30cm, sedimentation stability of the cement slurry is poor. To measure rheological parameters of the cement slurry, normal or high pressure viscometer shall be used when temperature is lower than 87℃; if the temperature is higher than 87℃, the high pressure viscometer shall be used.

(4) Thickening time of cement slurry.

Thickening time of cement slurry refers to the time duration that cement slurry loses its flowability during the flow process. According to GB 10238 Oil Well Cement, the time duration from start of slurry mixing to consistency reach to 100Bc could be taken as the pumpable limit time. However, for most cementing service companies，the standard of the thickening time or pumpable time for on-site operation is the time

（3）水泥浆流变性。

水泥浆流变性能是注水泥工艺设计的基本参数，其直接影响顶替效率。水泥浆的流变性能常用塑性黏度、动切力、稠度系数和流性指数表示，有时也用直观的流动度表示。当水泥浆的流动度达到20~30cm时，其流动性能较好；而当水泥浆流动度小于18cm时，其流动性能较差；流动度大于30cm时，水泥浆的沉降稳定性较差。对于水泥浆流变参数的测定而言，温度低于87℃时用常压或高压黏度计测定，而温度高于87℃时用高压黏度计测定。

（4）水泥浆稠化时间。

水泥浆的稠化时间是指水泥浆在流动过程中丧失流动能力的时间。GB 10238《油井水泥》规定：从开始混浆到稠度达100Bc所经历的时间为可泵送的极限时间。而多数固井服务公司都把稠度达到40Bc的时间定为现场施工的稠化时间或可泵送时间，这个时限是以实验室测试和被限定的现场经验为基础选定的。

duration that the consistency reach to 40Bc.This time standard was selected based on laboratory testing and the specific field experiences.

Thickening time of the cement slurry should be long enough to ensure safe pumping of the cement slurry under wells, but it should not be too long. This is because excessive long thickening time will prolong time of waiting-on-cement (WOC), increase the risk of oil/gas/water channeling and the non-working time of drilling. In general, the thickening time decreases with increasing temperature, hence it is important that the temperature conditions in the well are known.

(5) Water cement ratio of cement slurry.

The ratio between weights of water to dry cement is called water cement ratio, which affects density of the cement slurry. Small water cement ratio will result in a poor fluidity and large pumping resistance; but if the ratio is too large, the cement particles will sink and a large amount of free water will be appeared. The free water will merge and rise, destroy sealing property and reduce strength of the cement pastes. The frequent used cement slurry on-site is with a water cement ratio around 0.5. After being prepared into cement slurry, its density is in range from 1.78 to 1.98g/cm^3.

(6) Performance of cement slurry after solidification.

Compressive strength of cement paste: compressive strength refers to the pressure exerted per unit area when cement sample is destroyed. Under normal conditions, the solidified cement paste must withstand horizontal loads caused by the formation pore pressure and the axial load caused by weight of the casing, as well as the various pressures resulting from completion operations and reservoir stimulation.

Time of WOC: under normal circumstances, time of WOC for surface casing is 12h (sometimes 18 ~ 24h); time of WOC for the intermediate casing is 12 ~ 14h; and time of WOC for the production casing is mainly 24h.

Strength degradation of cement paste at high

temperature: under normal conditions, cement will solidify under the well and its strength will increase during continuous hydration. However, when well temperature exceeds 110℃, its strength value will decrease after a certain period of time. The higher the temperature is, the faster the strength regression will be. When temperature is in range of 110-120°C, the strength decays slowly, and will be destroyed within one month at 230°C. At 310°C, the strength will be destroyed within a few days. If the cement belongs to bentonite system with high water cement ratio, its high temperature stability will be worse.

(7) Stability of cement slurry.

Stability is one of the important properties of cement slurry. The column formed by cement slurry with poor stability is very uneven in density from top to bottom. In highly deviated wells and horizontal wells, this unevenness is more prominent. From low side to high side of the wellbore, the compactness and cementation of the cement paste are getting weakened, which will influence adversely on sealing quality of the cement sheath. Generally, cement slurries with poor stability are with many free liquids, which will also form channels for oil, gas and water in the cement column. This will affect sealing quality of the cement sheath also.

4.4 Cement additive

Additives and admixtures for oil well cement are materials used to adjust physical and chemical properties of the cement. They are auxiliary materials for oil well cement, and are also the important raw materials of cementing. The difference between additive and admixture is generally defined by the amount that added. If the added amount is more than 5% of cement mass, it will be called an admixture, if the amount is less than 5%, it is called an additive. Cement additive can be mainly classified into the following categories: retarders, accelerants, fluid loss control agents, weighting agents, lightening agents, and dispersants, etc.

4.4.1 Important additives for cementing

(1) Retarders.

During cementing operation in the high temperature and high pressure environments of deep wells, it is necessary to add an additive to prolong thickening time of the cement slurry, this additive is called retarder. Function of the retarder is to increase the thickening time of the cement slurry, reduce the viscosity, extend the pumpable time and prevent rapid setting. Good retarders should be able to work in a certain temperature range. In general, dosage of the retarder is quite small, sometimes only a few thousandths, but dosage error will influence the cement thickening time greatly. If the dosage is too large, the cement slurry will be super retarded.

①Types of retarder.

There are several types of oil well cement retarders: lignosulfonate, hydroxyl carboxylic acid and its salt, tannic acid, carbohydrates, cellulose derivatives, phosphates，inorganic compound boric acid and its salt, etc. Among them, the lignosulfonate and its modified products are common retarders，and their molecular formulas are as follows.

4.4.1 常用油井水泥外加剂

（1）缓凝剂。

高温高压深井固井施工中需要在水泥浆中添加延长稠化时间的外加剂，即缓凝剂，以增加水泥浆的稠化时间、降低黏度、延长可泵送时间、阻止水泥迅速固化等。好的缓凝剂应在一定温度区间内都具有缓凝作用。缓凝剂的加量一般较小，有时只有千分之几，缓凝剂加量的误差对水泥稠化时间的影响很大，加量过大会造成水泥浆超缓凝。

①缓凝剂类型。

油井水泥缓凝剂有以下几种类型：木质素磺酸盐类、羟基羧酸及其盐类、单宁酸类、糖类化合物、纤维素衍生物、磷酸盐类、无机化合物硼酸及其盐类缓凝剂等。其中，常用缓凝剂木质素磺酸盐及其改性产物的分子式如下所示。

M'：Na^+或$1/2Ca^{2+}$
木质素磺酸钠或木质素磺酸钙

M'：Na^+ or $1/2Ca^{2+}$
Sodium lignosulfonate or calcium lignosulfonate

M：Fe或Cr
铁铬木质素磺酸盐

M：Fe or Cr
Fe-Cr lignosulphonate

The common retarders and their application conditions are shown in Table 4-4.

Table 4-4 Common retarder models and application conditions

Model	Temperature/°C	Dosage/%	Components and physical properties	Manufacturer
GH-1	<110	0.1-1.0	Composed of various compounds such as cellulose derivatives and hydroxyl carboxylic acids; light yellow powder; suitable for dry mixing generally	Weihui, Henan
GH-7	110-140	0.5-2.0	Composed of various compounds such as gluconate and hydroxyl carboxylic acid; brown liquid; suitable for wet mixing	Weihui, Henan
SWH-2	100-150	0.2-1.5	Odorless solid white granular powder; Can be used for dry or wet mixing	Shandong World Group
DZH2	70-140	0.1-2.0	Light pink liquid	Sinopec Research Institute of Petroleum Engineering
DZH3	70-180	0.5-4.0	Light brown liquid	Sinopec Research Institute of Petroleum Engineering

目前，常用的缓凝剂及使用条件见表4-4。

表4-4 常用缓凝剂型号及使用条件

产品代号	适用温度/°C	加量范围/%	主要成分及物理特征	生产厂家
GH-1	<110	0.1~1.0	由纤维素衍生物、羟基羧酸等多种化合物组成，外观为浅黄色粉末，一般适于干混	河南卫辉
GH-7	110~140	0.5~2.0	由葡萄糖酸盐、羟基羧酸等多种化合物组成，外观为棕色液体，适于湿混	河南卫辉
SWH-2	100~150	0.2~1.5	白色无气味颗粒状固体粉末，既可干混也可湿混	山东沃尔德
DZH2	70~140	0.1~2.0	浅粉色液体	中国石化工程院
DZH3	70~180	0.5~4.0	棕褐色液体	中国石化工程院

② Mechanism of retarder.

Main mechanism of the retarder is to inhibit hydration speed of the cement minerals. There are two mechanisms in total:

a. Selective adsorption mechanism of cement minerals.

The order of organic retarders selectively adsorbed by cement minerals is:

$C_3A-C_4AT \rightarrow C_3S-C_2S$

When a retarder is added to cement slurry, it will selectively adsorb on surface of cement minerals and form a protective film, which will inhibit the hydration speed of aluminate component. While inhibiting hydration of C_3A, it will also inhibit hydration of silicate components (C_3S, C_2S) causing retarding effect.

②缓凝剂的作用机理。

缓凝剂起缓凝作用主要是抑制了水泥矿物的水化速度，综合起来有以下两种机理：

a. 水泥矿物选择性吸附机理。

一般水泥矿物选择性吸附有机缓凝剂的顺序为：

$C_3A-C_4AT \rightarrow C_3S-C_2S$

水泥浆中加入缓凝剂时，缓凝剂会选择性地吸附在水泥的矿物上，形成一层保护膜，抑制了铝酸盐组分的水化速度。在抑制C_3A水化的同时对硅酸盐组分(C_3S、C_2S)的水化也起抑制作用，并可起到缓凝作用。

b. Activity of groups in retarder molecules.

The organic molecules can cause retarding effects resulting from the structure of retarders. These organic molecules include cellulose, calcium gluconate (sodium), tannins and hydroxy acids containing retarding reactive groups, such as hydroxyl groups (—OH), ketone groups (—C=O), and the like which will adsorb on surface of cement minerals and inhibit the hydration.

(2) Accelerants.

Accelerants are chemicals used to shorten the thickening time of cement slurries. To add cement accelerants can accelerate setting and hardening of the cement or can alleviate the excessive retardation caused by addition of other additives (such as dispersants, fluid loss control agents, etc.).

①Types of accelerant.

Common accelerants are chloride accelerants (such as NaCl, $CaCl_2$, KCl, etc.), chlorine-free accelerants (sodium carbonate, sodium silicate, gypsum, formamide, triethanolamine, etc.), composite accelerants, and suspending accelerants, etc. The common accelerants and their application conditions are shown in Table 4-5.

Table 4-5 Common accelerant models and application conditions

Model	Temperature/℃	Dosage/%	Components and physical properties	Manufacturer
G201	30-60	0.3-2.0	Composed of alkaline inorganic salts; white granule or powder; can be used for dry or wet mixing	Weihui, Henan
G202	30-90	0.8-1.5	Modified from inorganic early strength material; white or slightly gray powder; non-toxic and odorless	Weihui, Henan
SWC-1	30-120	0.5-2.0	Prepared by many chemical materials; grayish white solid powder; odorless	Shandong World Group
DZC	30-120	0.5-2.0	Prepared by many chemical materials; bright white solid	Sinopec Research Institute of Petroleum Engineering

表4-5 常用促凝剂型号及使用条件

产品代号	适用温度/℃	加量范围/%	主要成分及物理特征	生产厂家
G201	30~60	0.3~2.0	由碱性无机盐组成，外观为白色颗粒或粉末，干混或湿混均可	河南卫辉
G202	30~90	0.8~1.5	由无机早强材料改性而成，外观为白色或略带灰色粉末，无毒、无臭	河南卫辉
SWC-1	30~120	0.5~2.0	由多种化学材料组成，外观为灰白色固体粉末，无气味	山东沃尔德
DZC	30~120	0.5~2.0	由多种化学材料组成，外观为亮白色固体	中国石化工程院

② Mechanism of accelerant.

Taking calcium chloride as an example, the mechanism of accelerant is as follows:

a. Impact on hydration of oil well cement.

$CaCl_2$ can accelerate hydration rate of aluminate or gypsum system to promote coagulation. The chloride ions will promote formation of ettringite until the gypsum is consumed. Growth of micro acicular ettringite will accelerate condensation of the cement.

b. Change on C—S—H structure of hydrated silicate gel.

$CaCl_2$ can convert the C—S—H gel into a flocculent loose structure, which will significantly increase the hydration rate.

c. Diffusion effect of chloride ions.

Since diffusion coefficient of chloride ions is much larger than that of calcium ions, the chloride ions will enter stern layer of hydrated calcium silicate gel firstly. This will inevitably cause hydroxide ions diffusing inversely to maintain the electrical balance, thus forming deposit of portlandite. The process will also lead to an end of the hydration induction period prematurely.

d. Change in composition of liquid phase.

Presence of $CaCl_2$ will strongly change ion distribution in liquid phase of the cement slurry. Chloride ion will not participate in composition of hydration products during cement hydration induction period, but due to intervention of chloride ions, the concentration of hydroxide and sulfate ions

② 促凝剂的作用机理。

以氯化钙为例，介绍促凝剂的作用机理如下：

a. 对油井水泥水化作用的影响。

$CaCl_2$ 可以加速铝酸盐或石膏体系的水化速度而起到促凝作用，氯离子促进钙矾石的形成，而且直到石膏被消耗完为止。微针状结晶钙矾石的生长将加快水泥凝结。

b. 水化硅酸盐凝胶 C—S—H 结构变化。

$CaCl_2$ 的存在可使 C—S—H 凝胶转化为絮状松散结构，明显提高水化速率。

c. 氯离子的扩散作用。

氯离子的扩散系数比钙离子大得多，因此氯离子扩散进入水化硅酸钙凝胶屏蔽层的速度大于钙离子的速度，这样势必引起氢氧根离子向外反扩散来保持电平衡，因而形成了羟钙石沉积，进而使水化诱导期提前结束。

d. 液相组成的变化。

$CaCl_2$ 的存在强烈地改变了水泥浆液相中的离子分布。氯离子在水泥水化诱导期中并不参与水化产物的组成，但由于氯离子的介入，降低了氢氧根和硫酸根离

is reduced, thus concentration of calcium ions is increased.

The coagulation accelerating effect of $CaCl_2$ on oil well cement is very complex, which includes many physical and chemical effects. In recent years, new mechanisms have been proposed for coagulation accelerating effect of oil well cement accelerants, including: common ion effect, form precipitates or crystals, catalysis.

(3) Fluid loss control agents.

In order to prevent and reduce water loss while cement slurry flowing through permeable formations, a filtration control additive is generally added. In primary cementing, the loss of fluid, i.e., water, to permeable subterranean formations can result in the premature gelation of the cement composition.

① Types of fluid loss control agent.

There are two types of material that can be used as fluid loss control agents for oil well cement: solid particulate materials (bentonite) and water soluble polymers (cellulose derivatives, nonionic synthetic polymers, anionic synthetic polymers and cationic polymers, etc.). Some of water soluble polymers that can be used as fluid loss control additives in cementing operations are shown below.

子的浓度，提高了钙离子的浓度。

$CaCl_2$ 对油井水泥促凝作用是错综复杂的，其中包括了许多物理、化学作用。近年来，对油井水泥促凝剂的作用机理又提出了新的理论，主要包括：同离子效应，形成沉淀物或者结晶物，催化作用。

（3）降失水剂。

为了防止和减少水泥浆在流经渗透地层时产生的失水，一般要加入降失水剂。在初次固井过程中，随着流体（即水）的损失而进入渗透性地层，这会导致水泥成分的提前固化。

① 降失水剂类型。

可作为油井水泥降失水剂的有两种材料：固体颗粒材料（膨润土）和水溶性高分子聚合物（纤维素衍生物、非离子人工合成聚合物、阴离子人工合成聚合物、阳离子聚合物等），可作为水泥浆降失水剂的部分水溶性聚合物如下所示。

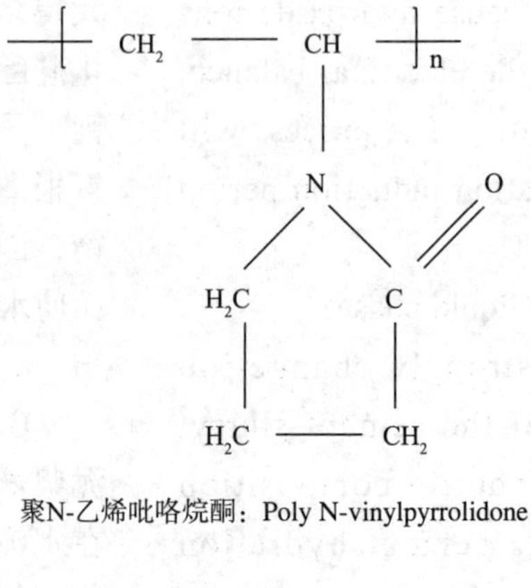

聚N-乙烯吡咯烷酮：Poly N-vinylpyrrolidone

Cement slurry chemistry
水泥浆化学 第4章

钠羧甲基纤维素: Sodium carboxymethyl cellulose

羟乙基纤维素: Hydroxyethyl cellulose

X: CH₂COONa 或 CH₂CH₂OH
X: CH$_2$COONa or CH$_2$CH$_2$OH

钠羧甲基羟乙基纤维素: Sodium carboxymethyl hydroxyethyl cellulose

The common fluid loss control additives and their application conditions are shown in Table 4-6.

目前，常用的降失水剂及使用条件见表4-6。

Table 4-6 Model and application condition of common fluid loss control additives

Model	temperature /℃	Dosage / %	Components and physical properties	Manufacturer
G301	40-120	0.8-2.0	Composed of various cellulose derivatives and other related promoters; brown powder; water loss of API less than 250mL; can be used for dry or wet mixing	Weihui, Henan
G315	40-120	5-7	Composed of a variety of water soluble polymers and other promoters; colorless transparent viscous liquid; suitable for wet mixing	Weihui, Henan

续表

Model	temperature /℃	Dosage / %	Components and physical properties	Manufacturer
SWJ-1	70-120	0.4-2.0	Composed of polymer and specific auxiliary material; brown powder; suitable for dry mixing	Shandong World Group
SWJ-2	90-110	0.4-2.5	Blended by polyene polymer and specific auxiliary materials; white or reddish solid powder; suitable for dry mixing	Shandong World Group
SWJ-3	90-150	3.0-5.0	Composed by a water soluble polymer; colorless transparent viscous liquid; with salt resistant effect; suitable for wet mixing	Shandong World Group
FSAM	30-90	3-8	Mainly composed by polyvinyl alcohol; colorless viscous liquid; suitable for wet mixing	Sinopec Research Institute of Petroleum Engineering
SCF	30-180	2-6	Composed by quinary copolymers of acrylic acid, acrylamide and other components; colorless viscous liquid; suitable for wet mixing	Sinopec Research Institute of Petroleum Engineering

表4-6 常用降失水剂型号及使用条件

产品代号	适用温度/℃	加量范围/%	主要成分及物理特征	生产厂家
G301	40~120	0.8~2.0	由多种纤维素衍生物及其他相关助剂组成，外观为棕色粉末，可使API失水小于250mL，干混或者湿混均可	河南卫辉
G315	40~120	5~7	由多种水溶性聚合物和其他助剂复合而成，外观为无色透明黏稠液体，适于湿混	河南卫辉
SWJ-1	70~120	0.4~2.0	由高分子聚合物掺以特定的辅助材料组成，外观为棕褐色粉末，适于干混	山东沃尔德
SWJ-2	90~110	0.4~2.5	由聚烯类的高分子聚合物并掺以特定的辅助材料组成，外观为白色或微红固体粉末，适于干混	山东沃尔德
SWJ-3	90~150	3.0~5.0	是一种水溶性聚合物，外观为无色透明黏稠液体，具有抗盐作用，适于湿混	山东沃尔德
FSAM	30~90	3~8	主要成分聚乙烯醇，外观为无色黏性液体，适于湿混	中国石化工程院
SCF	30~180	2~6	主要成分丙烯酸、丙烯酰胺等的五元共聚物，外观为无色黏性液体，适于湿混	中国石化工程院

②Mechanism of fluid loss control additive.

The mechanism of fluid loss control additive is very complicated and can be summarized as follows:

It can improve viscosity of cement slurry, which make the cement slurry difficult to dehydrate.

It can increase static shear force of cement slurry.

②降失水剂的作用机理。

降失水剂的作用机理十分复杂，可概括为以下几点：

提高水泥浆黏度，使之不易脱水。

提高水泥浆静切力，一旦静

Once it is still, the gelation will occur. During this process, there will be neither static pressure generated nor external pressure transmitted.

It can provide granular materials with different particle sizes, which will plug pores or micropores in the formation.

It can help water soluble polymer adsorb on surface of cement particles，which lead to a formation of adsorption hydration layer. The layer can make cement particles bridging to form a network structure, which will bind more free water, plug internal pores of the cement, and reduce permeability of the cement cake.

During application of fluid loss control additives, the following aspects shall be noted: each one is with its own characteristics. At different temperature and fixed dosage, the thickening times for same fluid loss control additive are different. And no regular pattern was observed between temperature and thickening time. Besides, many fluid loss control addictives are with retarding effect.

(4) Dispersants.

Dispersant, also known as drag reducer, is mainly used to reduce plastic viscosity and yield value of the cement slurry by adjusting surface charges of cement particles，which will result in the cement slurry with the best rheological parameters. Main functions of dispersants are: ① improve pumpability of cement slurry, reduce pump pressure under a certain flow rate, and ensure the cementing operation carried out smoothly; ② can help to form turbulent flow at low displacement and hence improve displacement efficiency of the drilling fluid and cementing quality; ③ can reduce water consumption without damaging rheology of the cement slurry and prepare cement slurry with higher density; ④ can adjust the thickening time curve tend to right angle, and hence improve strength and penetration resistance of the cement paste.

止即发生胶凝，既不产生静压，又不传递外压。

粒度大小分布不同的颗粒材料，堵塞地层孔隙或者微孔。

使水溶性聚合物吸附于水泥颗粒表面，形成吸附水化层，造成水泥颗粒桥接进而生成网状结构，束缚更多自由水，堵塞水泥内部孔隙，降低水泥滤饼的渗透性。

在应用降失水剂时应该注意：每一种降失水剂都有自己的特性，同一种降失水剂在同一种加量不同温度下的稠化时间都不同，而且没有一定规律可以遵循，同时很多降失水剂具有缓凝作用。

（4）分散剂。

分散剂又称减阻剂，其主要通过调节水泥颗粒表面电荷来降低水泥浆的塑性黏度和屈服值，使水泥浆获得最佳流变参数。分散剂的主要作用为：①提高水泥浆的可泵送性，降低一定流速下的泵压，使注水泥施工顺利；②能实现低排量下的紊流，提高对钻井液的顶替效率和固井质量；③能在不破坏水泥浆流变性的条件下，减少水的用量，配出较高密度的水泥浆；④可使稠化时间曲线趋于直角，提高水泥石强度和抗渗透能力。

① Types of dispersant.

According to chemical structure and performance of the dispersant, it can be divided into three categories: sulfonates (lignosulfonate, polynaphthalene sulfonate and aldehyde ketone polycondensate, etc.), non-sulfonates (polymaleic anhydride, acrylamide and acrylic acid copolymer, styrene and maleic anhydride sulfonated copolymer) and cement conditioner. Application conditions for different dispersants are different. The common dispersants and their application conditions are shown in Table 4-7.

① 分散剂类型。

根据分散剂的化学结构和使用性能可以分为三类：磺酸盐类（木质素磺酸盐、聚萘磺酸盐、醛酮缩聚物等）、非磺酸盐类（聚马来酸酐、丙烯酰胺-丙烯酸、苯乙烯-马来酸酐磺化共聚物）和水泥调节剂。不同分散剂的使用条件并不相同，常用分散剂型号与使用条件见表4-7。

Table 4-7 Common dispersant models and application conditions

Model	temperature /℃	Dosage / %	Components and physical properties	Manufacturer
USZ	30-150	0.3-0.8	Composed by polymerization of formaldehyde, acetone and other raw materials; orange color powder; with a certain retarding effect; can be used for dry or wet mixing.	Weihui, Henan
SWJZ-1	30-90	0.3-1.0	Sulfonated ketone aldehyde polycondensation; brownish black powder; suitable for dry mixing.	Shandong World Group
DZS	20-150	0.5-1.6	Acetone formaldehyde sulfonated polycondensation; brown liquid.	Sinopec Research Institute of Petroleum Engineering

表4-7 常用分散剂型号与使用条件

产品代号	适用温度/℃	加量范围/%	主要成分及物理特征	生产厂家
USZ	30~150	0.3~0.8	由甲醛、丙酮等原料聚合改性而成，为桔黄色粉末，具有一定的缓凝作用，可干混或湿混	河南卫辉
SWJZ-1	30~90	0.3~1.0	是一种有机合成的磺化酮醛缩合物，外观为棕黑色粉末，适于干混	山东沃尔德
DZS	20~150	0.5~1.6	丙酮甲醛磺化缩聚物，棕褐色液体	中国石化工程院

② Mechanism of dispersant.

The dispersants of oil well cement are generally surfactants, which can be roughly classified into anionic, cationic and nonionic dispersants. Their main mechanism is adsorption and dispersion.

When cement and water are mixed, there are positive and negative charges on surface of the cement particles. If solid phase maintains a high concentration, the cement particles will form a continuous network structure by the effect between positive and negative charges. As cement slurry is pumped, this network structure will be destroyed and viscosity of the cement

② 分散剂的作用机理。

油井水泥分散剂一般为表面活性剂，大体上可分为阴离子型、阳离子型和非离子型，其主要作用机理为吸附分散作用。

当水泥和水混合拌浆时，水泥颗粒表面存在正电荷和负电荷。当固相浓度较高时，通过正、负电荷间的作用使水泥颗粒形成一种连续的网状结构。当水泥浆被泵注时，此网状结构遭到破坏，水泥浆黏度下降。当向水泥浆中

slurry is decreased. When a certain amount of polyanion admixture is added into cement slurry, macromolecules in polyanion will adsorb on positive charge positions at surface of cement particles, so that surface of the particles is charged with the same charge. Therefore, aggregation of particles is inhibited by the electric repulsive force, and flocculation structure formed at the initial stage of cement hydration will collapse. Moreover, the cement slurry system can be maintained in a relatively stable suspension state.

Mechanism of the cationic dispersant is similar to that of the anionic dispersant. The difference is that molecules of cationic dispensing agent will adsorb on negative charge positions at surface of cement particles, and destroy or inhibit particle aggregation. By doing this, solid phase dispersion of the cement will be suspended in water phase, and the purpose of cement slurry fluidity improvement and rheological properties enhancement will be achieved.

(5) Weighting agents.

Additives or admixtures that could increase density of cement slurry is called weighting agents, typically to combat high bottom hole pressures. At present, the common weighting agents are barite, ilmenite and hematite, etc. Among them, barite is the most common cement weighting agent. Ilmenite powder can be used to increase the density of a mixture up to $2.4g/cm^3$ and does not require the addition of water when added to the slurry. Compare to barite, hematite requires less additional water, hence could increase the density up to $2.4g/cm^3$ or higher. At the same time, strength reduction of the cement paste is lower.

It is generally required that the particle size distribution of the weighted material should be similar to that of the cement. If the particles are too large, they will be easy to settle from cement slurry and wear the dry mixing device severely; however, if the particles are too small, the viscosity of the cement slurry will be increased easily. Water consumption of the weighted material should be low, and during process of the cement hydration, it shall be inertness and with good compatibility with other admixtures.

(6) Lightening agents.

Lightening agent, also known as filling agent, is mainly used to decrease density of the cement slurry, reduce pressure of static fluid column during cementing, avoid the loss of cement slurry through fractured formations, porous formations, high permeable formations and fissured-cavern leakage formations. Besides, it can increase slurry-making rate and reduce the cement consumption.

Generally, the lightening agents are classified into three types: water absorbing materials, light-weight materials, and foam cement. According to requirements of the actual conditions, several lightening agents are often used in the same cement slurry to ensure low density and stable performance.

In addition to above cement admixtures, defoamers, heat stabilizers, anti-gas-migration agents and plugging agents can also be used as admixtures for oil well cement.

4.4.2　Selection of the admixtures

(1) Select cement and admixture according to well conditions and formation features.

In order to carry out cementing operations better, the formation characteristics of working area, the related geological and drilling data must be fully understood. Then, selection of admixture and design of the cement slurry shall be made accordingly.

(2) Select oil well cements and admixtures according to temperature sensitivity of the admixtures.

During cementing, performance of admixtures will be affected by downhole temperature. Some admixtures are effective at low temperatures (10~80℃), while some other admixtures are effective at high temperatures

（6）减轻剂。

减轻剂也称填充剂，主要用于降低水泥浆密度，减小固井时静液柱压力，避免水泥浆通过裂缝性地层、多孔隙地层、高渗透地层和溶洞漏失地层时的流失，同时又可增加水泥造浆率，降低水泥用量。

通常，减轻剂分为三类：吸水性材料、轻质材料和泡沫水泥。有时在同一种水泥浆中经常使用几种减轻剂，以此达到密度低而性能又稳定的目的。

除了上述水泥外加剂外，消泡剂、热稳定性剂、防气窜剂、堵漏剂等均可成为油井水泥的外加剂。

4.4.2　外加剂的选择

（1）根据井况和地层特点选择水泥和外加剂。

为更好地进行固井作业，首先要充分了解工作地段的地层特点和有关地质、钻井资料，并据此进行外加剂的选择和水泥浆配方设计。

（2）根据外加剂对温度的敏感性选择油井水泥和外加剂。

在固井期间，外加剂的使用效果会受到井下温度的影响。有些外加剂在低温(10~80℃)时是有效的，而有些外加剂在高温

(80~200℃). Hence temperature limitation for admixtures must be considered before usage. To ensure performance of the admixtures, when temperature is close to the limitation, the admixture shall not be used.

(3) A compatible admixture system should be selected.

Since there are compatibility problems between different types of admixtures, a lot of experiments are often required during formula design of the cement slurry. In order to improve performance of the admixture, a compatible admixture system shall be selected. For special wells or under conditions that no proper admixtures could be found, an admixture screening test should be taken.

(4) Design method of the optimal formula.

It is often difficult to select a cement slurry formula which could ensure all of its performance parameters to meet the design requirements. The general operation is to adjust dosage of the fluid loss control additive and the dispersant under the premise of ensuring certain thickening time duration, so a lower water loss amount and better pumpability could be maintained.

(5) Mixing order of the admixtures.

Admixtures are usually added according to the following order: defoamers shall be added firstly to ensure the maximum agitation, followed by dispersants, and then retarders.

To ensure the best performance, if the cement slurry contains bentonite, it must be pre-hydrated in water for 30 minutes before adding other admixtures (including defoamers).

If the formula contains fluid loss control additives, the admixtures should be added in the following order: defoamers, fluid loss control additives, dispersants and, retarders.

This is the order recommended by most admixture service companies. However, this order can also be changed. If this order is intended to change, the formula

（80~200℃）时才是有效的，使用时必须认真考虑外加剂的使用温度界限。在温度接近它的限度时，就不能使用，否则将失去效果。

（3）选择外加剂时应选用配套的外加剂体系。

由于不同类别的外加剂之间存在相容性问题，在水泥浆配方设计时，往往要做大量的实验工作。为提高工作效率，最好选用配套的外加剂体系。对于特殊井或无配套外加剂供应时，应逐一进行外加剂配套筛选实验。

（4）最佳配方设计方法。

要确定一个水泥浆配方，使它的各项性能参数都达到设计要求通常是比较困难的。一般是在保证稠化时间不太长的前提下，调节降失水剂和分散剂的加量，以便具有较低的失水量和较好的可泵性。

（5）外加剂的混合顺序。

外加剂通常按下列顺序加入：首先是消泡剂，以保证其最大的搅拌效力；然后是分散剂；最后是缓凝剂。

如果水泥浆含有膨润土，必须在其他外加剂（包括消泡剂）加入之前，在水中预水化30min，以增加效果。

如果配方中含有降失水剂，应按如下顺序加入：消泡剂、降失水剂、分散剂和缓凝剂。

这个顺序是大多数外加剂服务公司推荐的。当然，这个顺序也是可以改变的，但事前必须在实验室对配方进行测试，并把确

must be investigated in the laboratory beforehand. Operators on-site must be notified of the new order according to result of the test.

4.5 Cement slurry system and its application

The cement slurry system refers to various types of cement slurries used for cementing in general and special formations.

4.5.1 Conventional cement slurry system

Conventional cement slurry system refers to the system composed of cement (including 9 kinds of oil well cements in API standard), fresh water and general cement slurry admixtures and additives, which is suitable for general formation. Both preparation and construction of such cement slurries are relatively simple.

4.5.2 Special cement types

(1) Ultra-low density cement slurry system.

The cement slurry systems with density lower than $1.20 g/cm^3$, some even lower than $1.00 g/cm^3$, were required for development of under-balanced drilling and certain low fracture pressure formations.

Ultra-low density cement slurry consists of cement, lightening materials, admixtures and water. By adjusting material dosage and optimizing the system, an ultra-low density cement slurry system can be designed to adapt different conditions. At present, ultra-low density cement slurry has been used for cementing in Kuwait, Venezuela and other countries.

(2) Ultra-high density cement slurry system.

In actual application, an ultra-high density cement slurry technology that capable of adapting complicated environments is still a research subject. This is because ultra-high density cement slurry is generally with poor

4.5 水泥浆体系及应用

水泥浆体系是指一般地层和特殊地层固井用的各类水泥浆。

4.5.1 常规水泥浆

常规水泥浆体系是指由水泥（包括API标准中的9种油井水泥）、淡水及一般水泥浆外加剂与外掺料配成，适用于一般地层。这类水泥浆的配制和施工都较为简单。

4.5.2 特种水泥浆体系

（1）超低密度水泥浆体系。

欠平衡钻井的发展和某些低破裂压力地层要求固井水泥浆密度低于$1.20 g/cm^3$，有些甚至要求水泥浆密度低于$1.00 g/cm^3$。

超低密度水泥浆由水泥、减轻材料、外加剂和水等组成，通过调整材料的加量和体系优化，可设计出适应不同条件的超低密度水泥浆体系。目前，超低密度水泥浆已在科威特、委内瑞拉等地进行了固井应用。

（2）超高密度水泥浆体系。

在使用超高密度水泥浆时，水泥浆的流动性差，固井质量差，能够适应复杂环境下的超高密度水泥浆技术仍是一个难题。目前，

fluidity and cementing performance. At present, the ultra-high density cement slurry with density of 2.82 g/cm³ has been developed, and successfully applied in cementing operation of Wen 72-421 well in Zhongyuan Oilfield. Both of its cementing success ratio at once time and fineness ratio success rate is 100%.

(3) Expansive cement slurry system.

Expansive cement slurry is a cement slurry that taking expansive agent as the main admixture. Such cement slurry will slightly expand during setting, which could combat the shrinkage of the conventional cement slurry in solidification. It can also improve connections between cement paste and well-bore as well as cement paste and casing, thus preventing occurrence of gas channeling.

Common expansion additives of cement slurry are semi-hydrated gypsum, aluminum powder, magnesium oxide and the like. Cement slurries with addition of these expansion additives are called semi-hydrated gypsum cement slurry, aluminum powder cement slurry, magnesium oxide cement slurry, etc.

(4) Thixotropy cement slurry system.

Thixotropy of cement slurry refers to the property that the cement slurry becomes thin after being stirred and thick after being static. When the cement slurry is in static, it looks like solidified slurry. Once it is stirred or shaken, the solidified slurry regains fluidity, and if it is still, it will be solidified again. In process of cementing displacement, the cement slurry is a thin fluid. After pumping stops, it can quickly form a rigid, self-supporting gel structure.

The thixotropic cement slurry system is a special cement slurry system with shear thinning behavior usually obtained by adding thixotropic agent to an ordinary silicate cement slurry system. In this system, performance of the thixotropic agent will influence performance of the cement slurry system importantly. Thixotropic cement slurry is considered to be an essential technical solution to solve the problem of severe lost circulation. At present, research on thixotropic cement slurry is still in its infancy. There are already some thixotropic cement slurry systems with relatively good performance in foreign countries, and their practical application has achieved certain effects. However, there are few domestic research and applications in this field.

(5) Latex cement slurry system.

Latex is a generic name used to describe a group of emulsified polymer. These emulsified polymers are with colloidal particle size in range from 0.05 to 0.5 μm, and usually the latex suspension contains about 50% solid phase. After being added latex, the cement is called latex cement. Its representative system is mainly styrene-butadiene latex system. With proper stabilizer, latex can be used as an admixture of cement slurry. By adding this admixture, the latex cement slurry will be with properties of anti-gas migration and toughness. Through admixture adjusting and system designing, latex cement slurry system could be suitable for cementing operations of many different well types such as slim hole wells, sidetracking wells and multilateral wells, etc.

The latex cement slurry system has been applied in liner cementing operation of the Jiudong block (Yumen oilfield), and its cementing quality has been greatly improved compared with conventional operation.

(6) Foam cement slurry system.

Foam cement slurry is composed of water, cement, gas, foaming agent and foam stabilizer.

The gas available is nitrogen or air.

The foaming agents available are water soluble surfactants, and foam stabilizers available are water soluble polymers. The most prominent advantages of foam cement slurry are low density and high strength, which are suitable for cementing of high permeable reservoirs fractured reservoirs and fissured-cavern reservoirs.

Recently, domestic scholars have proposed an aerated foam cement slurry cementing technology

决恶性井漏问题的一项重要技术手段，目前触变水泥浆的研究还处于起步阶段，国外已有性能相对较好的触变水泥浆体系，且实际应用也取得了一定效果，而国内研究应用很少。

（5）胶乳水泥浆体系。

胶乳是用于描述一种乳化聚合物的通用名称，其胶粒的粒径为0.05~0.5μm，通常胶乳悬浮液含有大约50%的固相。加入胶乳后的水泥被称为胶乳水泥，其代表体系主要是苯乙烯－丁二烯胶乳体系。胶乳辅以合适的稳定剂，可作为水泥浆的外加剂，配制出具有防气窜和韧性性能的胶乳水泥浆。通过外加剂调节和体系设计，可开发出适用于小井眼井、侧钻井、分支井等井固井作业的胶乳水泥浆体系。

胶乳水泥浆体系目前已在酒东区块尾管固井中进行了应用，其固井质量较以往得到了大幅度提高。

（6）泡沫水泥浆体系。

泡沫水泥浆由水、水泥、气体、起泡剂和稳泡剂配成。

可用的气体为氮气或者空气。

可用的起泡剂为水溶性表面活性剂，可用的稳泡剂为水溶性聚合物。泡沫水泥浆最突出的优点是密度低、强度高，适用于高渗透层、裂缝层、溶洞层的固井。

目前，国内研究学者已采用低密度泡沫水泥浆体系，利用高压气体混合发泡方法，通过合理

based on low density foam cement slurry system, by using high pressure gas mixed foaming method and designing gas injection rate reasonably. The technology has been applied in cementing operation of the Jiaoye 9 sidetracking well (Fuling shale gas field). It had successfully complete cementing by injecting cement slurry directly and the cementing quality is obviously better than that of floating bead cement system. The technology of aerated foam cement near-balanced cementing has good effects on solving cementing severe lost circulation, preventing shallow gas channeling and improving cementing displacement efficiency.

(7) Toughening and lost circulation control cement slurry system.

Lost circulation control is not the only requirement of cement slurry system during cementing operation. Certain toughness is also necessary to maintain the long term stability of mechanical properties of the cement paste. Therefore, it is necessary to develop a toughening and lost circulation control cement slurry system. Toughening and lost circulation control cement slurry system has applied in more than 60 wells in Lamadian, Saertu, Xingqu and other blocks of the Daqing oilfield with good results. This system is mainly composed of fiber, fluid loss control additive, retarding agent, dispersant, etc. Its applicable temperature range is from 40 to 93℃.

Since fibers of different size are with different bridging and filling characteristics, addition of the fiber material can improve lost circulation control of the cement slurry as well as compressive strength and impermeability of the cement paste. In addition, since fiber cement is with certain plasticity, it could enhance toughness of the cement paste.

There are many special cement slurry systems. In addition to the above systems, there are slight expansion anti-gas migration cement slurry systems, anti-corrosion cement slurry systems, salt-water stable cement slurry systems, super-retarding cement slurry systems, etc.

设计注气量，形成了充气泡沫水泥浆固井工艺技术。该项技术在焦页9侧钻井固井中得以应用，一次性注水泥浆成功实现全井封固，其固井胶结质量明显优于漂珠水泥体系。充气泡沫水泥近平衡固井技术，对解决固井恶性漏失、防止浅层气窜和提高固井顶替效率等具有好的效果。

（7）增韧防漏水泥浆体系。

在固井中，往往要求水泥浆不仅具有防漏作用，还要具有一定的韧性，进而满足水泥石力学性能的长期稳定性，因此需要开发增韧防漏水泥浆体系。增韧防漏水泥浆体系已在大庆油田喇嘛甸、萨尔图、杏区等区块应用60多口井，取得了较好的效果。该水泥浆体系主要由纤维、降失水剂、调凝剂、分散剂等组成，适用温度范围为40~93℃。

由于不同尺寸的纤维材料自身所具有的搭桥成网和填充特性，在水泥中加入纤维材料，可以改善水泥浆的防漏性能，提高水泥石的抗压强度和抗渗性能。此外，纤维水泥具有一定的塑性，可以增强水泥石的韧性。

特种水泥浆体系有很多，除了以上几种外，还有微膨胀防窜水泥浆体系、抗腐蚀水泥浆体系、盐水水泥浆体系、超缓凝水泥浆体系等。

第5章 Fracturing Fluid Chemistry 压裂液化学

Hydraulic fracturing is a well-established practice to enhance productivity of oil and gas wells. In process of hydraulic fracturing, a fracture is generated by pumping gel (fracturing fluid) and it remain open due to proppants, which can reduce fluid flow resistance for the hydrocarbon production (Figure 5-1).

水力压裂是利用水力作用在油气层中形成人工裂缝，并用支撑剂将裂缝支撑起来，以减小流体流动阻力的一种增产、增注措施（图5-1）。

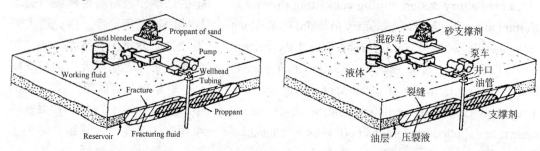

Figure 5-1 Scheme for hydraulic fracturing of reservoirs

图5-1 油气层水力压裂示意图

Fracturing effect of reservoirs is related to reservoir geological factors, fracturing fluid, proppant, fracturing process design and fracturing construction parameters. Among them, the fracturing fluid plays the role of pressure transmitting, fractures forming and extending, and proppants carrying. Since performance of fracturing fluid will directly affect size and diversion of fractures, whether the final effect of fracturing is satisfied or not is mainly depending on performance of the fracturing fluid.

油气层的压裂效果与油气层地质因素、压裂液、支撑剂、压裂工艺设计及压裂施工参数等有关。其中，压裂液起着传递压力、形成和延伸裂缝、携带支撑剂的作用。由于压裂液的性能直接影响形成裂缝的大小和裂缝的导流能力，因此压裂的效果主要取决于压裂液性能的好坏。

5.1 Definition and performance requirements of fracturing fluid

5.1 压裂液的定义及性能要求

Fracturing fluid refers to the working fluid used

压裂液指水力压裂改造油

in process of hydraulic fracturing. According to role of fracturing fluid in different stages of fracturing process, it can be divided into pad fluid, sand-laden fluid and displacement liquid.

(1) Pad fluid.

It is used for formation rupturing and geometric size fractures initiating. The fluid could help to cool the formation too. In order to improve its working efficiency, filtrate loss reducer shall be added, especially when it is used in high permeable formations. If it is necessary fine sand or ceramic powders (particle size from 100 to 320 mesh, sand ratio about 10%) or 5% diesel oil could also be added to plug small gaps in the formation and reduce filtrate-loss.

(2) Sand-laden fluid.

It is used to transport proppant (usually ceramsite or quartz sand) into the fractures and place sand to predetermined positions. In total amount of fracturing fluid, sand-laden fluid accounts for a large proportion. Like other fracturing fluids, it could initiate the fracture and cool the formation.

(3) Displacement fluid.

It is used for replacing all sand-laden fluid in the wellbore into fractures.

According to different design process requirements, pad fluid and sand-laden fluid which account for majority of the total liquid volume should possess certain ability of fractures initiation. After fracturing, the fluid flowed back for clean up while the proppant keeps an open channel for the hydrocarbon production, which shall lead to fracture walls and sand-filling fractures remaining sufficient diversion capacity. Therefore, an ideal fracturing fluid should meet the key requirements including strong sand suspension capacity, small filtrate-loss, low friction, low residue, good thermal stability and shear resistance, simple preparation, low cost and so on. For example, the proper viscosity of fracturing fluid is generally from 50 to 150mPa·s and filtration coefficient shall not be larger than 1×10^{-3}m/min$^{1/2}$ (Table 5-1).

气层过程中的工作液。根据压裂液在压裂过程中不同阶段的作用，可分为前置液、携砂液和顶替液。

（1）前置液。

其作用是破裂地层并造成一定几何尺寸的裂缝，同时还起到一定的降温作用。为提高其工作效率，特别是对高渗透层，前置液中需加入降滤失剂，必要时还需加入细砂或粉陶（粒径100～320目，砂比10%左右）或5%柴油，以堵塞地层中的微小缝隙，减少液体的滤失。

（2）携砂液。

它起到将支撑剂（一般是陶粒或石英砂）带入裂缝中并将砂子置于预定位置上的作用。在压裂液的总量中，这部分占的比例很大。携砂液和其他压裂液一样，都有造缝及冷却地层的作用。

（3）顶替液。

其作用是将井筒中的携砂液全部替入裂缝中。

根据不同的设计工艺要求，对于占总液量绝大多数的前置液及携砂液，都应具备一定的造缝能力并使压裂后的裂缝壁面及填砂裂缝有足够的导流能力。因此，一种好的压裂液应满足悬砂能力强、滤失小、摩阻低、低残渣、热稳定和抗剪切性能好、配制简单、成本低等条件。例如，压裂液的适宜黏度一般要求为50～150mPa·s，综合滤失系数不大于1×10^{-3}m/min$^{1/2}$（表5-1）。

Table 5–1 General technical standards of water base fracturing fluids (SYT 6376—2008)

S.N.	Item		Technical standard
1	Apparent viscosity of base fluid/mPa·s	20℃≤T<60℃	10-40
		60℃≤T<120℃	20-80
		120℃≤T<180℃	30-100
2	Crosslinking time/s	20℃≤T<60℃	15-60
		60℃≤T<120℃	30-120
		120℃≤T<180℃	60-300
3	Temperature and shear resistance	Apparent viscosity/mPa·s	≥50
4	Viscoelasticity	Storage modulus/Pa	≥1.5
		Loss modulus/Pa	≥0.3
5	Static filtration	Filtration coefficient/(m/min$^{1/2}$)	≤1×10^{-3}
6	Core matrix permeability damage rate/%		≤30
7	Dynamic filtration	Filtration coefficient/(m/min$^{1/2}$)	≤9×10^{-3}
8	Dynamic filtrate-loss permeability damage rate/%		≤60
9	Gel breaking properties	Breaking time/min	≤720
		Apparent viscosity of gel breaking/mPa·s	≤5.0
		Surface tension of gel breaking/(mN/m)	≤28.0
		Interfacial tension between gel breaking liquid and kerosene/(mN/m)	≤2.0
10	Residue content/(mg/L)		≤600
11	Demulsification rate/%		≥95
12	Compatibility of fracturing fluid filtrate and formation water		No precipitation, no flocculation
13	Friction reduction rate/%		≥50

表5–1 水基压裂液通用技术标准（SYT 6376—2008）

序号	项目		指标
1	基液表观黏度/mPa·s	20℃≤T<60℃	10~40
		60℃≤T<120℃	20~80
		120℃≤T<180℃	30~100
2	交联时间/s	20℃≤T<60℃	15~60
		60℃≤T<120℃	30~120
		120℃≤T<180℃	60~300
3	耐温耐剪切能力	表观黏度/mPa·s	≥50
4	黏弹性	储能模量/Pa	≥1.5
		损耗模量/Pa	≥0.3
5	静态滤失性	滤失系数/(m/min$^{1/2}$)	≤1×10^{-3}
6	岩心基质渗透率损害率/%		≤30
7	动态滤失性	滤失系数/(m/min$^{1/2}$)	≤9×10^{-3}
8	动态滤失渗透率损害率/%		≤60
9	破胶性能	破胶时间/min	≤720
		破胶液表观黏度/mPa·s	≤5.0
		破胶液表面张力/(mN/m)	≤28.0
		破胶液与煤油界面张力/(mN/m)	≤2.0
10	残渣含量/(mg/L)		≤600
11	破乳率/%		≥95
12	压裂液滤液与地层水配伍性		无沉淀，无絮凝
13	降阻率/%		≥50

Low viscosity of fracturing fluid will cause poor sand suspension capacity, while high viscosity will lead to small filtrate-loss, which is also not desirable because small filtrate-loss is tending to generate short and narrow fractures (Figure 5-2). In this case height of fractures will be large and no good for forming wide and long fractures.

若压裂液黏度过小会造成悬砂能力差，黏度过高则滤失小，不易造长缝、宽缝（图5-2）。可使产生的裂缝高度大，不利于产生宽而长的裂缝。

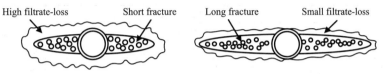

Figure 5-2 Effect of fracturing fluid filtrate-loss on fracture geometry

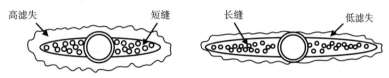

图5-2 压裂液滤失对裂缝几何形态的影响

5.2 Classification of fracturing fluid system

Fracturing fluid systems used widely in oilfields at home and abroad can be divided into water base fracturing fluid, oil base fracturing fluid, emulsified fracturing fluid, foam fracturing fluid, alcohol based fracturing fluid, clean fracturing fluid, and slickwater fracturing fluid, etc. Among them, water base and oil base fracturing fluids are widely used. A summary of all fluid systems is given in Table 5-2.

5.2 压裂液体系分类

国内外油田常用压裂液体系可分为水基压裂液、油基压裂液、乳化压裂液、泡沫压裂液、醇基压裂液、清洁压裂液及清水压裂液等类型。其中，水基压裂液和油基压裂液应用比较广泛，常用各种类型压裂液体系见表5-2。

Table 5-2 Different fracturing fluids and their application conditions

Base fluid	Type	Main ingredient	Application
Water base	Linear type	HPG、TQ、CMC、HEC、PAM	Short fractures, low temperature
	Crosslinking type	Crosslinking agent + HPG or HEC	Long fractures, high temperature
Oil base	Linear type	Oil, thickened oil	Water-sensitive formation
	Crosslinking type	Crosslinking agent + oil	Water-sensitive formation, long fractures
	O/W emulsion	Emulsifier + oil + water	Suitable for filtrate-loss control
Foam base	Acid base foam	Acid + foaming agent + N_2	Low pressure, water-sensitive formation
	Water base foam	Water + foaming agent + N_2 or CO_2	Low pressure formation
	Alcohol base foam	Methanol + foaming agent + N_2	Low pressure formation with water lock

续表

Base fluid	Type	Main ingredient	Application
Alcohol base	Linear system	Gelatinized water + alcohol	Eliminate water lock
	Crosslinking system	Crosslinking system + alcohol	
Surfactant base	Clean fracturing fluid	Viscoelastic surfactant + brine	Low permeability shallow reservoir

Note: HPG: hydroxypropylguar gum; CMC: carboxymethyl cellulose; HEC: hydroxyethyl cellulose; TQ: sesbania gum; PAM: methylene polyacrylamide.

表5-2　各类压裂液及其应用条件

压裂液基液	压裂液类型	主要成分	应用对象
水基	线型	HPG、TQ、CMC、HEC、PAM	短裂缝、低温
	交联型	交联剂+HPG或HEC	长裂缝、高温
油基	线型	油、稠化油	水敏性地层
	交联型	交联剂+油	水敏性地层、长裂缝
	O/W乳状液	乳化剂+油+水	适用于控制滤失
泡沫基	酸基泡沫	酸+起泡剂+N_2	低压、水敏性地层
	水基泡沫	水+起泡剂+N_2或CO_2	低压地层
	醇基泡沫	甲醇+起泡剂+N_2	低压存在水锁的地层
醇基	线性体系	胶化水+醇	消除水锁
	交联体系	交联体系+醇	
表面活性剂基	清洁压裂液	黏弹性表面活性剂+盐水	低渗透浅薄油层

注：HPG:羟丙基胍胶；CMC:羧甲基纤维素；HEC:羟乙基纤维素；TQ:田菁胶；PAM:聚丙烯酰胺。

5.2.1　Water base fracturing fluid

Water base fracturing fluid is prepared by water used as solvent or dispersion medium with thickening agent and addition agent. Thickening agent used in this fracturing fluid is mainly water soluble polymers. These polymers will swell into sol in water, and form a highly viscous gel after cross-linking, which have advantages of high viscosity, strong sand suspension capacity, low filtrate-loss, low friction and so on. Presently, application of water base fracturing fluid accounts for more than 70% of all fracturing operations.

Preparation process of water base fracturing fluid:

Water + Addition agent + Thickening agent⟶Sol solution.

Water + Addition agent + Crosslinking agent⟶Crosslinking fluid.

Sol solution + Crosslinking fluid⟶Water base gelling fracturing fluid.

5.2.1　水基压裂液

水基压裂液是以水作溶剂或分散介质，向其中加入稠化剂、添加剂配制而成的。该类压裂液使用稠化剂主要使用水溶性聚合物作为稠化剂，这些高分子聚合物在水中溶胀成溶胶，交联后形成黏度极高的冻胶，具有黏度高、悬砂能力强、滤失低、摩阻低等优点。目前，其应用约占压裂施工的70%以上。

水基压裂液的配液过程是：

水+添加剂+稠化剂⟶溶胶液。

水+添加剂+交联剂⟶交联液。

溶胶液+交联液⟶水基冻胶压裂液。

The volume ratio of crosslinking fluid and sol solution is from 0.01 to 0.12.

There are mainly three types of thickening agents being used in water base fracturing fluid:

(1) Nature plant gum, such as guar gum, fenugreek gum, sesbania gum, sophora bean gum, konjac gum, alginate gel and its derivatives, etc.;

(2) Cellulose derivatives, such as carboxymethyl cellulose, hydroxyethyl cellulose, carboxymethyl-hydroxyethyl cellulose, etc.;

(3) Synthetic polymers, such as polyacrylamide, partially hydrolyzed polyacrylamide, methylene polyacrylamide, carboxymethyl polyacrylamide and so on.

Guar gum (GG) is a long-chain polymer composed of mannose and galactose. The main raw material of guar gum is guar beans grown in Pakistan and India. There are 15%-18% insoluble residues in guar gum powder, and dosage of usage is from 0.4% to 0.7%. Guar gum could be modified to hydroxypropyl guar gum (HPG) by using propylene oxide, in this process, a large number of plant elements in the polymer could be removed and water insoluble residue could be reduced to 2%-4% at minimum. Chemical structure of hydroxypropyl guar gum is shown in Figure 5-3. Modified HPG has advantages of low residue, good thermal stability and strong biodegradability (Table 5-3), which is the main type of thickening agent in fracturing fluid.

溶胶液：交联液＝１００：（1~12）。

水基压裂液中使用的稠化剂主要有三种类型：

（1）天然植物胶，如胍胶、香豆胶、田菁胶、槐豆胶、魔芋胶和海藻胶及其衍生物等；

（2）纤维素衍生物，如羧甲基纤维素，羟乙基纤维素，羧甲基－羟乙基纤维素等；

（3）合成聚合物，如聚丙烯酰胺、部分水解聚丙烯酰胺、甲叉基聚丙烯酰胺及羧甲基聚丙烯酰胺等。

胍胶（GG）是一种由甘露糖和半乳糖组成的长链高分子聚合物，原料主要是生长在巴基斯坦和印度的胍胶豆。胍胶粉有15%~18%的不溶残余物，使用时加量为0.4%~0.7%。利用环氧丙烷将胍胶改性为羟丙基胍胶（HPG，其化学结构如图5-3所示），可去除聚合物中大量的植物成分，水不溶残余物最小时仅含2%~4%。经过改性的HPG具有低残渣、好的热稳定性和较强的耐生物降解性能等优点（表5-3），是目前压裂液稠化剂应用的主要类型。

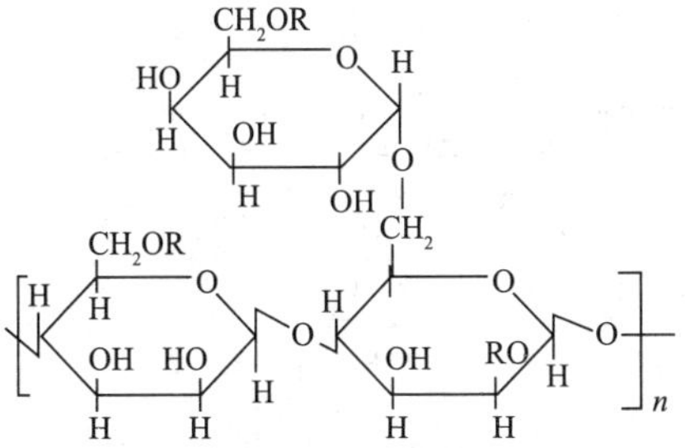

Figure 5-3　Unit structure of HPG, R=CH$_2$—CHOH—CH$_3$

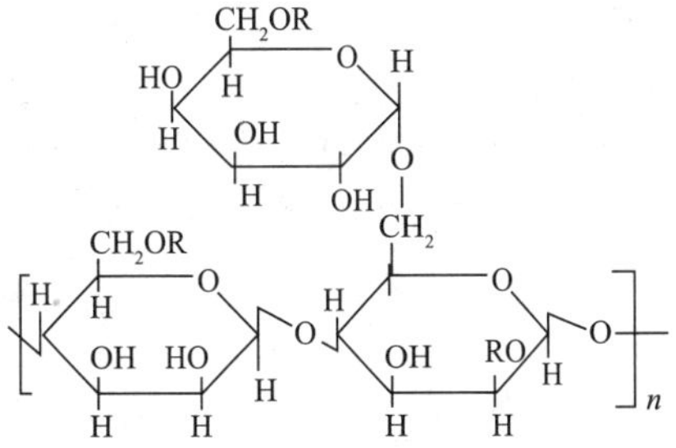

图5-3　HPG, R=CH$_2$—CHOH—CH$_3$单元结构

Table 5-3 Properties of thickening agents commonly used in fracturing fluid

Thickening agent	Water content/%	Water insoluble content/%	Viscosity of 1% solution/ (mPa·s, 30℃, 170s^{-1})
Fenugreek gum	4.5-9.0	6.0-13.0	160-210
Sesbania gum	8.0-14.0	20.0-35.0	120-160
Guar gum (Pakistan)	9.5	20.0	309
Hydroxypropyl sesbania gum	6.0-11.0	7.5-16.0	100-160
Hydroxypropyl guar gum (China)	7.0-12.0	8.0-15.0	170-280
Hydroxypropyl guar gum (USA)	8.2	4.4	297

表5-3 压裂液常用稠化剂的性能

稠化剂	含水量/%	水不溶物含量/%	1%溶液黏度/ (mPa·s, 30℃、170s^{-1})
香豆胶	4.5~9.0	6.0~13.0	160~210
田菁胶	8.0~14.0	20.0~35.0	120~160
胍尔胶（巴基斯坦）	9.5	20.0	309
羟丙基田菁胶	6.0~11.0	7.5~16.0	100~160
羟丙基胍尔胶（国内）	7.0~12.0	8.0~15.0	170~280
羟丙基胍尔胶（美国）	8.2	4.4	297

5.2.2 Oil base fracturing fluid

Oil based fracturing fluid is prepared by oil used as solvent or dispersion medium with various addition agents. For example, dissolve phosphate in hydrocarbon (diesel or light crude oil) and add small amount of aluminates, an oil base gelling system will be formed by Al^{3+} crosslinking(Figure 5-4). This kind of fracturing fluid is suitable for low pressure, partial oil wetting, and strong water-sensitive formation. However, it is with disadvantages of high cost, easy to cause fire and stain workers, equipment and sites.

5.2.2 油基压裂液

油基压裂液是以油作为溶剂或分散介质，与各种添加剂配制成的压裂液。如将磷酸酯溶解于烃类（柴油或轻质原油），添加少量铝酸盐后，通过Al^{3+}的交联作用（图5-4），可形成油基冻胶体系。这类压裂液适用于低压、偏油润湿、强水敏地层，但存在成本高、容易引起火灾、易使作业人员、设备及场地受到油污等缺点。

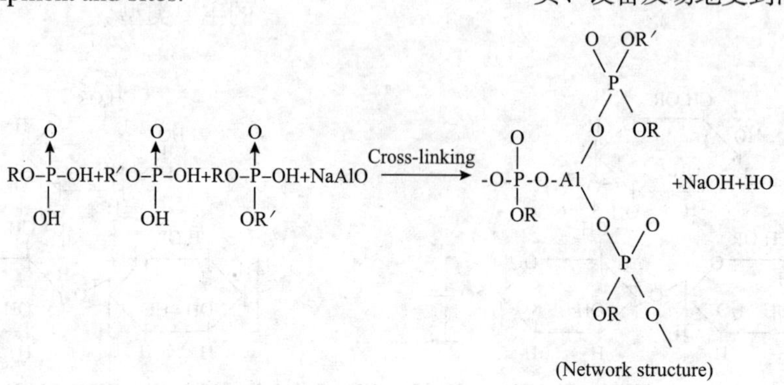

Figure 5-4 Cross-linking of oil based fracturing fluid
(R is a alkyl group from C_1 to C_8, R′ is an alkyl group from C_8 to C_{18}.)

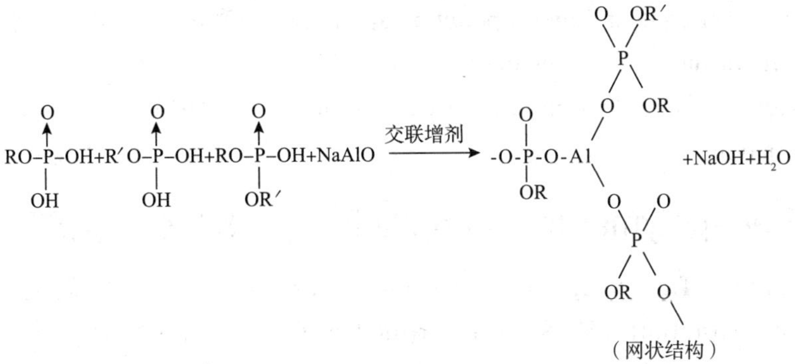

$$RO-\underset{\underset{OH}{|}}{\overset{\overset{O}{\uparrow}}{P}}-OH + R'O-\underset{\underset{OH}{|}}{\overset{\overset{O}{\uparrow}}{P}}-OH + RO-\underset{\underset{OR'}{|}}{\overset{\overset{O}{\uparrow}}{P}}-OH + NaAlO \xrightarrow{\text{交联增剂}} \text{(网状结构)} + NaOH + H_2O$$

图5-4　油基冻胶压裂液交联增稠机理
（图中R为$C_1 \sim C_8$的烃基，R′为$C_8 \sim C_{18}$的烃基。）

5.2.3 Emulsified fracturing fluid

Emulsified fracturing fluid is prepared by adding surfactant with emulsification to aqueous and oil phase. According to the different dispersion medium, it is divided into oil-in-water fracturing fluid and water-in-oil fracturing fluid. With characteristics of less damage to the formation, lower cost than oil based fracturing fluid and poor temperature resistance, emulsified fracturing fluid is suitable for fracturing of shallow wells, low temperature and low sand ratio water sensitive formations.

5.2.4 Foam fracturing fluid

Foam fracturing fluid is a dispersion system in which nitrogen, carbon dioxide or air dispersed in water (or oil) containing foaming agent. The foaming agent could be alkylsulfonate, alkylbenzene sulfonate, quaternary ammonium salts, OP surfactants, etc. In the process of mixing gas and liquid, the surfactant is used to form foams by making gas into bubble and evenly dispersed in liquid. In water, mass fraction of the foaming agent is generally in range from 0.005 to 0.02. In order to improve foam stability under high temperature, it is often necessary to add polymer foam stabilizer. Since with characteristics of high suspension capacity, low density, low friction, small filtrate-loss,

5.2.3 乳化压裂液

乳化压裂液是在水相和油相中加入具有乳化作用的表面活性剂配制而成。根据分散介质的不同，分为水包油状压裂液和油包水状压裂液。该类压裂液具有对地层伤害小，成本比油基压裂液低，耐温性差的特点，适应于浅井、低温、低砂比的水敏性地层压裂。

5.2.4 泡沫压裂液

泡沫压裂液是氮气、二氧化碳或空气分散于含起泡剂的水（或油）中的分散体系。其中，起泡剂可用烷基磺酸盐、烷基苯磺酸盐、季铵盐、OP型表面活性剂等，这些表面活性剂的作用是在气、液混合后，使气体成气泡状均匀分散在液体中形成泡沫。在水中，起泡剂的质量分数一般是0.005~0.02。为了改善高温条件下泡沫的稳定性，常需要加入高分子化合物类的泡沫稳定剂。该类压裂液具有高悬浮能力、低

relatively low water content, low oil pollution and easy discharge after fracturing, this type of fracturing fluid is suitable for low pressure, low permeability and water-sensitive formation.

密度、摩阻小、滤失量小、含水量相对少、压裂后易排出及对油层污染小等特点，适应于低压、低渗、水敏性地层。

5.2.5 Free-polymer fracturing fluid

5.2.5 清洁压裂液

Free-polymer fracturing fluid, also known as viscoelastic surfactant（VES）fracturing fluid, is primarily prepared by water, long carbon chain surfactants, water soluble salts and (or) alcohols.

清洁压裂液也称为黏弹性表面活性剂压裂液，主要由水、长碳链表面活性剂、水溶性盐和（或）醇配成。

Long chain surfactants can be cationic (such as alkyl quaternary ammonium salts, etc.), anionic (such as alkyl sulfates, alkylsulfonate, etc.), nonionic (such as polyoxyethylene polyoxypropylene alkyl ether, etc.) and zwitterionic surfactants, as shown in Figure 5-5.

长碳链表面活性剂可采用阳离子型（如烷基季铵盐等）、阴离子型（如烷基硫酸盐、烷基磺酸盐等）、非离子型（如聚氧乙烯聚氧丙烯烷基醇醚等）和两性型的长碳链表面活性剂，如图5-5所示。

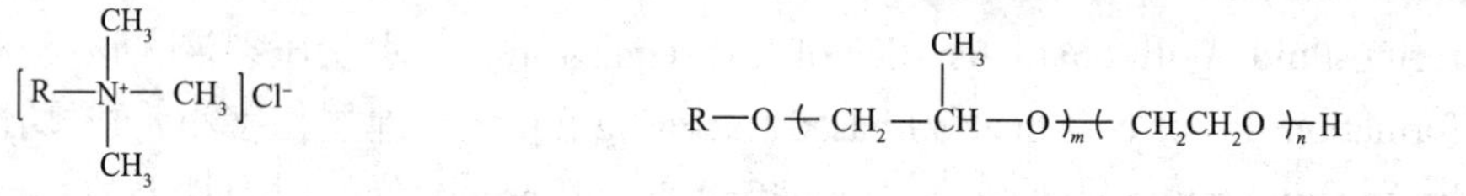

(a) Alkyl trimethylammonium chloride ($R=C_{16}-C_{30}$)　(b) Polyoxyethylene polyoxypropylene alkyl ether ($R=C_{16}-C_{30}$)

Figure 5-5　Long chain surfactants

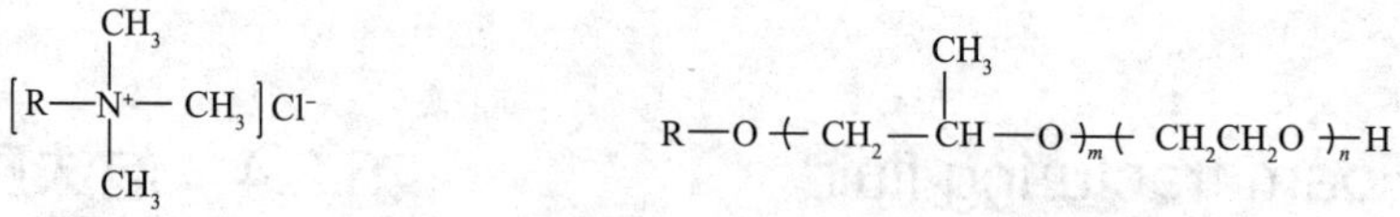

(a) 氯化烷基三甲基铵($R=C_{16} \sim C_{30}$)　(b) 聚氧乙烯聚氧丙烯烷基醇醚($R=C_{16} \sim C_{30}$)

图5-5　长碳链表面活性剂

Water soluble salt can be inorganic salt or organic salt such as potassium chloride, magnesium chloride, ammonium chloride or sodium salicylate.

水溶性盐可采用无机盐与有机盐，如氯化钾、氯化镁、氯化铵或水杨酸钠等。

Water soluble alcohol can be ethanol, isopropanol, etc.

水溶性醇可采用乙醇、异丙醇等。

When long chain surfactant dissolves in salt and (or) alcohol solution of certain concentration, the salt and (or) alcohol will promote growth of the surfactant micelle. Spherical micelles will transform to rod micelles or linear micelles that will form three dimensional network structure (Figure 5-6), often accompanied by

当长碳链表面活性剂溶于一定浓度的盐和（或）醇溶液中时，盐和（或）醇会促进表面活性剂胶束生长，球形胶束转变为棒状或线状胶束，线状胶束相互缠绕可形成三维空间网状结构

high viscoelasticity and other rheological properties (such as shear thinning, thixotropy, etc.). In addition, if contacting with hydrocarbons or dilution of formation water, the entangled wormlike micelle structure of VES fracturing fluid will be destroyed into single spherical micelle. After gel breaking, viscosity and surface tension of the aqueous solution become low. Therefore, flowback of the fracturing fluid is comparatively easy to clean up. In general, Free-polymer fracturing fluid is with advantages of easy preparation, no residue, low damage, shear stability and so on.

（图5-6），常伴随高黏弹性和其他流变特性出现（如剪切稀释、触变性等）。此外，VES压裂液在遇到油层烃类物质或地层水稀释条件下，该胶束结构会受到破坏变为表面活性剂单个胶束，且破胶后的水溶液黏度和表面张力低，因而压裂液返排较为彻底。总体来看，清洁压裂液具有配制容易、无残渣、低伤害、剪切稳定等优点。

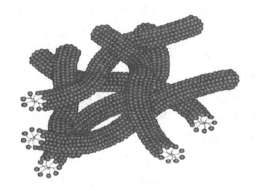

Figure 5-6 Network structure formed by linear micelles

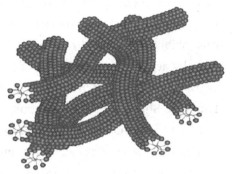

图5-6 表面活性剂棒状胶束相互缠绕形成的蠕虫状网络结构

Since free-polymer fracturing fluid cannot form filter cake on surface of reservoir rock, it shows full filtrate-loss characteristic and suitable for conditions that formation permeability is less than $5 \times 10^{-3} \mu m^2$. Besides, there are many deficiencies on currently developed viscoelastic surfactant types in China, especially on temperature resistance and costs comparing with guar polymer fracturing fluids. In order to expand application of free-polymer fracturing fluid, it is necessary to conduct research deeply on the fracturing fluid.

由于清洁压裂液在储层岩石表面不能形成滤饼，表现为全滤失特征，适合在地层渗透率小于$5 \times 10^{-3} \mu m^2$的状况下使用。此外，国内目前开发的黏弹性表面活性剂产品种类有限，在耐温性、成本等方面存在诸多不足，有待进一步深入研究、完善，扩大其应用领域。

5.2.6 Slickwater fracturing fluid

Since permeability of shale matrix is very low (generally less than $1 \times 10^{-3} \mu m^2$), only a few shale gas wells with fully developed natural fractures can be directly put into production after drilling. To obtain ideal

5.2.6 清水压裂液

页岩基质渗透率很低（一般小于$1 \times 10^{-3} \mu m^2$），因此仅有少数天然裂缝特别发育的页岩气井完钻后可直接投入生产，而90%

productivity, more than 90% of wells shall go through proper measures including acidification, fracturing and so on. Presently, the fracturing fluids used frequently in hydraulic fracture of shale gas wells are slickwater fracturing fluid, fiber based fracturing fluid and VES fracturing fluid. With characteristics of low cost and small damage on formation, the slickwater fracturing fluid becomes the main fracturing technology for shale gas development.

Composition of slickwater fracturing fluid is mainly water and sand. From 98.0% to 99.5% of slickwater fracturing fluid is water containing sand. Addition agents including friction reducer, surfactant, scale inhibitor, clay stabilizer, bactericide and so on, generally account for 0.5%-2.0% of the total volume. For Halliburton's Water Frac system, Baker Hughes' HydroCare Slickwater system (working temperature up to 150℃) and Schlumberger's OpenFRAC SW system, all of these fluids are slickwater fracturing fluids.

Friction reducer is the core of addition agents in slickwater fracturing fluid. It is mainly used to solve the problem of high friction caused by high pump speed (to carry sand by flow rate) in operation of slickwater fracturing fluid. Acrylamide polymers, polyethylene oxide (PEO), guar gum and its derivatives, cellulose derivatives, and viscoelastic surfactants can be used as friction reducer. Among them, since it could meet mixing requirements on site and with characteristics of low cost and fast dissolution rate, polyacrylamide reducer plays an important role in formula of slickwater fracturing fluid.

The mechanism of polymer reducer is quite complicate. Some researches indicate that this mechanism is to reduce flow resistance of fracturing fluid by storing turbulent energy. It will be introduced when explaining drag reduction mechanism of pipeline crude oil drag reducer.

以上的井需要经过酸化、压裂等储层改造才能获得比较理想的产量。目前，页岩气井水力压裂常用的压裂液类型有清水压裂液、纤维压裂液和清洁压裂液。清水压裂液成本低，地层伤害小，是目前页岩气开发最主要的压裂技术。

清水压裂液，又称减阻水或滑溜水压裂液，其组成以水和砂为主。清水压裂液中98.0%~99.5%是混砂水，添加剂一般占压裂液总体积的0.5%~2.0%，包括降阻剂、表面活性剂、阻垢剂、黏土稳定剂以及杀菌剂等。哈里伯顿公司的Water Frac体系、贝克休斯公司的HydroCare Slickwater 体系（使用温度达150℃）及斯伦贝谢公司的OpenFRAC SW 体系都属于清水压裂液。

降阻剂是清水压裂液的核心添加剂，主要解决由于清水压裂施工中要求泵速较大（利用流速携砂）导致的摩阻较高的问题。丙烯酰胺类聚合物、聚氧化乙烯（PEO）、胍胶及其衍生物、纤维素衍生物以及黏弹性表面活性剂等均可作为降阻剂使用。其中，聚丙烯酰胺降阻剂具有低成本、溶解速度快、能够适用于现场施工混配要求等特点，是目前页岩气清水压裂液配方中的主角。

聚合物降阻剂的降阻机理较为复杂，研究认为它是通过储藏紊流能量的机理减少压裂液的流动阻力。该机理将在管输原油减阻剂的减阻机理中进行介绍。

5.2.7 LPG fracturing fluid

Liquid Petroleum Gas (LPG) anhydrous fracturing fluid refers to a LPG gel system prepared by liquefied petroleum gas as base liquid and special thickening agent. The main component of LPG is propane that is a lower hydrocarbon. The fracturing fluid system has good rheological property and sand carrying capability, and can be directly put into production without flowback after fracturing.

Compared with the traditional water base fracturing fluid, LPG fracturing fluid not only has better fracture geometry control and sand carrying performance, but also has features of no damage, no water lock, no polymer residue and no clay swelling. After fracturing, there will be only proppant remaining in the formation. Therefore, it can effectively improve reservoir fracture conductivity and single well production. At the same time, it will improve gas release efficiency in the formation and enhance stimulation effect of fracturing. Hence the fracturing fluid system is suitable for most oil and gas reservoirs, especially water-sensitive tight reservoirs.

At present, new anhydrous fracturing technologies also includes nitrogen anhydrous fracturing, liquid CO_2 anhydrous fracturing and liquid nitrogen anhydrous fracturing. In order to expand application scale and level of anhydrous fracturing technology in China's unconventional oil and gas reservoirs, it is necessary to enhance further research on method selection, mechanism of anhydrous fracturing and development of related equipment.

5.3 Addition agents of fracturing liquid

In order to improve fracturing effect, many addition agents are added in fracturing fluid, such as proppant, crosslinking agent, breakers, clay stabilizers, cleanup

5.2.7 LPG无水压裂液

LPG无水压裂液指采用液化石油气（主要成分为丙烷）作为基液，结合专用的稠化剂，稠化后形成LPG凝胶体系。该压裂液体系具有较好的流变及携砂性能，压后无须反排直接投产。

与传统水基压裂液相比，LPG压裂液不仅具有较好的裂缝形态控制性能和携砂性能，而且兼具无伤害、无水锁、无聚合物残留、无黏土膨胀及压后仅有支撑剂留在地层中的特点，因此能够有效提高储层裂缝导流能力和单井产量，同时能够改善地层内气体的释放效率，提高压裂增产效果。该压裂液体系适合大部分油气储层，特别是致密易水敏储层。

目前，新型的无水压裂技术还包括氮气无水压裂、液态CO_2无水压裂和液氮无水压裂。为了扩大无水压裂技术在我国非常规油气藏的现场应用规模，尚需要在无水压裂方法选择、无水压裂机理及相关配套设备研发等方面进一步加强研究。

5.3 压裂液添加剂

为了提高压裂效果，在压裂液中用到许多添加剂，如支撑剂、交联剂、破坏剂、黏土稳定剂、助排剂、防乳化剂、降滤失

additive, emulsion inhibitor, filtrate loss reducer, etc. Here we will focus on the first three addition agents.

5.3.1 Proppant

Proppant is a substance that being carried into fractures by fracturing fluid for supporting the fracture after pressure is released. A good proppant should be with low density, high strength, good chemical stability, cheap and easy to be obtained.

Particle size of proppant is generally in range from 0.4 to 1.2 mm.

Natural proppants include quartz sand, bauxite, alumina, zircon, walnut shells and so on.

High strength proppants include sintered bauxite (ceramsite), aluminum alloy balls, and plastic balls, etc.

Low density proppants include microporous sintered bauxite, walnut shells and so on.

Proppants that are coated with resins (such as phenolic resins) or silicone proppants generally possess good chemical stability.

A certain proportion of solid particles with special purpose could also be mixed into proppant. Such as swellable particle, paraffin inhibitor, scale inhibitor, demulsifier, corrosion inhibitor and other particles added during fracturing of oil wells; And clay stabilizer, bactericide and other particles that being added during fracturing process of water wells. After fracturing, these solid particles can play corresponding roles in oil recovery and water injection.

5.3.2 Crosslinking agent

Crosslinking agent is a chemical agent that can chemically bond the active groups on polymer chain dissolved in water by crosslinking ions (groups) and forming a jelly with three dimensional network structure. When the above water soluble polymer thickening agent dissolves in water, viscosity of the solution will increase. This solution is also referred as

剂等，这里重点介绍前三种添加剂。

5.3.1 支撑剂

支撑剂是指用压裂液带入裂缝，在压力释放后用于支撑裂缝的物质。一种好的支撑剂应密度低、强度高、化学稳定性好、便宜易得。

支撑剂的粒径一般为0.4~1.2mm。

天然支撑剂有石英砂、铝矾土、氧化铝、锆石和核桃壳等。

高强度的支撑剂有烧结铝矾石（陶粒）、铝合金球和塑料球等。

低密度的支撑剂有微孔烧结铝矾石及核桃壳等。

化学稳定性好的支撑剂一般为树脂（如酚醛树脂）或有机硅覆盖的支撑剂。

在支撑剂中还可混入一定比例的有特殊用途的固体颗粒，如在油井压裂时加入水膨体、防蜡剂、防垢剂、破乳剂、缓蚀剂等；在水井压裂时加入黏土稳定剂、杀菌剂等。压裂后，这些固体颗粒可在采油和注水中起相应的作用。

5.3.2 交联剂

交联剂是能通过交联离子（基团）将溶解于水中的高分子链上的活性基团以化学键连接起来形成三维网状冻胶的化学剂。前面所述的水溶性聚合物稠化剂溶于水后可提高溶液黏度，通常称为线性胶。但是，线性胶增黏

linear gel. However, concentration of the polymer used for linear gel thickening is large, and viscosity of the solution decreases rapidly with increase of temperature, which causes many application problems for high temperature and deep well fracturing. Crosslinking agent can effectively reduce concentration of the polymer, significantly increase relative molecular mass of the polymer, enhance viscosity of the solution, and help to improve temperature stability of the original polymer. The crosslinking agents commonly used in water base fracturing fluid are shown in Table 5-4.

所用聚合物浓度较大，且溶液黏度随温度增加快速下降，在高温深井压裂施工中存在许多应用难题。使用交联剂可以有效降低聚合物使用浓度，明显增加聚合物的有效相对分子质量，提高溶液的黏度，有助于增加原聚合物的温度稳定性。常用的水基压裂液的交联剂见表5-4。

Table 5-4 Crosslinking groups and crosslinking agents

Crosslinking group	Code of thickening agent	Crosslinking agent	Crosslinking conditions
—COO$^-$	HPAM, CMC	$BaCl_2$, $AlCl_3$, $K_2Cr_2O_7+Na_2SO_3$, $KMnO_4+KI$	Acid crosslinking
Ortho cis hydroxyl	GG	Borax, boric acid, sodium diborate, sodium pentaborate, titanorganic, organic zirconium	Alkaline crosslinking
Ortho trans hydroxyl	HEC, CMC	Aldehyde, dialdehyde	Acid crosslinking
—$CONH_2$	HPAM, PAM	Aldehyde, dialdehyde, Zr^{4+}, Ti^{4+}	Acid crosslinking
CH_2CH_2O	PEO (polyethylene oxide)	Lignin, calcium sulfonate phenolic resin	Alkaline crosslinking

表5-4 交联基团和交联剂

交联基团	稠化剂代号	交联剂	交联条件
—COO$^-$	HPAM、CMC	$BaCl_2$、$AlCl_3$、$K_2Cr_2O_7+Na_2SO_3$、$KMnO_4+KI$	酸性交联
邻位顺式羟基	GG	硼砂、硼酸、二硼酸钠、五硼酸钠、有机钛、有机锆	碱性交联
邻位反式羟基	HEC、CMC	醛、二醛	酸性交联
—$CONH_2$	HPAM、PAM	醛、二醛、Zr^{4+}、Ti^{4+}	酸性交联
CH_2CH_2O	PEO (聚环氧乙烷)	木质素、磺酸钙酚醛树脂	碱性交联

With ortho cis hydroxyl structure, the HPG molecule is able to crosslink with multivalent ions and form jelly. The following is an example of crosslinking reaction of boron crosslinking agent, in which borax react with HPG:

The optimal pH value of crosslinking condition is from 9 to 10. It is suitable for oil and gas reservoirs fracturing when temperatures below 150℃.

(1) Sodium borate dissociates into boric acid and sodium hydroxide in water:

HPG分子具有邻位顺式羟基结构，它可以和多价离子交联生成冻胶，下面以硼交联剂为例介绍硼砂与HPG的交联反应：

交联条件以pH值为9~10最佳。适用于温度低于150℃油气层压裂。

（1）硼酸钠在水中离解成硼酸和氢氧化钠：

$$Na_2B_4O_7 + 7H_2O \rightleftharpoons 4H_3BO_3 + 2NaOH$$

(2) Boric acid is further hydrolyzed to tetrahydroxyborate ion:

（2）硼酸进一步水解形成四羟基合硼酸根离子：

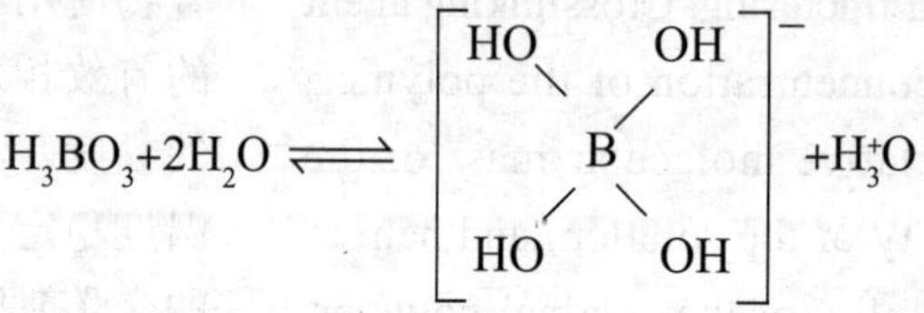

$$H_3BO_3 + 2H_2O \rightleftharpoons \left[\begin{array}{c} HO \quad OH \\ B \\ HO \quad OH \end{array}\right]^- + H_3O^+$$

(3) Borate ion combined with ortho cis hydroxy group:

（3）硼酸根离子与邻位顺式羟基结合：

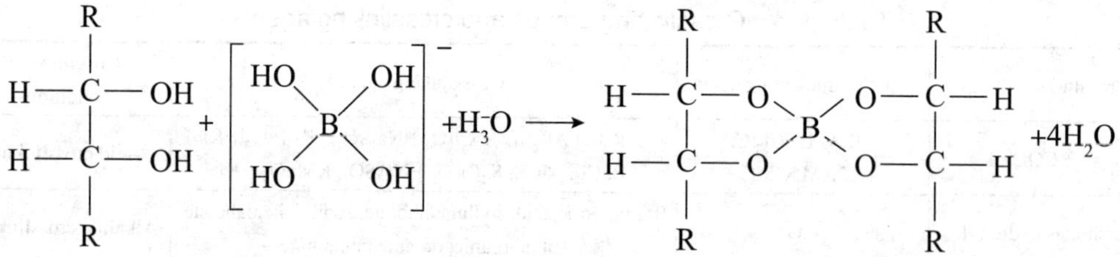

With advantages of non toxicity, low cost and good viscoelasticity, boron crosslinking fracturing fluid can be pumped into high temperature and deep reservoirs. However, during the process of fracturing operation, if the fracturing fluid crosslinks rapidly, jelly will be formed instantaneously, causing increase of fracturing fluid friction. Subjected by high speed shearing, the jelly will become thinning in fracturing strings, which can result in the decrease of fracture initiating and sand carrying capacity; If crosslinking speed of fracturing fluid is too slow to make the crosslinking reaction occurred after entering the formation, it will also lead to screen out and the failure of operation. Therefore, only when crosslinking time of fracturing fluid is consistent with the time fracturing fluid flows through the fracturing strings, performance of the fracturing fluid will be with the best. To solve this problem, a retarding crosslinking technique of fracturing fluid has been proposed. Retarding crosslinking is helpful for dispersion of crosslinking agent, resulting in higher viscosity and better temperature stability of the

硼交联压裂液具有无毒、价廉、黏弹性好的优点，能够泵入高温、深层储油层。但是，在压裂施工过程中，如果压裂液交联进行得很快，瞬间形成冻胶，使压裂液的摩阻升高，冻胶在压裂管柱中受到高速剪切而变稀，造缝与携砂能力变差；如果压裂液交联进行得很慢，进入地层后仍未交联，也会造成脱砂，使施工失败。因此，只有压裂液的交联时间与压裂液流经压裂管柱的时间一致，才能使压裂液性能最佳。针对此问题，人们提出了压裂液延迟交联技术。延迟交联有利于交联剂的分散，产生更高的黏度并改善压裂液的温度稳定性。同时，延迟交联使管路中流动的大部分液体是线性胶，而能形成低的泵送摩阻。

fracturing fluid. At the same time, retarding crosslinking can ensure that majority of fluid in pipeline is linear gel, hence keeping low pumping friction.

Crosslinking-time is the time of linear gel solution required from liquid to rigid structure after adding crosslinking agent. Crosslinking time, temperature resistance and gel breaking ability are three important parameters of crosslinking agent performance evaluation.

There are three ways to retard boron crosslinking speed:

First, wrap borax by polymer and form solid dry particles, which will slow dissolution rate of the borax resulting in the delay of jelly formation;

Second, reduce pH value of sol solution and inhibit borax hydrolysis to delay formation of jelly;

Third, add complexing agent (such as glycerin) into boric acid solution for forming complex (Figure 5-7) that can mask boric acid; The higher the pH value will lead to the more binding stability between borate and ligand. In this case, reaction time of boric acid and HPG can be greatly retarded due to the weak dissociation of borate ion.

交联时间为线性胶溶液加入交联剂至呈现刚性结构所需要的时间。交联时间、耐温能力和破胶能力是评价交联剂性能的三个重要指标。

延迟硼交联速度有三种途径：

一是采用聚合物将硼砂包裹，形成固体干颗粒，减缓硼砂的溶解速率来延迟冻胶形成时间；

二是降低溶胶液的pH值，抑制硼砂的水解来延迟冻胶形成时间；

三是在硼酸溶液中加入一种络合剂（如甘油），使之先与硼酸形成络合物（图5-7），将硼酸掩蔽起来；如果pH值愈高，硼酸盐与配体结合愈牢固，离解出的硼酸盐离子少，可以大大延缓硼酸与HPG的反应时间。

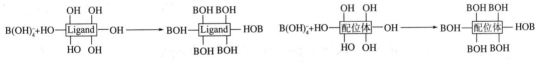

Figure 5-7 Synthetic reaction of organic boric acid

图5-7 有机硼交联剂的合成反应式

5.3.3 Breaker

Fracturing fluid breaker refers to chemical agents that can reduce viscosity of fracturing fluid to a sufficiently low level within a certain time. In the whole process of fracture generating and proppant carrying, the fracturing fluid containing ideal breaker should maintain an ideal high viscosity. Once pumping is completed, the liquid shall be immediately broke from gel or jelly to water under effect of the breaker (Figure 5-8).

5.3.3 破坏剂

压裂液破坏剂是指在一定时间内能将压裂液的黏度降低到足够低的化学剂。理想的破坏剂在整个造缝和携砂过程中，应保持理想高黏度，一旦泵送完毕，液体立刻破胶化水（图5-8）。

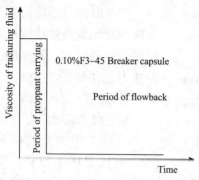

Figure 5-8 Viscosity of fracturing liquid added ideal breaker in process of fracture operations

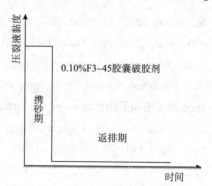

图5-8 在不同压裂时期加入理想破胶剂的压裂液黏度变化

The agent that breaks gelling fracturing fluid is called a breaker.

There are three types of breaker for water base gelling fracturing fluid: oxidizing agent, enzyme and latent acid.

(1) Oxidizing agent.

The oxidizing agents used in gelling fracturing fluid breaking are mainly including ammonium persulfate, potassium persulfate, potassium permanganate (sodium permanganate), tert-butyl hydroperoxide, hydrogen peroxide, potassium dichromate and other compounds. By polymer oxidative degradation, they will break structure of gel or jelly.

(2) Enzyme.

Enzyme is a kind of special protein. The commonly used enzyme breakers are amylase, cellulase, pancreatin and protease. They break structure of gel or jelly by catalyzing hydrolysis and degradation of polysaccharides. Enzyme breaker can only be used under conditions that temperature below 65℃ and pH value range from 3.5 to 8.0.

(3) Latent acid.

Latent acid is a substance that can be converted into acid under certain conditions, such as organic esters like methyl formate, ethyl acetate, and triethyl phosphate. Some compounds such as benzotrichloride, dichlorotoluene, and chlorobenzene can be used also since they will release acid under high temperature

冻胶压裂液的破坏剂称为破胶剂。

水基冻胶压裂液的破胶剂有三类：氧化剂、酶和酸。

（1）氧化剂。

用于冻胶压裂液破胶的氧化剂主要有过硫酸铵、过硫酸钾、高锰酸钾（钠）、叔丁基过氧化氢、过氧化氢、重铬酸钾等化合物。它们通过聚合物氧化降解，破坏冻胶结构。

（2）酶。

酶是一种特殊蛋白质，常用的酶破胶剂有淀粉酶、纤维素酶、胰酶、蛋白酶，它们对聚糖水解降解起催化作用，破坏冻胶结构。酶破胶剂只能用于温度低于65℃和pH值为3.5~8.0的条件下。

（3）潜在酸。

潜在酸是一定条件下能转变为酸的物质，如甲酸甲酯、乙酸乙酯、磷酸三乙酯等有机酯，以及三氯甲苯、二氯甲苯、氯化苯等化合物在较高温度条件下能放出酸。潜在酸不是通过破坏聚合物，而是通过改变pH值条件使冻胶结构被破坏而起作用的。

conditions. The latent acid will not destroy structure of polymer, but breaks the gel or jelly structure by changing pH conditions.

In addition, the latest technology of breaker application is oxidants encapsulation. Wrapped solid oxidant with inert membrane, then the oxidant will be released into fracturing fluid as the membrane degrades, or slowly penetrated by its laden fluid. The results show that breaker capsule technology can avoid the problem of "gradual gel breaking" under high breaker concentration condition. It greatly improves applicability and effectiveness of oxidative gel breaking.

此外，破胶剂应用的最新发展是氧化剂中的胶囊包制技术。在胶囊包制的过程中，固体氧化剂用一种惰性膜包起来，然后膜层降解或慢慢地被其携带液所渗透，而将氧化剂释放到压裂液中。研究表明，使用胶囊破胶剂避免了"逐渐破胶"的问题，使用浓度高，大大地提高了氧化破胶的适用性和有效性。

第6章 Acidification and acid additives
酸化及酸液添加剂

Acidification is an important technical measure for oil and gas production. It is also essential to increase production and injection. Its basic principle is to inject formulated acid composed by addition agents, acid of certain type and concentration into formation according to a certain order. In this case, some minerals or plugs in pores and fractures of the formation will be dissolved. Therefore, permeability and seepage conditions of the formation will be improved, and production of oil and gas wells (or injection capacity of injection wells) will be restored or improved.

As an oil and gas well stimulation measure, acidification was begun to use in the last century. Due to its great contribution to production increase of oil and gas wells, acidification has been highly valued, promoted, applied and developed widely. At present, acidification technology is not only successfully applied to conventional reservoir stimulation, but also be used on special oil and gas wells (such as high temperature deep wells, low pressure and low permeable oil wells, high sulfur wells, high porosity and low permeable reservoirs, etc.) and complex structural wells effectively. With these contributions, it plays an important role in production increase of oil and gas fields.

6.1 Acid treatment process

Acidification is a general term for a type of process in which acid is used to increase production

酸化是油气井投产、增产和注入井增注的一项重要技术措施。其基本原理是按照一定顺序向地层中注入一定类型、浓度的酸液和添加剂组成的配方酸液，溶蚀地层岩石部分矿物或孔隙、裂缝内的堵塞物，提高地层或裂缝渗透性，改善渗流条件，达到恢复或提高油气井产能（或注入井注入能力）的目的。

酸化作为一种油气井增产措施始于20世纪。由于酸化措施对油气井增产的巨大贡献，因而受到油田的高度重视并得以推广应用和广泛发展。目前，酸化技术不但成功地应用于常规油气层增产改造，还可对特殊油气井(如高温深井、低压低渗油井、高含硫井、高孔低渗储层等)及复杂结构井等进行有效的作业，为油气田增产发挥重要作用。

6.1 酸处理工艺

酸化是利用酸液增产增注的一类工艺方法的统称。根据酸

and stimulation. According to method and purpose of acidification construction, process of acidification can be divided into acid washing, matrix acidzing and acid fracturing.

6.1.1 Acid washing

Acid washing is a process that removes acid-soluble scales from mine shaft or dredges perforations. It refers to injecting a small amount of acid into the predetermined well section and dissolving the well wall scaling or plugs in perforation.

6.1.2 Matrix acidizing

Matrix acidification (matrix acidizing) is to inject formulated acid into the reservoir pores of sandstone or carbonate rock when pressure is lower than formation fracture pressure. To sandstone reservoirs, the acid will generally penetrate in radial direction, dissolving particles and plugs (mud, cement and rock debris, etc.) in pores, thereby pollution near the mine shaft will be removed. By doing this, the matrix permeability will be restored or improved. However, under certain conditions, this process may also form high permeable etched channels [Figure 6-1(a)] and the contamination zones could be bypassed by these channels.

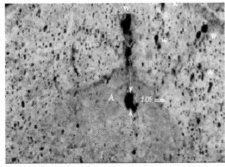
(a) Channel etched by acid from sandstone

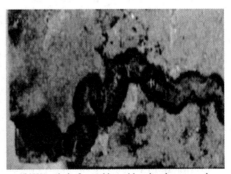
(b) Wormhole formed by acid and carbonate rock

Figure 6-1　Reaction of acid and rock

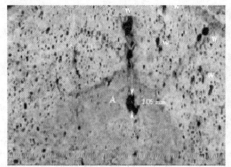
(a)酸液与砂岩作用形成的孔道

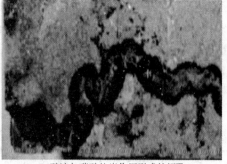
(b)酸液与碳酸盐岩作用形成的蚓孔

图6-1 酸液与岩石作用

To carbonate reservoirs, the acid will mainly expand fractures by dissolving the fracture walls or plugs in microfractures or forming a pore similar to holes made by earth worms, which is called acid-etched wormhole [Figure 6-1(b)]. The contamination zones will be bypassed by these wormholes and the seepage conditions of the formation will be improved.

Since the shale is quite fragile, in order to maintain natural fluid flow boundaries and reduce or prevent water and gas production, fracturing acidification is an operation that not shall be taken riskily. In this case, the most effective stimulation measure is matrix acidification. In most cases, the purpose of matrix acidification is primarily to remove contaminations and bypass contamination zones. Generally speaking, acidification and plug removal measures can greatly improve productivity of contaminated oil wells; however effect of plug removal and acidification is not obvious for uncontaminated wells.

6.1.3 Acid fracturing

Acid fracturing is a measure which acid is squeezed into reservoir and makes fractures in it under pressure condition higher than reservoir breakdown pressure or closure pressure of natural fractures. The acid will then react with rock on fracture walls, etching non-uniformly and forming grooved or rugged etch fractures (Figure 6-2). Unlike fractures formed by hydraulic fracturing

对于碳酸盐岩储层，酸液则主要通过溶解微裂缝中堵塞物或溶蚀裂缝壁面，扩大裂缝；或者形成类似于蚯蚓的孔道，简称为酸蚀蚓孔［图6-1（b）］而旁通污染带，从而改善地层渗流条件。

由于页岩的易碎性，或者为了保持天然液流边界以减少或防止水、气采出，而不能冒险进行压裂酸化时，一般最有效的增产措施就是基质酸化。大多数情况下，基质酸化的目的主要在于解除污染物和旁通污染带。一般来讲，对于受污染的油井，采用酸化解堵措施，可以大大提高油井产能，而对于未受到污染的井，解堵酸化效果不明显。

6.1.3 酸化压裂

酸化压裂是指在高于储层破裂压力或天然裂缝的闭合压力下，将酸液挤入储层，在储层中形成裂缝，同时酸液与裂缝壁面岩石发生反应，非均匀刻蚀缝壁岩石，形成沟槽状或凹凸不平的刻蚀裂缝（图6-2）。不同于水力

(Figure 6-3), such fractures will not completely close at the end of construction but will eventually form some artificial fractures with certain geometrical dimensions and conductivity. They will improve the seepage conditions of oil and gas wells, thus enable oil and gas wells production increase.

压裂形成的裂缝（图6-3），这种裂缝在施工结束时不完全闭合，最终形成具有一定几何尺寸和导流能力的人工裂缝，改善油气井的渗流状况，从而使油气井获得增产。

Figure 6-2　Non-uniform etching of the fracture wall

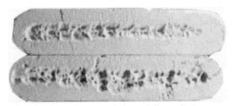

图6-2　酸液非均匀刻蚀裂缝壁面

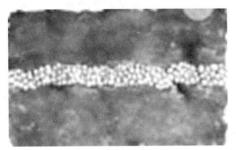

Figure 6-3　Sand packed fracture through hydraulic fracturing

图6-3　水力压裂填砂裂缝

Acid fracturing is mainly suitable for low permeable carbonate reservoirs but not for sandstone formations. This is because acid will dissolve cementing materials in the sandstone, resulting sand grains evenly detached. These grains will be carried away by the acid. In this process, there will be no dissolution groove formed. After pressure released, fractures will completely close. In addition, acid fracturing will easily destroy the screened layers which are with weak intrinsic vertical permeability, making it connect with the adjacent formation that does not need to be fractured.

酸压主要适用于低渗透碳酸盐岩储层，而不适用于砂岩地层。因为酸液溶蚀了砂岩中的胶结物，砂粒均匀脱落并被酸液带走，不会形成溶蚀沟槽，卸压后裂缝会完全闭合。此外，酸压容易破坏天然垂直且渗透性较差的遮挡层，使之与邻近不需要压开的地层连通。

6.2　Acid type and additives

The rational use of acid and additive systems plays a major role in final effect of acidification. The key of selection is to understand how various acids and

6.2　酸液类型及添加剂

酸液及添加剂体系的合理使用对酸化效果起着主要作用，其选择的关键在于了解各类酸液及

additives work and their scope of application.

6.2.1 Requirements for acids

Acids being used for acidification must be selected according to specific conditions of the construction well. Selected acid should meet the following requirements:

(1) It must have strong dissolution ability, and the product of reaction can be dissolved in residual acidic water. Besides, it must have good compatibility with reservoir fluid and will not pollute the reservoir;

(2) After chemical additives were added, the physical and chemical properties of the acid shall meet construction requirements;

(3) Transportation and construction are convenient and safe;

(4) It should be of low cost and with extensive supply sources.

With development of acidification technology, more and more acids are used in acidification at home and abroad. Currently, the commonly used acids can be classified into inorganic acids, liquid organic acids, powdered organic acids, multi-component (or mixed) acids and retarded acid, etc. The commonly used acids are shown in Table 6-1.

添加剂的作用及其适用范围。

6.2.1 酸液应满足的要求

酸化设计必须针对施工井的具体情况选择适当的酸液，选用的酸液应满足以下要求：

（1）溶蚀能力强，生成的产物能够溶解于残酸水中，与储层流体配伍性好，对储层不产生污染；

（2）加入化学添加剂后所配制成的酸液其物理、化学性质能够满足施工要求；

（3）运输、施工方便，安全；

（4）价格便宜，货源广。

随着酸化工艺技术的发展，国内外酸化用酸液越来越多，目前常用的酸可分为无机酸、液体有机酸、粉状有机酸、多组分（或混合）酸或缓速酸等类型。每类酸的常用品种见表6-1。

Table 6-1 Commonly used acid types for acidification

Acid type	Item	Properties	Application
Inorganic acids	Hydrochloric acid	Strong dissolution, low price and widely supplied; fast reaction speed and serious corrosion	Widely used in acidification of carbonate rock reservoirs and sandstone reservoirs with high carbonate content
	Hydrochloric acid-hydrofluoric acid (mud acid)	Strong dissolution, fast reaction speed, serious reaction, easy to produce secondary pollution	Sandstone reservoir matrix acidification
	Fluoboric acid	The reaction is slow and the hydrolysis rate is greatly affected by temperature. Area affected is large	Unplugging and acidification of deep sandstone reservoirs
	Phosphoric acid	Used to remove sulfides, corrosion residues and carbonate plugs, as well as clay minerals dissolved by HF. Reaction rate is slow	High carbonate and shale content reservoirs with water or acid sensitive clay minerals. Heavily polluted sandstone reservoirs that could not be processed by mud acid can be treated with phosphoric acid / HF

Acid type	Item	Properties	Application
Organic acids	Formic acid	Slow reaction, weak corrosion	Acidification of high temperature carbonate rock reservoirs
	Acetic acid		
Powdered acids	Sulfamic acid	Slow reaction and weak corrosion; convenient transportation and low dissolution; easy to produce hydrolysis insoluble matter under high temperature	Unplugging and acidification of carbonate rock reservoirs under temperature not higher than 70℃
	Monoehloroacetic acid	Slow reaction and weak corrosion; Convenient transportation and low dissolution; it is stronger and more stable than sulfamic acid	Unplugging and acidification of carbonate rock reservoirs
Multi-component acid	Acetic acid-hydrochloric acid	Can ensure strong dissolution while conduct thorough deep acidification	Deep acidification of high temperature carbonate rock reservoirs
	Mixed formic acid-hydrochloric acid		
Retarded acid	Viscous acid	Good retarding effect and small filtrate-loss volume; poor stability at high temperature, and residual acid is not easy to flowback	Acidification of medium and low temperature carbonate rock reservoirs
	Emulsified acid	Good retarding effect and weak corrosion; large friction and limited discharge capacity	Carbonate rock reservoirs
	Gelatinized acid	Good retarding effect and small filtrate-loss volume; poor stability at high temperature. It will cause severe pollution to reservoir without gel breaking	Carbonate rock reservoirs
	Chemical retarded acid	Good retarding effect but construction is quite difficult	Carbonate rock reservoirs
	Foamed acid	Good retarding effect and small filtrate-loss volume. It will not pollute reservoir but cost is quite high. The construction is difficult	Acid fracturing for low pressure, low permeable water sensitive carbonate reservoir

表6-1 酸化常用酸型

酸类	名称	特点	适用条件
无机酸	盐酸	溶解力强，价廉货源广；反应速度快，腐蚀严重	广泛用于碳酸盐岩储层酸化和碳酸盐含量高的砂岩储层酸化
	盐酸-氢氟酸(土酸)	溶解力强，反应速度快，反应严重，易产生二次污染	砂岩储层基质酸化
	氟硼酸	反应慢，水解速度受温度影响较大。处理范围大	砂岩储层深部解堵酸化
	磷酸	反应速度慢，用以解除硫化物、腐蚀产物及碳酸盐类堵塞物，HF溶解黏土矿物	碳酸盐含量和泥质含量高，含有水敏及酸敏性黏土矿物，污染较重，又不易用土酸处理的砂岩储层，可用磷酸/HF处理
有机酸	甲酸（蚁酸）	反应慢，腐蚀性弱	高温碳酸盐岩储层酸化
	乙酸（冰醋酸）		
粉状酸	氨基磺酸	反应慢，腐蚀性弱，运输方便；溶蚀能力低，在高温下易产生水解不溶物	温度不高于70℃的碳酸盐岩储层解堵酸化
	氯醋酸	反应慢，腐蚀性弱，运输方便；溶蚀能力低，较氨基磺酸酸性强而稳定	碳酸盐岩储层解堵酸化
多组分酸	乙酸-盐酸混合酸	可保证较强的溶解力，又可较好地实现深部酸化	高温碳酸盐岩储层的深部酸化
	甲酸-盐酸混合酸		

续表

酸 类	名 称	特 点	适用条件
缓速酸	稠化酸	缓速效果好，滤失量小；高温下稳定性差，残酸不易返排	中、低温碳酸盐岩储层的酸化
	乳化酸	缓速酸效果好，腐蚀性弱；摩阻大，排量受限	碳酸盐岩储层
	胶化酸	缓速效果好，滤失量小；高温下稳定性差，未破胶对储层污染严重	碳酸盐岩储层
	化学缓速酸	缓速效果好，施工难度大	碳酸盐岩储层
	泡沫酸	缓速效果好，滤失量小，对储层污染小；成本高，施工困难	低压、低渗水敏性碳酸盐岩储层酸压

Except hydrochloric acid-hydrofluoric acid, formic acid-hydrofluoric acid and other mixed acids, acids above can be used for production stimulation of carbonate rock reservoirs. Characteristics and applications of hydrochloric acid, hydrofluoric acid, mud acid and organic acid will be introduced below.

除盐酸-氢氟酸、甲酸-氢氟酸等混合酸外，上述其他酸都可用于碳酸盐岩储层的增产措施。下面主要介绍盐酸、氢氟酸、土酸和有机酸特性及其用途。

6.2.2 Commonly used acid type

6.2.2.1 Hydrochloric acid

Hydrochloric acid is a strong inorganic acid. It is aqueous solution of hydrogen chloride and is a strong acid reductant with strong corrosivity. Hydrochloric acid can dissolve corrosion residues that plug in oil and water well and restore formation permeability.

6.2.2 常见的酸液类型

6.2.2.1 盐酸

盐酸系无机强酸，它是氯化氢的水溶液，是一种具有强腐蚀性的强酸还原剂。盐酸能溶解堵塞油水井的腐蚀产物，恢复地层的渗透性。

$$Fe_2O_3 + 6HCl \longrightarrow 2FeCl_3 + 3H_2O$$
$$FeS + 2HCl \longrightarrow FeCl_2 + H_2S \uparrow$$

Hydrochloric acid could also dissolve limestones (such as dolomite) and improve formation permeability.

盐酸还能够溶解灰岩（石灰岩、白云岩），改善地层渗透性。

$$CaCO_3 + 2HCl \longrightarrow CaCl_2 + CO_2 \uparrow + H_2O$$
$$CaMg(CO_3)_2 + 4HCl \longrightarrow CaCl_2 + MgCl_2 + 2CO_2 \uparrow + 2H_2O$$

Concentration of hydrochloric acid used for oil well acidification is generally 0.05~0.15 (mass fraction). High concentration acid with mass fraction reach to 0.25~0.35 is also commonly used. Generally, dilute acid is used in oil well acidification, but due to the following advantages, high concentration hydrochloric acid is also

用于油井酸化的盐酸的浓度一般为0.05~0.15（质量分数），也常用高浓度酸，其质量分数可达0.25~0.35。一般情况下，油井酸化使用稀酸，但使用高浓度盐酸酸化具有以下优势：

be used:

(1) Acid/rock reaction speed is relatively slow, and the effective radius will increase;

(2) Comparatively more carbon dioxide will be produced per each unit volume of hydrochloric acid, which is helpful for discharge of residual acid;

(3) More calcium chloride and magnesium chloride will be produced per unit volume of hydrochloric acid, which will improve viscosity of residual acid and control acid/rock reaction speed. Besides, it will help suspension and discharge of solid particles, carrying them away from the formation;

(4) Less affected by dilution of formation water.

6.2.2.2 Hydrofluoric acid

Industrial hydrofluoric acid is aqueous solution of hydrogen fluoride; normally its concentration is 40%.

Hydrofluoric acid can dissolve plugs or clays of cemented formation (main component is kaolinite or montmorillonite), and restore formation permeability:

(1) Montmorillonite:

$$Al_4Si_8O_{20}(OH)_4 + 72HF \longrightarrow 8H_2SiF_6 + 4H_3AlF_6 + 24H_2O$$

(2) Kaolinite:

$$Al_2Si_2O_5(OH)_4 + 24HF \longrightarrow 2H_2SiF_6 + 2H_3AlF_6 + 9H_2O$$

Hydrofluoric acid could also dissolve siliceous substances (such as quartz and feldspar) in sandstone, improving formation permeability:

(1) Quartz:

$$SiO_2 + 6HF \longrightarrow H_2SiF_6 + 2H_2O$$

(2) Albite:

$$Na_2O \cdot Al_2O_3 \cdot 6SiO_2 + 50HF \longrightarrow 2NaF + 6H_2SiF_6 + 2H_3AlF_6 + 16H_2O$$

During acid treatment of sandstones, hydrofluoric acid is usually used in combination with hydrochloric acid. The mixed acid of hydrochloric and hydrofluoric acid is called mud acid. In mud acid, mass fraction of hydrochloric acid is 6%-15% and hydrofluoric acid is 3%-15%. Conventional mud acid formula consist 3% hydrofluoric acid and 12% hydrochloric acid.

In actual construction, before acidification by hydrofluoric acid, the formation is usually pretreated by hydrochloric acid that containing ammonium chloride. Hydrochloric acid in pretreatment liquid is used to remove carbonate in the formation because carbonate can react with hydrofluoric acid, forming sediment and plug the formation:

氟酸的混合酸称为土酸。在土酸中，盐酸质量分数为6%~15%, 氢氟酸质量分数为3%~15%。常规土酸配方中含3%的氢氟酸和12%的盐酸。

在实际施工中，氢氟酸或土酸酸化前，常用含氯化铵的盐酸溶液对地层进行预处理。预处理液中的盐酸用于除去地层中的碳酸盐，因为碳酸盐可与氢氟酸反应生成沉淀物堵塞地层：

$$CaCO_3 + 2HF \longrightarrow CaF_2 \downarrow + CO_2 + H_2O$$
$$CaCO_3 \cdot MgCO_3 + 4HF \longrightarrow CaF_2 \downarrow + MgF_2 \downarrow + 2CO_2 \uparrow + 2H_2O$$

Even for sandstone formation, there are certain amount of carbonate can be found. That is why it is necessary to pretreat the formation with hydrochloric acid. So carbonate could be removed before acidification with hydrofluoric acid or mud acid, reducing adverse effects of above precipitation reaction.

Ammonium chloride in pretreatment liquid is used to prevent precipitation of fluorosilicate. Fluorosilicic acid can be formed by reaction of hydrofluoric acid with quartz, silicate or aluminosilicate. Although fluorosilicic acid is water soluble, its salt is mostly insoluble in water. After reacting with sodium and potassium, ions in the formation, fluorosilicic acid can form precipitates and plug the formation:

即使砂岩地层也会有一定数量的碳酸盐，所以用氢氟酸或土酸酸化地层前，必须用盐酸预处理，除去碳酸盐，以减小上述沉淀反应的不利影响。

预处理液中的氯化铵是用来防止氟硅酸盐沉淀生成的，因氢氟酸与石英、硅酸盐、硅铝酸盐反应都可生成氟硅酸，氟硅酸本身是水溶的，但它的盐大多不溶于水，所以氟硅酸与地层中的钠、钾离子可生成堵塞地层的沉淀：

$$H_2SiF_6 + \begin{matrix} 2Na^+ \\ 2K^+ \end{matrix} \longrightarrow \begin{matrix} Na_2SiF_6 \\ K_2SiF_6 \end{matrix} \downarrow + 2H^+$$

Since ammonium salt is the only water-soluble one in fluorosilicates, ammonium chloride is added to the pretreatment liquid so other metal ions could be replaced by ammonium ions. This will reduce the chances of plugging caused by fluorosilicate precipitation in the formation.

在氟硅酸盐中只有铵盐是水溶的，因此在预处理液中加入氯化铵，用铵离子取代其他金属离子，减少氟硅酸盐沉淀对地层造成的堵塞。

6.2.2.3 Low molecular organic acid

Formic acid, acetic acid, propionic acid, and mixture of them are the commonly used low molecular organic acids. They can react with limestone formations, but since the reaction rate is much lower than that of hydrochloric acid, they are suitable for high temperature formations:

$$CaCO_3 + 2RCOOH \longrightarrow (RCOO)_2Ca + CO_2 \uparrow + H_2O$$
$$CaCO_3 \cdot MgCO_3 + 4RCOOH \longrightarrow (RCOO)_2Ca + (RCOO)_2Mg + 2CO_2 \uparrow + 2H_2O$$

Since calcium carboxylate and (or) magnesium carboxylate produced by the reaction are nearly insoluble to water, mass fraction of the low molecular carboxylic acid should not be too high. For example, mass fraction of formic acid should not exceed 11%, acetic acid should not exceed 18%, and propionic acid should not exceed 28%.

6.2.2.4 Latent acid

Latent acid is a substance that could produce acid under formation conditions. Although it is not acid, it can be considered as a special form of acid.

Commonly used latent acids are:

(1) Halohydrocarbon.

Halohydrocarbon will be hydrolyzed at 120-370℃ and can be applied to formations with temperature ≥120℃.

$$CCl_4 + 2H_2O \longrightarrow 4HCl + CO_2$$
$$Cl_2CH\text{-}CHCl_2 + 2H_2O \longrightarrow 4HCl + OHC\text{-}CHO$$
$$F_2CH\text{-}CHF_2 + 2H_2O \longrightarrow 4HF + OHC\text{-}CHO$$
$$FClCH\text{-}CHClF + 2H_2O \longrightarrow 2HCl + HF + OHC\text{-}CHO$$

(2) Halogen salt.

With the effect of initiator, halogen salt will decompose at 80-120℃ and produce corresponding inorganic acid. For example, with formaldehyde as initiator, ammonium chloride and ammonium fluoride will react as following:

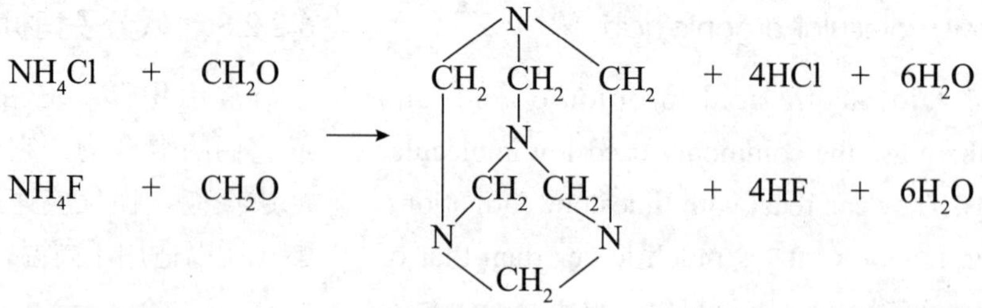

$$NH_4Cl + CH_2O \longrightarrow (CH_2)_6N_4 + 4HCl + 6H_2O$$

$$NH_4F + CH_2O \longrightarrow (CH_2)_6N_4 + 4HF + 6H_2O$$

In addition to hydrochloric acid and hydrofluoric acid, the above reaction produces hexamethylenetetramine which could help to corrosion inhibition.

上述反应除产生盐酸和氢氟酸外，还生成具有缓蚀作用的六次甲基四胺。

(3) Ester, acid halide and anhydride.

Low molecular esters, acid halides and acid anhydrides can be hydrolyzed at a certain temperature and form corresponding acid.

（3）酯、酰卤、酸酐。

低分子的酯、酰卤、酸酐在一定温度下都可用过水解反应生成相应的酸。

$$HCOOCH_3 + H_2O \longrightarrow HCOOH + CH_3OH$$
$$CH_3COOCH_3 + H_2O \longrightarrow CH_3COOH + CH_3OH$$
$$CH_3COCl + H_2O \longrightarrow CH_3COOH + HCl$$
$$(CH_3CO)_2 + H_2O \longrightarrow 2CH_3COOH$$

(4) Retarded hydrofluoric acid.

Both fluoroboric acid and fluorophosphoric acid are latent acids that could form hydrofluoric acid.

（4）缓速氢氟酸。

氟硼酸和氟磷酸都是生成氢氟酸的潜在酸。

$$HBF_4 + 3H_2O \longrightarrow H_3BO_3 + 4HF$$
$$HPF_6 + 4H_2O \longrightarrow H_3PO_4 + 6HF$$

By adding $AlCl_3$ into mud acid and forming fluoroaluminum complex, the amount of HF can also be controlled, achieving the purpose of retard.

将$AlCl_3$加到土酸中形成氟铝络合物也能控制HF的量，达到缓速的目的。

$$AlCl_3 + 4HF \longrightarrow HAlF_4 + 3HCl$$
$$HAlF_4 + 2HCl \longrightarrow AlFCl_2 + 3HF$$

Ammonium fluorotitanate is also a retarded acid that can control the forming of HF:

氟钛酸铵也是一种能控制HF生成的缓速酸：

$$(NH_4)_2TiF_6 \longrightarrow 2NH_4 + TiF_6^{2-}$$
$$TiF_6^{2-} + H_2O \longrightarrow TiOF_4^{2-} + 2HF$$

These reactions are characterized by a small equilibrium constant. They will slowly release as generated HF is gradually consumed, achieving the

这些反应的特点是平衡常数较小，随着生成的HF逐渐消耗而缓慢释放，达到缓速的目的。产

purpose of retard. The latent acid producing hydrofluoric acid needs to be treated with ammonium chloride hydrochloric acid as pad fluid before acidification.

6.2.3 Acid additives and selection

In order to improve the performance of acid and prevent adversely affecting in reservoir caused by acid, it is necessary to add certain chemicals to acid, which are collectively referred as acid additives. Common types of acid additives are: corrosion inhibitors, surfactants, ferric ion stabilizer, diverting agent and so on.

The general requirements for acid additives are:

(1) High efficiency and good effect;

(2) Good compatibility with acids, reservoir fluids and rocks;

(3) With extensive supply sources and low cost.

With development of acidification technology, more and more acid additives have been used at home and abroad, hence types and varieties of the acid additives are also being improved. In this section, the main types of additives which are commonly used will be introduced.

6.2.3.1 Corrosion inhibitor

Both hydrochloric acid and hydrofluoric acid have strong corrosive effect on metals. During acid treatment, the acid will cause corrosion problems since it will directly in contact with storage tank, fracturing equipment, downhole oil pipes and the casing. The equipment corrosion problem will become severe when temperature of deep well bottom hole and the acid concentration are both high. If effective corrosion inhibitor is not added, the acid will not only damage the equipment, but also shorten their service life, even cause accidents. Moreover, the reaction residue of acid and steel will be squeezed into the reservoir, causing plugging and reducing acidification effect. Therefore, corrosion rate of injected acid and steel must be

生氢氟酸的潜在酸在酸化前，也需要用氯化铵的盐酸溶液作前置液对地层进行处理。

6.2.3 酸液添加剂及选择

为了改善酸液性能，防止酸液在储层中产生不利影响，需要在酸液中加入某些化学物质，这些化学物质统称为酸液添加剂。常用酸液添加剂的种类有：缓蚀剂、表面活性剂、铁离子稳定剂、分流剂等。

对酸液添加剂的总的要求是：

（1）效能高，处理效果好；

（2）与酸液、储层流体及岩石配伍性好；

（3）来源广，价格便宜。

随着酸化工艺技术的发展，国内外采用的酸液添加剂越来越多，类型和品种也在不断改进，本节就常用的主要添加剂类型作简单介绍。

6.2.3.1 缓蚀剂

无论是盐酸还是氢氟酸对金属都有很强的腐蚀作用。酸处理时，由于酸直接与储罐、压裂设备、井下油管、套管接触，特别是深井井底温度很高，而所用的酸浓度较高时，便会给这些金属设备带来严重腐蚀。如果不加入有效的缓蚀剂，不但会损坏设备、缩短使用寿命，甚至会造成事故，而且因酸液和钢铁的反应产物被挤入储层，造成储层堵塞而降低酸化处理效果。因此，必须将注入酸液对钢材的腐蚀速度控制在允许的安全标准范围

controlled within the permissible safety standards.

Corrosion inhibitors used in acids include inorganic corrosion inhibitors (such as arsenic compounds) and organic corrosion inhibitors (such as pyridines, alkynols, aldehydes, thioureas, amines, etc.).

Research and application practices show that: Organic corrosion inhibitors are more effective than inorganic ones; In application, corrosion inhibitor shall be used under proper dosage because effect of excessive dosage is not good, and the best dosage should be determined by experiment; Effect of single corrosion inhibitor is not as good as that of composite formulation, and the best compounding formula should be selected by experiment.

6.2.3.2 Surfactants

By adding surfactants, the acid will show a lot of functions. Surfactants can be divided into the following categories according to these functions:

(1) Cleanup additive.

These surfactants are acid and salt resistant, and can effectively reduce interfacial tension between acid and crude oil even under concentrated acid and high salt conditions. They could reduce the Jamin's effect and make residual acid easily discharged from the formation.

Usable surfactants are amine salt surfactants, quaternary ammonium salt surfactants, pyridine surfactants, nonionic anionic surfactants, and fluorosurfactants. Since they are excellent on reducing interfacial tension under acidification conditions, fluorosurfactants are the ideal cleanup additive.

(2) Emulsion inhibitor.

Adding specific surfactant to acid can offset effect of original natural emulsifier (such as petroleum acid and similar substances) in crude oil and prevent the acid from emulsification with crude oil in reservoir. Commonly used emulsion inhibitors are surfactants with branched structure, such as polyoxyethylene-polyoxypropylene-propylene-glycol-ether or the like.

(3) Acid sludge inhibitor.

Adding combination of anionic alkylaryl sulfonates and nonionic surfactant in liquid acid, then aromatic solvent and complexing agent, the complexing agent could complex ferric ions under acidic conditions. Put the mixture into the acid or pad fluid could prevent plugging caused by acid sludge which may form during the acidification of asphaltene crude oil.

The commonly used acid sludge inhibitors are alkylaryl sulfonates, aromatic mutual solvents, glycol ethers and the like. Among them, alkylaryl sulfonates are usually used with nonionic surfactants since they are with very small solubility. Besides, an excellent emulsion inhibitor is also necessary to prevent alkylaryl sulfonates and the crude oil from producing emulsion.

(4) Mutual solvent.

Ethylene glycols are the mainly used mutual solvent. The commonly used mutual solvents include ethylene glycol monobutyl ether (EGMBE), diglycol monbutyl ether (EGMEB) and butoxy triethanol (BOTP), etc. After being added into pad fluid or post-pad fluid, the mutual solvent can maintain water wettability of rock, reduce adsorption loss of surfactant to solid phase particles, and enhance compatibility of various additives in the acid.

Mutual solvents are mostly used for sandstone acidification, also for carbonate strata. Before squeezing hydrochloric acid, ethylene glycol monobutyl ether solution with certain concentration shall be used to pre-wash the limestone reservoir. This process could achieve the effect of both cleaning agent and deoiling agent and improve the effect of the acid treatment.

It must be emphasized that surfactant is multifunctional. If surfactants are added to acid without analysis, it may not work well or cause negative effect. Therefore, compatibility among the added surfactant, the corrosion inhibitor and other additives should always be paid special attention to.

6.2.3.3 Ferric ion stabilizer

During reservoir acidification, since the acid will dissoluate corrosion products (such as iron oxide, ferrous sulfide) of iron equipments and iron-bearing minerals (such as siderite, hematite), it could always find ferric ions (Fe^{2+} and Fe^{3+}) in residual acid.

At mass fraction of 0.6%, Fe^{3+} and Fe^{2+} will hydrolyze when pH is greater than 2.2 and 7.7 respectively:

$$Fe^{2+} + H_2O \longrightarrow Fe(OH)_2 \downarrow + 2H^+$$
$$Fe^{3+} + H_2O \longrightarrow Fe(OH)_3 \downarrow + 2H^+$$

They will then form precipitate (or secondary precipitation) and plug the formation.

Since pH of residual acid is generally 4~6, under this condition, Fe^{2+} is hard to be hydrolyzed but Fe^{3+} is easy to be hydrolyzed. Therefore, stability of Fe^{3+} in residual acid must be considered. In practical applications, whether it is necessary to add ferric ion stabilizer in acid should be determined by the content of Fe^{3+} in reservoir rock.

Chemical agent that is capable of stabilizing Fe^{3+} in residual acid is called ferric ion stabilizer.

In order to prevent iron precipitation in residual acid, complexing agent (such as acetic acid, lactic acid, citric acid, sodium ethylenediaminetetraacetate, etc.) or reductant (such as isoascorbic acid) can be added to acid. Mechanism of the former is to form water soluble complex by Fe^{3+} and multivalent complex ion (that is with large stability constant); while the latter is to reduce Fe^{3+} to Fe^{2+} and preventing iron hydroxide precipitation. In addition, the pH control agent (usually acetic acid) can be added to acid to maintain residual acid with low pH value. This will prevent secondary

加入的表面活性剂与缓蚀剂及其他添加剂的配伍性。

6.2.3.3 铁离子稳定剂

在油气层酸化处理过程中，由于酸液对钢铁腐蚀产物（如氧化铁、硫化亚铁）和含铁矿物（如菱铁矿、赤铁矿）的溶解作用，在残酸中可产生铁离子（Fe^{2+}和Fe^{3+}）。

Fe^{3+}和Fe^{2+}在质量分数0.6%条件下，分别在pH值大于2.2和pH值大于7.7时水解：

生成沉淀（或称为二次沉淀），堵塞地层。

由于残酸的pH值一般为4~6，在此条件下，Fe^{2+}很难发生水解，而Fe^{3+}容易发生水解。因此，必须考虑Fe^{3+}在残酸中的稳定问题。实际应用中，应根据储层岩石中Fe^{3+}的含量来确定是否需要在酸液中添加铁离子稳定剂。

能将Fe^{3+}稳定在残酸中的化学剂称为铁离子稳定剂。

为了防止残酸中产生铁沉淀，可以采用在酸液中加入络合剂（如乙酸、乳酸、柠檬酸、乙二胺四乙酸钠盐等）或还原剂（如异抗坏血酸等）的方法避免。前者是通过稳定常数大得多价络离子与Fe^{3+}生成可溶于水的络合物；而后者使Fe^{3+}还原成Fe^{2+}防止氢氧化铁沉淀产生。此外，在酸液中加入pH值控制剂（一般使用的是乙酸），可使残酸维持

iron precipitation.

6.2.3.4 Diverting agent

Diverting agent is also called temporary plugging agent or steering agent. Adding appropriate volume of diverting agent to the acidification pad fluid or acid is aimed to temporarily plug the acidified layer (or high permeable layer). This will make the subsequent acid divert to a different layer or low permeable layer (severe pollution layer). In this case, the acid is uniformly fed and helpful for homogeneous acidification. The key of diversion acidizing (also known as temporary plugging and acidization) is to find a diverting agent that is compatible with the formation and acid phase.

Diverting agent can be made of water soluble, oil soluble or acid soluble solid. It shall be injected after mixing with laden fluid (commonly use water soluble polymer solution). The plugs made by diverting agent shall be removed spontaneously (or manually) after a certain period of time. Generally, diverting agent shall be slightly soluble in water and acid, easily dissolved in oil or organic solvent, and shall be discharged with residual acid or crude oil after acidification.

At present, the frequently used diverting agent mainly includes water soluble polymers (which will expand with acid, including polyethylene, polyoxymethylene, polyacrylamide and guar gum, etc.), inert solids (silica fume, rock salt, oil soluble resin, etc.), naphthalene, benzoate granules and so on. Among them, the most widely used is benzoates. For example, mix water soluble ammonium benzoate with acid will form benzoate, which is slightly soluble in water.

These diverting agents can also reduce the filtrate-loss along with fracture walls during acid fracturing of carbonate rock reservoirs. Therefore, it can also be used as filtrate reducer during the acid fracturing.

低pH值，防止铁的二次沉淀。

6.2.3.4 分流剂

分流剂，有时也称暂堵剂或转向剂。在酸化前置液或酸液中加入适当的分流剂，暂时封堵已酸化层（或高渗透层），使后续的酸液转向到另外一层或低渗层（污染严重层），达到均匀进酸、最终实现均匀酸化的目的。要实现分流酸化（也称暂堵酸化），关键是寻找与地层和酸液相配伍的分流剂。

分流剂可用水溶性、油溶性或酸溶性的固体制成，与携带液（常用水溶性聚合物溶液）混合后注入，经过一定时间后可自行（或人工）解堵。一般要求分流剂在水、酸中微溶，在油或有机溶剂中易溶，酸化后随产出残酸或原油等排出。

目前采用的分流剂主要有遇酸膨胀的水溶性聚合物（聚乙烯、聚甲醛、聚丙烯酰胺、瓜胶等）、惰性固体（硅粉、岩盐、油溶性树脂等）、萘、苯甲酸颗粒等。其中，应用最广的是苯甲酸系列，例如苯甲酸铵溶于水，与酸液混合后可生成微溶于水的苯甲酸。

这些分流剂也可降低碳酸盐岩储层酸压时酸液沿裂缝壁面的滤失的。所以，也可以作为酸压时的降滤剂。

6.3 Acidification technology with retarded acid

6.3.1 Principle of acidification technology with retarded acid

Due to the fast speed of acid/rock reaction and the short acid penetration distance, the conventional acidification can only eliminate damage near the well. There is a common problem for both mud acid and hydrochloric acid acidification: production increasing period is too short.

If we solve this problem by increasing acid concentration, the acid penetration distance will be increased, but a new problem will be raised by severe plugging and corrosion of mud, sand and emulsion (especially in case of high temperature deep wells). Therefore, in order to improve acid treatment effect, it is necessary to conduct deep acidification in the formation through retarded acid technique.

Since reaction between formation and acid is a kind of heterogeneous reaction, it is often possible to increase its time and reduce the rate of acid/rock reaction by studying its process (Figure 6-4).

6.3 缓速酸酸化技术

6.3.1 缓速酸酸化技术原理

在常规酸化施工中，由于酸岩反应速度快，酸的穿透距离短，只能消除近井地带的伤害。在土酸酸化或盐酸酸化作业中，都普遍存在酸化增产有效期较短的问题。

若采用提高酸的浓度，可增加酸穿透距离，但又产生严重的泥砂、乳化液堵塞及腐蚀问题（尤其是高温深井）。因此，必须运用缓速酸技术对地层进行深部酸化以改善酸处理效果。

鉴于酸与地层的反应是多相反应，往往可以通过研究酸岩反应过程来增加酸岩反应时间，降低酸岩反应速度（图6-4）。

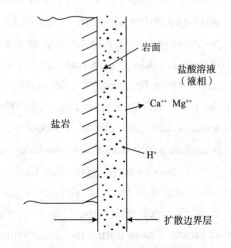

Figure 6-4　Process of acid/rock reaction　　图6-4　酸岩反应过程示意图

(1) Generation of active acid:

In most cases, the active acid is pumped in from ground. There is a large group called "latent acid" in retarded acid, they will produce active acid under formation conditions, and the process of active acid generation is a slow reaction.

(2) Transfer of acid to reaction wall:

This step is carried out under the influence of diffusion, convection mixing and mixing caused by density gradient or formation leakage.

(3) Acid reacts with surface of rocks.

(4) Reaction product diffuses from rock surface to liquid phase.

Only if the above steps belong to slow reaction, effecting time of the active acid will be prolonged.

6.3.2 Process of acidification with retarded acid

There is a general process of acidification in the major oilfields of home and abroad to ensure the following advantages: preventing precipitation of Fe^{3+}, reducing damage caused by secondary precipitation and clay mineral migration, improving relevance of the formula and conducting deep acidification to the formation, as well as acidification effect improvement. The order of the general acidification process by using retarded acid is: Inject pretreatment solution → inject pad fluid → inject treatment liquid → inject post-pad fluid → inject displacement fluid → shut-in reaction → flowback. Specific process steps should be designed according to actual conditions of the well and acid. Functions of the above mentioned fluids are as follows:

(1) Pretreatment solution:

Mainly used to remove organic plugging and clean up crude oil on surface of the rock. It is helpful for reaction of acid and the formation.

(2) Pad fluid:

To relieve some of plugs or filtrate-loss that may be

（1）活性酸的生成：

在多数情况下，活性酸由地面泵入。缓速酸中有一大类"潜在酸"，即在地层条件下产生活性酸，其生成属慢反应。

（2）酸至反应壁面的传递：

该步骤是在扩散、对流混合、由密度梯度引起的混合或地层漏失等作用下进行的。

（3）酸与岩石表面反应。

（4）应产物从岩石表面扩散到液相。

上述步骤中，有任何一步是慢反应，都能延长活性酸的作用时间。

6.3.2 缓速酸酸化工艺步骤

为了防止 Fe^{3+} 沉淀，减少因二次沉淀、黏土矿物运移等造成的伤害，通过提高配方的针对性，并对地层进行深部酸化，以增加酸化效果。国内外各大油田的缓速酸酸化工艺步骤一般为：注预处理液→注前置液→注处理液→注后置液→注顶替液→关井反应→返排。具体的工艺步骤要根据井的实际情况和所用的酸液来设计。各种处理液作用如下：

（1）预处理液：

解除有机物堵塞，清除岩石表面的原油，有利于酸液与地层接触反应。

（2）前置液：

解除地层可能与处理液产生的某些堵塞或滤失。

（3）处理液：

解除地层主要损害，提高近

produced by treatment fluid with the formation.

(3) Treatment liquid:

To relieve major damages in the formation and improve permeability of near wellbore formation.

(4) Post-pad fluid:

Displacing the main acid into deep formation for conducting deep acidification. It will prevent secondary precipitation at the near wellbore area.

(5) Displacement fluid:

Displacing treatment fluid into the formation to prevent acid from corroding pipe strings.

6.3.3 Commonly used acidification technology by retarded acid

6.3.3.1 Deep formation acidification by latent acid

Acidification of latent acid refers to carry out acid/rock reaction by active acid generated from chemical reaction under formation conditions. It is helpful for increasing permeability of the deep formation. At present, acidifications in research and on site are frequently carried out by hydrochloric acid or hydrofluoric acid that generated from halogen salts, halohydrocarbons, esters of low molecular weight organic acids or fluoboric acid.

6.3.3.2 Acidification by foamed acid

The process of acidification by foamed acid is to reduce the acid/rock reaction speed by retarding the diffusion of H^+ to rock surface under effect of the foams and high viscosity of the acid. Compared with other liquid acids, foam acid has a lot of advantages including low fluid column pressure, low filtrate-loss rate, high viscosity, strong suspension force, small dosage, good backflow, long acid effective distance, low overall cost and high economic efficiency, etc. Besides, it is easy for construction and brings little damage to the formation. Therefore, acidification with foam acid has been widely

井地层渗透率。

（4）后置液：

把主体酸替入地层深部，达到深部酸化的目的，并防止近井地带产生二次沉淀。

（5）顶替液：

把处理液替入地层，防止酸液对管柱造成腐蚀。

6.3.3 常用的缓速酸酸化技术

6.3.3.1 潜在酸地层深部酸化

潜在酸酸化是指在地层条件下，通过化学反应产生活性酸进行酸岩反应，以提高地层深部的渗透率。目前，研究和现场应用较多的是利用卤盐、卤代烃、低分子有机酸的酯以及氟硼酸生成盐酸或氢氟酸进行酸化。

6.3.3.2 泡沫酸酸化

泡沫酸酸化是通过泡沫及其高黏性对H^+向岩面的扩散的阻碍作用，降低酸岩反应速度。与其他酸液相比，泡沫酸具有液柱压力低、滤失率低、黏度较高、悬浮力强、用量小、对地层伤害小、返排性好、酸液有效作用距离长、施工比较简便、综合成本较低、经济效益高等优点。因此，用泡沫酸进行酸化作业受到油田工作者的普遍重视，并在各

used by oilfield workers and has achieved remarkable results in major oil fields.

6.3.3.3 Acidification by viscous acid (gelled acid)

Viscous acid, also known as gelled acid, refers to the acid with increased viscosity by adding thickening agent (such as polyacrylamide, hydroxyethyl cellulose, and guar gum, etc.). During the process of acid fracturing, it can help to increase viscosity and reduce speed of filtrate-loss. Viscous acid could also be used in matrix acidizing of fractured or erosion type low primary porosity formations, mainly to clean preferential paths and make less liquid enter low permeable layer. With adsorption effect of thickening agent on the rock surface, it can reduce diffusion rate of H^+ to rock surface, retarding the whole process by its high viscosity and low filtrate-loss. Because acidification by viscous acid can save corrosion inhibitors, reduce pumping friction and decrease formation damage, it has been highly valued at home and abroad since it was developed and constructed in the 1970s.

6.3.3.4 Acidification by emulsified acid

Emulsified acid is an emulsified dispersion system of oil and acid. The commonly used emulsified acid is oil-external emulsion, in which the oil phase can be crude or petroleum fraction such as diesel, kerosene, gasoline, or combination of crude oil and other light hydrocarbon oils. The acid mainly refers to hydrochloric acid and hydrofluoric acid. Other mixed acids can also be used for acidification with emulsified acid.

Since emulsified acid is with oil external phase, it will form oil film on the rock surface after entering the formation, so the acid will not contact with the rock wall directly. After a certain period of time, the oil film will be destroyed by high temperature of the formation, and then the acid is able to react with the rock wall.

High friction is also a main obstruction on field application of emulsified acid. In order to solve this

大油田取得了显著效果。

6.3.3.3 稠化酸（胶凝酸）酸化

稠化酸又称胶凝酸，是指通过加入稠化剂（如聚丙烯酰胺、羟乙基纤维素和胍胶等），提高了黏度的酸。稠化酸用于酸压是为了增加黏度和降低滤失速度。而用于裂缝型或溶蚀型低原生孔隙度地层的基质酸化的作用主要是为了清洗高渗通道和较少液体进入低渗透层中。由于其黏度高，滤失性低，以及稠化剂在岩石表面的吸附，降低了H^+向岩面的扩散速度，起到缓速作用。稠化酸酸化能节省部分缓蚀剂，降低泵送摩阻，减轻地层伤害，因而自20世纪70年代研制并施工以来受到国内外重视。

6.3.3.4 乳化酸酸化

乳化酸是油和酸的乳化分散体系。通常使用的乳化酸是油外相乳状液。油相可用原油或石油馏分。如：柴油、煤油、汽油等，也可将原油同其他轻烃油混合使用。酸液主要是盐酸、氢氟酸，其他混合酸也能用于乳化酸酸化。

由于乳化酸为油外相，进入地层后乳化液在岩石表面形成油膜，酸液不会直接与岩壁接触。当油膜在地层高温下经过一定时间破坏后，酸才能与岩石壁面反应。

高摩阻是妨碍乳化酸现场应用的一个重要问题。对此，可

problem, low friction microemulsified acids shall be developed, such as taking organic hydrocarbons instead of high-viscosity crude oil or filling gas into oleic acid emulsion to generate three-phase emulsified acid.

6.3.3.5 Chemical retarded acid

Chemical retarded acid refers to acid with surfactant addition. After adsorbing on the rock surface, surfactant in the acid will form a protective film: it will let the rock surface wets by oil, which will make reaction speed of acid/rock reaction slow. At present, this kind of retarded acid can prolong acid/rock reaction by 5~10 times, its working temperature can reach to 150℃.

The choice of surfactant will be determined by properties of the rock surface. If the rock surfaces are with negative charges, cationic or nonionic surfactants shall be chosen. Most cationic surfactants have the property of adsorbing sand and clay, while anionic surfactants are commonly used for carbonate rock formations. Nonionic surfactants can be used for both rock types.

Commonly used cationic surfactants are amines and quaternary ammonium salts, anionic surfactants are alkylsulfonates and alkylbenzene sulfonates, and nonionic surfactants is polyoxyethylene polyoxypropylene ether. In particular, combination of anionic and nonionic surfactants works very well with carbonate rocks.

6.3.3.6 Micellar acids

Micelle acid is a new type of retarded acid developed by overseas researchers in the 1980s. Its working mechanism is the principle of surfactant micellization in colloid chemistry, i.e. forming micelle dispersion system at a certain concentration and encapsulating molecules of acidizing fluid in the micelles, so speed of reaction will be retarded.

In micelle state, physical properties of surfactant

用有机烃类代替高黏度的原油或在油酸乳化液中充入气体，形成三相乳化酸，发展低摩阻的微乳化酸。

6.3.3.5 化学缓速酸

化学缓速酸是指加有表面活性剂的酸。酸中的活性剂在岩石表面吸附后形成保护薄膜。而且活性剂使岩石表面油润湿，黏附的油膜延缓了酸岩反应速度。目前，这类缓速酸能延长酸岩反应时间5~10倍，使用温度达到150℃。

选择表面活性剂要根据岩石表面性质而定。凡带有负电荷的岩石表面用阳离子或非离子型活性剂。多数阳离子型活性剂都有吸附砂粒及黏土的通性，而碳酸盐岩地层常用阴离子型活性剂。非离子型活性剂可在两种岩石中使用。

常用的阳离子型活性剂为胺和季铵盐，阴离子型活性剂为烷基磺酸盐和烷基苯磺酸盐，非离子型活性剂为聚氧乙烯聚氧丙烯醚。对于碳酸盐岩，把阴离子型和非离子型活性剂混合使用效果良好。

6.3.3.6 胶束酸

胶束酸是国外在20世纪80年代发展起来的新型缓速酸。它是利用胶体化学中表面活性剂的胶团化原理，在一定浓度下形成胶团分散体系，将酸化液分子包裹在胶团中而达到缓速的目的。

由于处在胶团状态的表面活性剂分子溶液的表/界面张力、电

molecules including surface/interface tension, conductivity, density, and ability of washing and solubilization are drastically changed. Therefore, compared to the conventional acids, the micellar acids are more suitable for acidification of oil and gas wells due to its features of good suspension carrying, emulsion inhibiting, demulsification and capillary resistance reduction.

导率、密度和洗涤与增溶能力等物理性质发生剧变。故与常规酸相比，胶束酸具有良好的悬浮携带、防乳破乳和降低毛细管阻力等特性，适合油气井酸化作业。

Chemical water plugging and profile control technology
第7章 化学堵水与调剖技术

Water production in oil and gas well is one of the inevitable major problems in middle and late stages of oil and gas field development. In the process of water flooding and polymer flooding, water and polymer solution injected into the formation are often absorbed by high permeable thin layer due to heterogeneity of the formation permeability, making water injection profile unconformable; At the same time, the water often invade oil production wells along the high permeable layer prematurely (Figure 7-1), resulting poor oil displacement in middle and low permeable layers and too fast water content increase in oil production wells.

油气井出水是油气田开发中后期不可避免的主要问题之一。由于地层渗透率的不均质，在水驱和聚合物驱过程中，注入地层的水和聚合物溶液常被厚度不大的高渗透层所吸收，吸水剖面很不均匀；同时，这些水常沿高渗透层过早侵入油井（图7-1），致使中低渗透层驱油效果差，油井产液中含水上升过快。

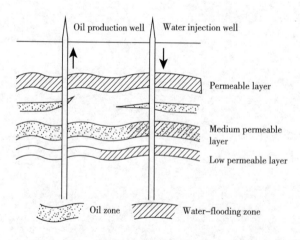

Figure 7-1　Schematic diagram of injection water advancing along a single layer

图7-1　注入水单层突进示意图

In addition, water production of oil production well will consume formation energy, reduce pumping efficiency, exacerbate corrosion and scaling of pipelines

此外，油井出水消耗地层能量，降低抽油井泵效，加剧管线、设备的腐蚀和结垢，增加脱水站

and equipment, increase load on dewatering stations, and, if the water is not injected back, will increase possibility of environmental pollution.

At present, profile control and water plugging technology is generally used for water channeling treatment of oil and water wells in oilfields. As a key technology in management of oil production and water injection wells during development, it can effectively improve the swept volume of injection water, reduce water cut, as well as increase the oil production and recovery.

7.1 Water plugging of oil production well

The produced water in oil production wells shall be controlled during production. On the one hand, the high permeable layer could be plugged by injecting profile control agent from injection well to reduce possibility of water advancing to oil production well along the high permeable layer; on the other hand, water layer of the oil production well shall be plugged, that is, to select an effective water plugging agent for oil production well water plugging.

7.1.1 Classification of water plugging agent

Water plugging agent of the oil production well(water plugging agent for short) refers to substance that is injected into the formation from oil production well to control water production. Most water plugging agents can also be used as profile control agents.

According to different plugging effect on oil and water layers, the water plugging agent could be divided into selective and non-selective water plugging agent.

Selective water plugging agent will only functioning with water but not oil. Hence it will cause plugging in water layer but affect little to oil layer. Or it may change interface characteristics between oil, water and rocks, reducing water phase permeability, thereby decrease the water production of the well.

Non-selective water plugging agent shows no difference in plugging oil, water or layers producing oil or water during the plugging process.

When non-selective water plugging method is used, water layers must be separated before plugging. Otherwise, plugging agent may cause side effects on the production layer. It is quite complicated in operation process, and necessary to re-open interlayer of production after plugging. Comparing with non-selective water plugging, the selective water plugging method is characterized by stop water production through reaction between plugging agent and formation water, which will not affect production of the production layer. However, selective water plugging is with disadvantages of large agent dosage and high cost.

Since selective water plugging agent is with better development prospects, it will be mainly introduced here. Besides, due to low flow resistance of the high permeable layer (high water content layer), the non-selective water plugging agent will preferentially enter the high permeable layer and functioning, so some non-selective water plugging agents will also be introduced below.

7.1.2 Selective water plugging agent

Selective water plugging agent is suitable for constructions in which oil layer and plugged water layer could not be separated easily by packers. It will mainly conduct plugging by the difference between oil and water or oil production layer and the water layer.

Selective water plugging agent is divided into water-based, oil-based and alcohol-based type according to solvent or dispersion medium used during preparation.

7.1.2.1 Water-based water plugging agent

Water-based plugging agents are the most widely used plugging agents in selective plugging agents. It

率，从而降低油井出水量。

非选择性堵水剂是指封堵过程中对油、水或产油、产水层的封堵没有差别的一类堵水剂。

采用非选择性堵水方法时必须分隔水层，再对水层进行封堵，否则堵剂可能对产层起副作用。这种方法在工艺上较复杂，封堵后还需要做再次打开生产夹层的善后工作。相比非选择性堵水，选择性堵水方法的特点在于堵剂通过与地层水的反应来阻止出水层段水的产出，而不阻碍产层的开采。但是选择性堵水存在着堵剂用量大、成本高的缺点。

鉴于选择性堵水剂具有较好的发展前途，下面着重介绍选择性堵剂。此外，由于高渗透层（高含水层）流动阻力小，非选择性堵水剂将优先进入高渗透层起选择性堵水作用，因此下面也将介绍一些非选择性堵水剂。

7.1.2 选择性堵水剂

选择性堵水剂适用于不易于用封隔器将油层与带封堵水层分隔开时的施工作业，主要利用油、水或产油层、水层之间的差异进行封堵。

选择性堵水剂按配制所用的溶剂或分散介质分为水基堵水剂、油基堵水剂和醇基堵水剂。

7.1.2.1 水基堵水剂

水基堵水剂是选择性堵剂中应用最广、品种最多、成本较低

is with the most varieties and the lowest costs, and includes various water soluble polymers, foams, and soaps. Among them, the most commonly used water plugging agent is water soluble polymers.

(1) Allyl polymers.

HPAM (Partially hydrolyzed polyacrylamide) is the most common water-based polymer plugging agent.

HPAM has obvious selectivity for oil and water. It could reduce at most 10% of the oil-phase permeability, and this figure could be more than 90% for water-phase.

In oil production wells, water plugging selectivity of HPAM is mainly revealed on the following aspects:

① Preferentially enter formations with high water saturation.

② Since water scouring will expose surface of the formation, after entering the formation, HPAM will be adsorbed on the formation surface with help of hydrogen bonds (Figure 7-2).

③ The unabsorbed HPAM molecules will stretch in water, reducing water permeability of the formation [Figure7-3(a)].

④ HPAM can form a water film that reduces oil flow resistance [Figure 7-3(b)].

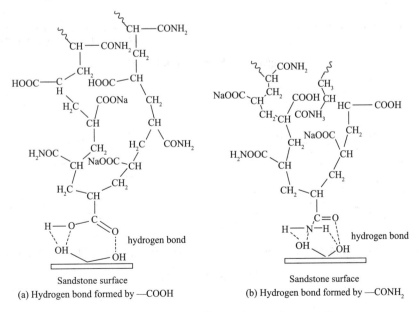

Figure 7-2　Adsorption of HPAM on sandstone surface

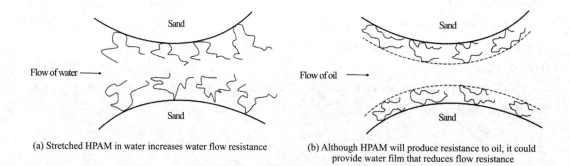

(a) Stretched HPAM in water increases water flow resistance
(b) Although HPAM will produce resistance to oil, it could provide water film that reduces flow resistance

Figure 7-3　Selectivity of HPAM

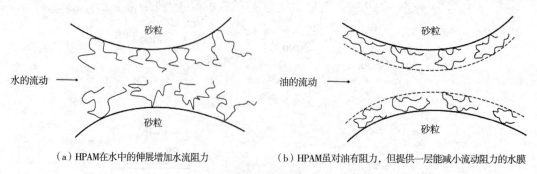

（a）HPAM在水中的伸展增加水流阻力　　（b）HPAM虽对油有阻力，但提供一层能减小流动阻力的水膜

图7-3　HPAM的选择性

HPAM with relative molecular mass ranging from 3.0×10^6 to 1.2×10^7 and hydrolysis degree in range of 10%~35% can be used for oil production well plugging.

In order to increase adsorption capacity of HPAM in the formation and improve water plugging capacity of HPAM, HPAM can be dissolved in saline solution and then injected into the formation. Since salts can increase adsorption amount of HPAM on rock surface. Crosslinker (such as aluminum sulfate or aluminum citrate) solution can also be used for formation pretreatment, aiming to reduce electronegativity or even convert the rock surface into electropositive, which will increase adsorption capacity of HPAM on the rock surface. Or HPAM with low hydrolysis degree could be injected first to improve adsorption capacity of HPAM on the rock surface by non-ionic property of —$CONH_2$ in HPAM; then inject alkali to increase hydrolysis degree of the unadsorbed HPAM to improve its water control capacity.

In addition, anionic, cationic and nonionic terpolymer (such as copolymer of partially hydrolyzed polyacrylamide-dimethyldiallylammonium chloride and similar copolymers) and cationic polymer (such as copolymer of acrylamide-acrylamide propylene trimethylammonium chloride and similar copolymers) are also with similar water plugging mechanism.

(2) Foams.

Foams with water dispersion medium can preferentially enter water layer and be there stably. Water will then be plugged by superimposed Jamin effect. In reservoir, oil can be emulsified to form three-phase foams in dispersion medium of the foam. Oil beads in dispersion medium will undergo the process shown in Figure 7-4, which will make the foams break. That is why foams entering reservoir will not act like plugging agent. Therefore, foams are also a selective water plugging agent.

相对分子质量在 3.0×10^6~1.2×10^7，水解度在 10%~35% 的 HPAM 均可用于油井堵水。

为了提高 HPAM 在地层中的吸附量，从而提高 HPAM 对水的封堵能力，可将 HPAM 溶于盐水中注入地层，因为盐可提高 HPAM 在岩石表面的吸附量；也可用交联剂（如硫酸铝或柠檬酸铝）溶液预处理地层，减小岩石表面的负电性，甚至可将岩石表面转变为正电性，提高 HPAM 在岩石表面的吸附量；还可先注入低水解度的 HPAM，利用 HPAM 中—$CONH_2$ 的非离子性质提高 HPAM 在岩石表面的吸附量，再注入碱，提高 HPAM 未吸附部分的水解度，以提高这部分 HPAM 的控水能力。

此外，阴阳非三元共聚物（如部分水解丙烯酰胺-二甲基二烯丙基氯化铵共聚物等）和阳离子聚合物（如丙烯酰胺-丙烯酰胺亚丙基三甲基氯化铵共聚物等）也具有类似堵水机理。

（2）泡沫。

以水作分散介质的泡沫可优先进入出水层，并在出水层稳定存在，通过叠加的贾敏效应，封堵来水。在油层中，油可乳化在泡沫的分散介质中形成三相泡沫。分散介质中的油珠，可经历如图 7-4 所示的过程，引起泡沫的破坏，所以进入油层的泡沫不堵塞油层。因此，泡沫也是一种选择性堵水剂。

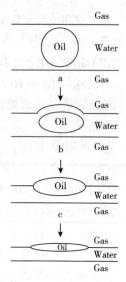 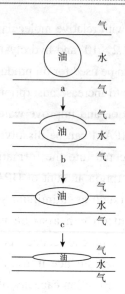

Figure 7-4 Dispersion medium film destruction process in three-phase foams

图7-4 三相泡沫中分散介质膜的破坏过程

Foaming agents are primarily sulfonate surfactants. Thickening agent such as sodium carboxymethyl cellulose, polyvinyl alcohol, polyvinylpyrrolidone or the like could be added into foaming agent to improve foam stability. Gas used for foam preparation could be nitrogen or carbon dioxide, which could be converted from liquid state. Nitrogen can also be produced by reaction. For example, inject NH_4 or NO_2 directly or other substances capable of producing NH_4 or NO_2 such as $NH_4Cl + NaNO_2$ or $NH_4NO_3+KNO_2$ into the formation, and control system with pH (such as $NaOH+CH_3COOCH_3$) to make the system alkalinized and then acidized, i.e. the system shall be alkaline at the beginning, so nitrogen generation will be inhibited. When the system enters the formation, pH changes to acidic, and then nitrogen can be generated by the following reaction and produce foams in foaming agent solution:

$$NH_4NO_2 \longrightarrow N_2 \uparrow + 2H_2O$$

(3) Water soluble soaps.

Water-soluble soaps refer to high carbon number organic acid salts which are soluble in water, such as sodium rosinate, sodium naphthenate, fatty acid sodium

泡沫的起泡剂主要是磺酸盐型表面活性剂。为了提高泡沫的稳定性，可在起泡剂中加入稠化剂如钠羧甲基纤维素、聚乙烯醇、聚乙烯吡咯烷酮等。制备泡沫用的气体可以是氮气或二氧化碳，它们可由液态转变而来。例如，向地层注 NH_4、NO_2 或能产生此物质的其他物质如 $NH_4Cl + NaNO_2$ 或 $NH_4NO_3+KNO_2$，用pH值控制系统（如 $NaOH+CH_3COOCH_3$）使体系先碱后酸，即开始时体系为碱性，抑制氮气产生；氮气也可通过反应产生。当体系进入地层时，pH值转变为酸性，即可通过下面反应产生氮气，在起泡剂溶液中产生泡沫：

（3）水溶性皂。

水溶性皂是指能溶于水中的高碳数有机酸盐，如松香酸钠、环烷酸钠、脂肪酸钠等。下面重

and so on. The selective water plugging mechanism of sodium rosinate is as following:

Sodium rosinate can be formed by the reaction of rosin (wherein mass fraction of rosin acid is 80%-90%) and sodium carbonate (or sodium hydroxide):

点介绍松香酸钠的选择性堵水机理。

松香酸钠可由松香（其中松香酸的质量分数在80%~90%）与碳酸钠（或氢氧化钠）反应生成：

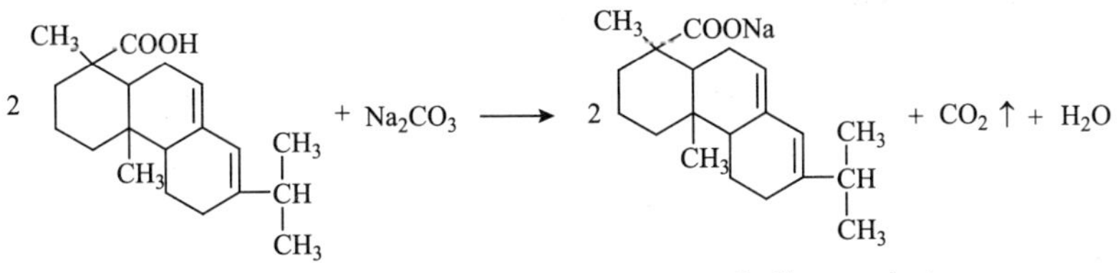

$$2\ \text{Rosin acid} + Na_2CO_3 \longrightarrow 2\ \text{Sodium rosinate} + CO_2\uparrow + H_2O$$

$$2\ \text{松香酸} + Na_2CO_3 \longrightarrow 2\ \text{松香酸钠} + CO_2\uparrow + H_2O$$

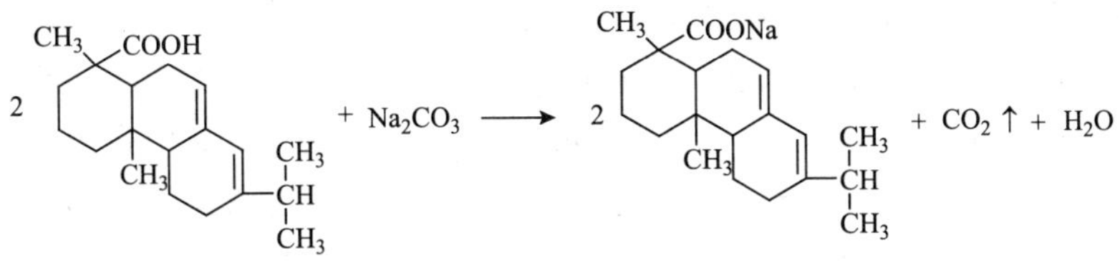

Since sodium rosinate reacts with calcium and magnesium ions will form calcium rosinate and magnesium rosinate precipitate that insoluble in water;

由于松香酸钠与钙、镁离子反应，生成不溶于水的松香酸钙、松香酸镁沉淀；

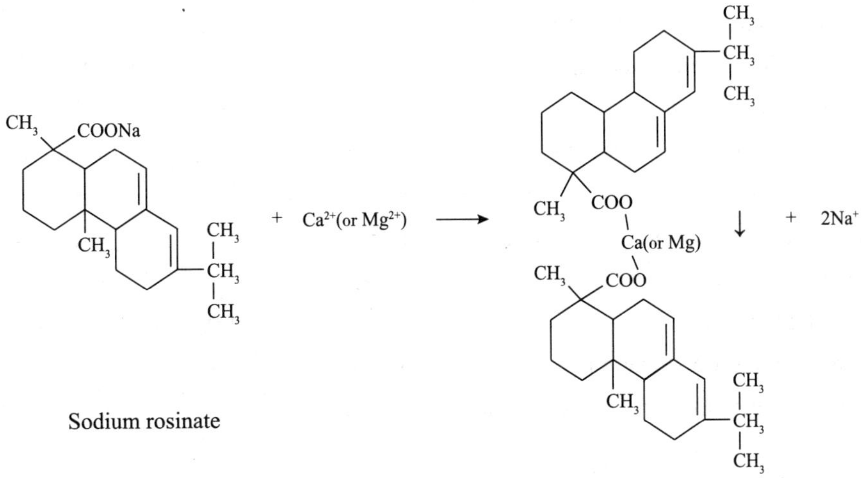

$$\text{Sodium rosinate} + Ca^{2+}(\text{or }Mg^{2+}) \longrightarrow \text{calcium(magnesium) rosinate}\downarrow + 2Na^+$$

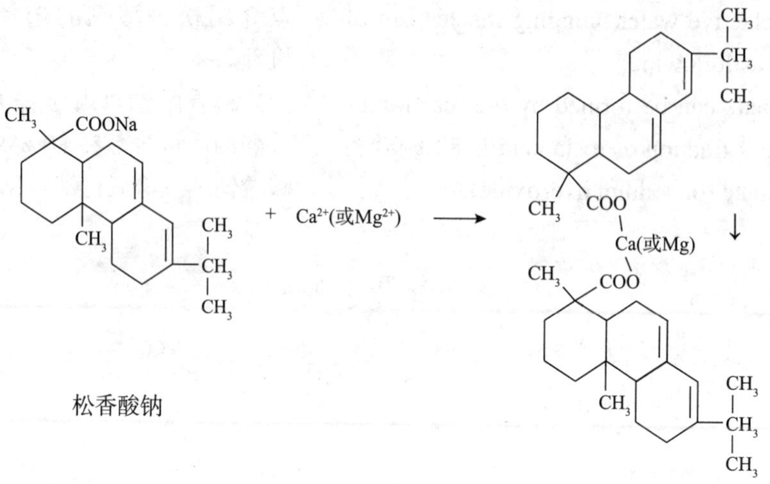

松香酸钠 + Ca²⁺(或Mg²⁺) ⟶ 松香酸钙(或镁)↓ + 2Na⁺

Sodium rosinate is suitable for oil production well water plugging in reservoir with high concentration of calcium and magnesium ions (for example, higher than 1×10^3 mg/L). In case oil in reservoir does not contain calcium or magnesium ions, sodium rosinate will not plug the reservoir.

7.1.2.2 Oil-based water plugging agent

(1) Organosilicones.

Most water plugging agents that suitable for selective water plugging is organosilicon compounds. In organosilicon compounds, hydrocarbyl halomethylsilane is the most widely used liquid that with low viscosity and easy to be hydrolyzed.

Hydrocarbyl halomethylsilane can be represented by the general formula R_nSiX_{4-n}, wherein R represents hydrocarbon radical, X represents halogen (i.e., fluorine, chlorine, bromine or iodine), and n represents any integer in range of 1-3.

Dimethyl dichloro silicane is a type of hydrocarbyl halomethylsilane made from silica fume and methyl chloride:

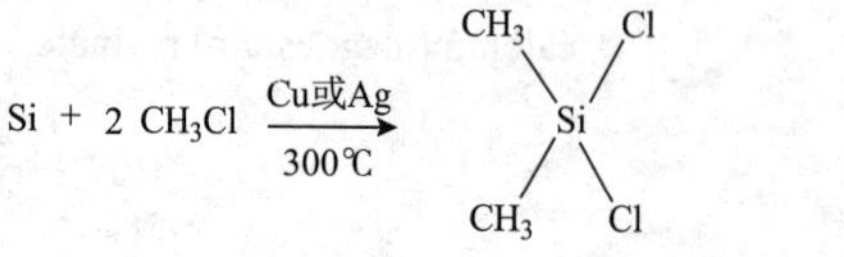

$$Si + 2CH_3Cl \xrightarrow[300℃]{Cu或Ag} (CH_3)_2SiCl_2$$

dimethyl dichloro silicaneand

所以，松香酸钠适用于油层水中钙、镁离子质量浓度高（例如高于 1×10^3 mg/L）的油井堵水。油层的油不含钙、镁离子，松香酸钠不堵塞油层。

7.1.2.2 油基堵水剂

（1）有机硅类。

适用于选择性堵水的大多为有机硅化合物。烃基卤代甲基硅烷是有机硅化合物中使用最广泛的一种易水解、低黏度的液体。

烃基卤代甲硅烷可用通式 R_nSiX_{4-n} 表示，式中，R 表示烃基，X 表示卤素（即氟、氯、溴或碘），n 表示 1~3 的整数。

二甲基二氯甲硅烷是一种烃基卤代甲硅烷，它由硅粉与一氯甲烷制成：

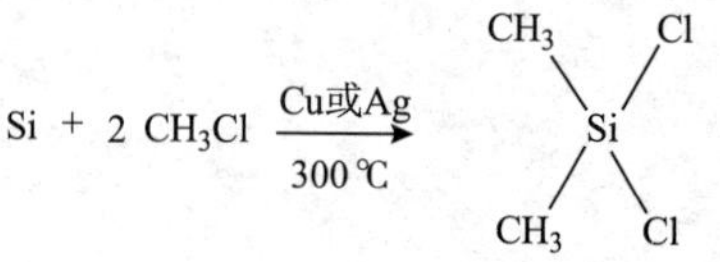

$$Si + 2CH_3Cl \xrightarrow[300℃]{Cu或Ag} (CH_3)_2SiCl_2$$

二甲基二氯甲硅烷

Chemical water plugging and profile control technology

Hydrocarbyl halomethylsilane reacts with water and forming corresponding silanol. Polyol in the silanol is easily to be polycondensedand formingpolysilanol precipitate, which will plug water layer. The following is an example of dimethyl dichloro silicaneand water reaction, illustrating how water plugging precipitates formed:

$$(CH_3)_2SiCl_2 + 2H_2O \longrightarrow (CH_3)_2Si(OH)_2 + 2HCl$$

Dimethyl methyl silanediol

$$n(CH_3)_2Si(OH)_2 \longrightarrow HO-[Si(CH_3)_2-O]_n-H\downarrow + (n-1)H_2O$$

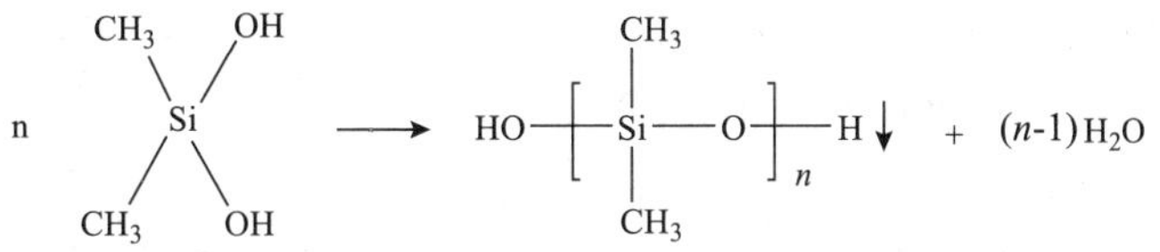

Poly dimethyl methyl silanediol

烃基卤代甲硅烷与水反应，生成相应的硅醇。硅醇中的多元醇很容易缩聚，生成聚硅醇沉淀，封堵出水层。下面以二甲基二氯甲硅烷与水反应为例，说明堵水沉淀的产生：

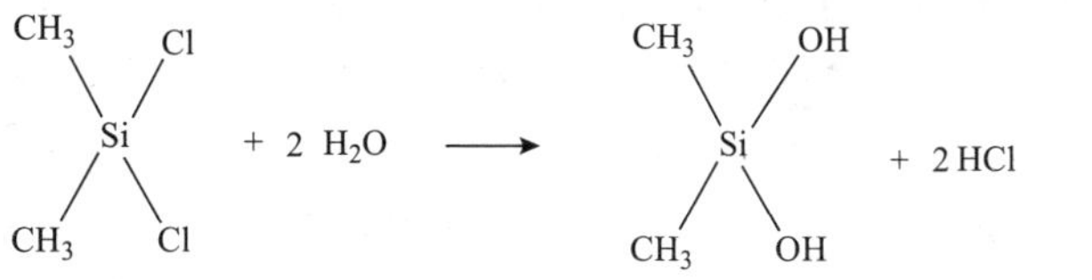

$$(CH_3)_2SiCl_2 + 2H_2O \longrightarrow (CH_3)_2Si(OH)_2 + 2HCl$$

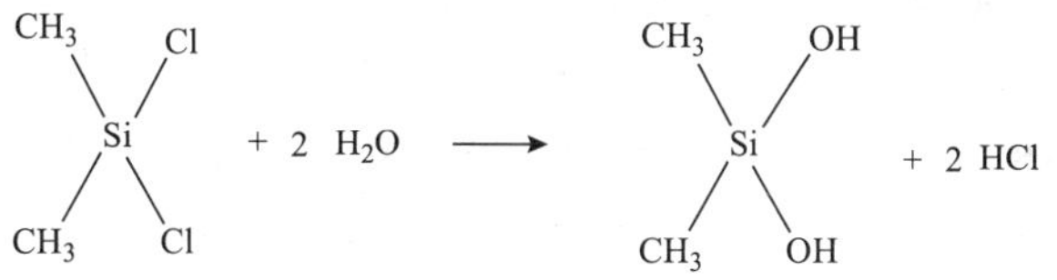

二甲基甲硅二醇

$$n(CH_3)_2Si(OH)_2 \longrightarrow HO-[Si(CH_3)_2-O]_n-H\downarrow + (n-1)H_2O$$

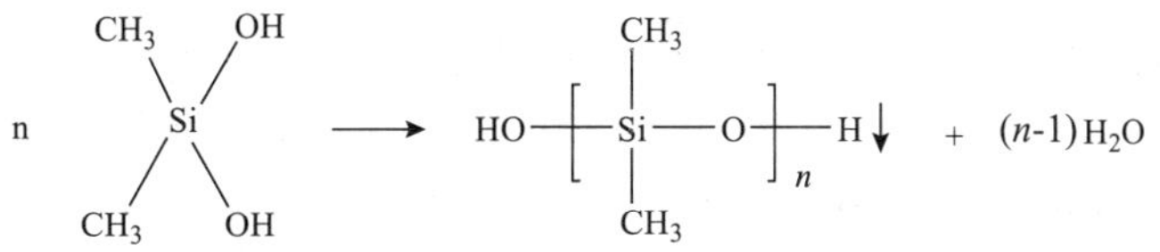

聚二甲基甲硅二醇

Since silane in hydrocarbyl halide is oil-soluble, it must be formulated into oil solution.

(2) Heavy oil.

①Active heavy oil.

This is a kind of heavy oil that added emulsifier. The emulsifier is water-in-oil type (such as Span 80). It will make water-in-oil emulsion with high viscosity when the heavy oil is exposed to water.

Heavy oil of this type can be used as selective

由于烃基卤代中硅烷是油溶性的，所以必须将它配成油溶液使用。

（2）稠油类。

①活性稠油。

这是一种溶有乳化剂的稠油。该乳化剂为油包水型乳化剂（如Span 80），它可使稠油遇水后产生高黏的油包水乳状液。

water plugging agent directly since it contains significant amounts of water-in-oil emulsifiers (such as naphthenic acid, resin, asphaltene, and so on). Active heavy oil can also be made by dissolving oxidized asphalt in oil. Asphalt of this type is both the water-in-oil emulsifier and the gelatinizer for oil.

② Heavy oil-solid powder.

Under effect of the emulsifier, the mixture of heavy oil and solid powder are pumped into the formation and making it form water-in-oil emulsion with formation water. The emulsion will then change the surface properties of the rock, impeding flow of formation water, hence reducing water phase permeability. Content of resin and asphaltene in heavy oil shall be greater than 45%, viscosity shall be greater than 500 mPa·s; particle size of solid powder such as shell powder, lime or cement shall be 150-200 mesh, and the surfactant shall be AS or ABS. The formula composition (mass fraction) shall be: heavy oil：powder：water = 100：3：230. This plugging agent can be used for water plugging in the sandstone oil reservoirs in which water are from the same layer. Before injection, the agent shall be heated to 50-70 ℃.

③ Couple heavy oil.

Water plugging agent of this type is prepared by dissolving low polymerization degree phenol formaldehyde resin, phenol furfural resin or their mixture as couplant in heavy oil. By reacting with formation surface, these resins will form chemisorption, which will strengthen combination (coupling) of formation surface of heavy oil, making discharging process of the water plugging agent difficult, hence prolonged the effective period.

④ Heavy oil-in-water.

Water plugging agent of this type is prepared by emulsifying heavy oil in water by oil-in-water emulsifier. It is easier to enter the water layer since water is the external phase and with low viscosity.

由于稠油中含有相当数量的油包水型乳化剂（如环烷酸、胶质、沥青质等），所以可将稠油直接用于选择性堵水。也可将氧化沥青溶于油中配成活性稠油。这种沥青既是油包水型乳化剂，也是油的稠化剂。

②稠油-固体粉末。

在乳化剂的作用下，稠油、固体粉末混合液泵入地层后与地层水形成油包水型乳状液，可改变岩石表面性质，使地层水的流动受阻并因此降低水相渗透率。其稠油中胶质和沥青含量应大于45%，黏度大于500 mPa·s，固体粉末贝壳粉、石灰或水泥的粒度为150~200目，表面活性剂为AS或ABS。配方组成（质量分数）为：稠油：粉末：水=100：3：230。该堵剂可用于出水类型为同层水的砂岩油层堵水，在注入地层前应加热至50~70℃。

③耦合稠油。

这种堵水剂是将低聚合度的苯酚甲醛树脂、苯酚糠醛树脂或它们的混合物作耦合剂溶于稠油中配成的。由于这些树脂可与地层表面反应，产生化学吸附，加强了地层表面与稠油的结合（耦合），使它不易排出，延长有效期。

④水包稠油。

这种堵水剂是用水包油型乳化剂将稠油乳化在水中配成的。因乳状液中水是外相，黏度低，所以易进入水层。在水层中，由于乳化剂在地层表面吸附，使乳状液破坏，油珠聚并为高黏度的稠油，产生很大的流动阻力，减

Because emulsifier adsorbs on formation surface in water layer, the emulsion is destroyed and the oil beads are aggregated into the heavy oil with high-viscosity, which will generate large flow resistance and reduces water production. Emulsifier for heavy oil-in-water shall preferably be cationic surfactant, because it will be easily adsorbed on negatively charged sandstone surface and causing emulsion destruction.

(3) Oil-based cements.

Oil-based cement is suspension cement in oil. Surface of cement is hydrophilic. When it enters water layer, water will replace oil on cement surface and react with the cement, making it to solidified and plug the water layer. In this process, the cement used shall be oil production well cement suitable for the corresponding well depth. And the oil shall be gasoline, kerosene, and diesel or low viscosity crude oil. In addition, a surfactant (such as carboxylateor sulfonate surfactant) shall also be added to change the suspension mobility. For example, add 300-800 kg oil production well cement and 0.1–1.0 kg surfactant to $1m^3$ oil could prepare oil-based cement with 1.05-1.65g/cm^3 density, which can be used for water plugging in oil production wells.

7.1.2.3 Alcohol-based water blocking agent

Alcohol solution of dimmer rosin is the most commonly used alcohol-based water plugging agent.

Under effect of sulfuric acid, rosin can be polymerized into dimmer rosin.

Dimmer rosin is soluble in low molecular alcohols (such as methanol, ethanol, n-propanol, etc.) and is insoluble in water. When alcohol solution of dimmer rosin is mixed with water, the water will dissolve in alcohol immediately, which will reduce solubility of dimmer rosin in low molecular alcohol, making it saturated and precipitated. Since softening point of dimmer rosin is quite high (at least 100 ℃), it will remain in solid state after precipitation, which could plug water layer very well.

The content of dimmer rosin in alcohol solution shall be around 40%-60% (mass fraction). If the content is too high, viscosity of the solution will be too large; if content is too low, effect of water plugging will not be good. Usage dose of the solution shall be around $1m^3$ per each meter of the formation on thickness.

7.1.3 Non-selective water plugging agent

Non-selective water plugging agent is suitable for plugging single water layers or layers with high water content in oil and gas wells. It is mainly divided into the following five types: resin, precipitation, gel, jelly and dispersion type water plugging agent.

7.1.3.1 Resin type water plugging agent

Resin type water plugging agent is with a space structure. It is polycondensed by low molecular substances, and is insoluble and infusible to polymer substance. Phenolic resins, urea-formaldehyde resins and epoxy resins are all belong to water plugging agent of this type.

When resin is mixed with curing agent (a catalyst capable of accelerating curing, such as oxalic acid, etc.) and injected into formation, it will be solidified with certain strength after a certain period of time under effect of curing agent and temperature. The solid state resin will then plug pores and water layers.

松香二聚物易溶于低分子醇（如甲醇、乙醇、正丙醇等）而难溶于水，当松香二聚物的醇溶液与水相遇，水即溶于醇中，减小了低分子醇对松香二聚物的溶解度，使松香二聚物饱和析出。由于松香二聚物软化点较高（至少100℃），所以松香二聚物析出后以固体状态存在，对于水层有较高的封堵能力。

在松香二聚物的醇溶液中，松香二聚物的含量为40%~60%（质量分数），含量太大，则黏度太高；含量太小，则堵水效果不好。其用量为每米厚地层 $1m^3$ 左右。

7.1.3 非选择性堵水剂

非选择性堵水剂适用于封堵油气井中单一含水层和高含水层，主要分为以下五种类型：树脂型堵水剂、沉淀型堵水剂、凝胶型堵水剂、冻胶型堵水剂和分散体型堵水剂。

7.1.3.1 树脂型堵水剂

树脂型堵水剂是由低分子物质通过缩聚反应产生的具有体型结构，不溶、不熔高分子物质的堵水剂。酚醛树脂、脲醛树脂、环氧树脂等都属于这类堵水剂。

当树脂液与固化剂（指能加速固化的催化剂，如草酸等）混合后挤入地层，在固化剂和温度的作用下，树脂液可在一定时间内转变为具有一定强度的固态树脂，达到堵塞孔隙、封堵水层的目的。

Resin plugging agent is mainly used for plugging high permeable layers, oil production wells with severe bottom water, channel flooding and sand production, as well as high temperature oil production wells. Water plugging technology with resin agent possesses characteristics of easy squeeze (into the formation), large plugging strength and good effects. However, since its cost is high and mistakenly plugged areas will be difficult to handle, application of this technology is not wide currently.

7.1.3.2 Precipitation type water plugging agent

Precipitation type water plugging agent is consisted by two substances that can react to form precipitation. Inject two kinds of inorganic chemical solutions separated by spacer liquid into the formation. During the injection process, let precipitation formed in pores and plug the formation physically. Since both reactants are aqueous solutions with low viscosity as the water, thus can preferentially enter the high water content layer and effectively plug high permeable layer. Precipitation type water plugging agent is with characteristics of high success rate, long expiration period, simple construction, low cost, easy plug removal and wide applicability, but it is easy to pollute the reservoir.

The most commonly used precipitation type water plugging agents is water glass-brine system.

As an important raw material in daily chemical and chemical industries, sodium silicate $x\text{Na}_2\text{O} \cdot y\text{SiO}_2$, also known as water glass, is a solid or viscous liquid with turquoise or brown color, sometimes no color. Its physical properties vary with ratio of sodium oxide and silicon dioxide. The molar ratio of SiO_2 to Na_2O in water glass is usually called modulus (M) of water glass. M is often in range of 2.7-3.3.

The brine system includes CaCl_2, FeCl_2, FeCl_3, FeSO_4, $\text{Al}_2(\text{SO}_4)_3$ and formaldehyde.

Generally speaking, the larger the water glass

树脂堵剂主要用于封堵高渗透地层，油井底水和窜槽水出砂严重及高温的油井。该技术具有堵剂易挤入地层，封堵强度大，效果好等特点，但所需费用高，误堵后很难处理，目前应用较少。

7.1.3.2 沉淀型堵水剂

沉淀型堵水剂是由两种能反应生成沉淀的物质组成的，向地层注入由隔离液隔开的两种无机化学剂溶液，在注入过程中，使其在地层孔道中形成沉淀，对被封堵地层形成物理堵塞，从而封堵地层孔道。由于这两种反应物均系水溶液且黏度较低，与水相近，因此，能优先进入高吸水层，有效地封堵高渗透层。沉淀型堵剂作业成功率高，有效期长，施工简单，价格较低，解堵容易，适用性强，但易污染油层。

最常用的沉淀型堵水剂为水玻璃-卤水体系。

硅酸钠 $x\text{Na}_2\text{O} \cdot y\text{SiO}_2$ 又名水玻璃、泡花碱，无色、青绿色或棕色的固体或黏稠液体，其物理性质随着成品内氧化钠和二氧化硅的比例不同而不同，是日用化工和化工工业的重要原料。通常将水玻璃中 SiO_2 与 Na_2O 的摩尔比称为水玻璃的模数（M）。M 通常为2.7~3.3。

卤水体系包括 CaCl_2、FeCl_2、FeCl_3、FeSO_4、$\text{Al}_2(\text{SO}_4)_3$、甲醛。

modulus is, the larger the precipitation amount will be, and plugging capacity is better.

7.1.3.3　Gel type water plugging agent

Gel type water plugging agent is formed by sol gelation.

The most commonly used gel type water plugging agent is silicic acid gel. To conduct water plugging with silicic acid gel, the water glass and an activator shall be mixed (by adding the water glass into the activator for forming acidic silicic acid sol or adding the activator into the water glass to form alkaline silicic acid sol) and then injected into the formation. They can also be divided into several slugs separated by a spacer fluid. By injected alternately into water layer to make sure they could mix after reaching to certain depth. Mixing water glass with activator, the silicic acid sol will be formed firstly, and it will then convert into silicic acid gel.

Saline gel is a new gel type water plugging agent developed in recent years. It is suitable for deep formation plugging. Its composition includes hydroxypropyl cellulose (HPC), sodium dodecyl sulfate (SDS) and saline, which are mixed to form the gel. Its advantage is metal salts such as chromium or aluminum is not necessary to be added as activators, gelation could be simply controlled by salinity in water. Viscosity of HPC/SDS fresh water solution is 80mPa·s. After mix with saline solution, the viscosity could reach to 70000mPa·s. This gel could reduce water permeability by 95% in the core flow test of sandstone. It is not necessary to specially design or handle reservoir during construction, and the validity period of plugging is up to six months. When there is no saline solution in the reservoir, the viscosity will decrease within a few days.

7.1.3.4　Jelly type water plugging agent

Jelly type water plugging agent is formed by polymer solution crosslinking reaction.

一般来说，水玻璃模数增大，沉淀量也增大，而沉淀量越大，堵塞能力就越大。

7.1.3.3　凝胶型堵水剂

凝胶型堵水剂是由溶胶胶凝产生的堵水剂。

最常用的凝胶型堵水剂是硅酸凝胶。当用硅酸凝胶封堵时，可将水玻璃和活化剂混合（将前者加入后者生成酸性硅酸溶胶或将后者加入前者生成碱性硅酸溶胶）后注入地层；也可将它们分成几个段塞，中间以隔离液隔开，交替地注入水层，让它们进入水层一定距离后才混合。水玻璃与活化剂混合后，首先生成硅酸溶胶，随后转变为硅酸凝胶。

盐水凝胶是近年来发展的新型凝胶堵剂，适用于深部地层封堵。它的组成包括羟丙基纤维素（HPC）、十二烷基硫酸钠（SDS）及盐水，三者混合后形成凝胶。优点是不需加入铬或铝等金属盐作活化剂，而是控制水的含盐度引发胶凝。HPC/SDS的淡水溶液黏度为80mPa·s，当与盐水混合后黏度可达70000mPa·s。该凝胶在砂岩的岩心流动实验中，可使水的渗透率降低95%。施工时不必对油藏进行特殊设计和处理，有效期达半年。当地层中不存在盐水时，几天内就会使其黏度降低。

7.1.3.4　冻胶型堵剂

冻胶型堵剂是由高分子溶液经交联反应形成的堵水剂。

Polymers can be crosslinked are mainly polyacrylamide (PAM), partially hydrolyzed polyacrylamide (HPAM), carboxymethyl cellulose (CMC), hydroxyethyl cellulose (HEC), hydroxypropyl cellulose (HPC), sodium lignosulfonate (Na-Ls), calcium lignosulfonate (Ca-Ls) and so on.

The crosslinker are mostly multi-nuclear hydroxy bridge complex ions (Cr^{3+}, Zr^{4+}, Ti^{3+}, Al^{3+}, etc.) that formed by high valent metal ions and aldehydes (formaldehyde, glyoxal, etc.).

There are many kinds of plugging agents in this type, such as aluminum jelly, chrome jelly, zirconium jelly, titanium jelly and aldehyde jelly.

7.1.3.5 Dispersion type water plugging agent

Dispersing plugging agent is mainly a solid dispersion for plugging ultra-high permeable layers. For example, solid dispersions such as clay/cement, calcium carbonate/cement and fly ash/cement can be used to plug ultra-high permeable layer in wells.

7.1.4 Selection of water plugging agent

Water plugging agent shall be selected according to the following principles preferentially:

(1) Water-based water plugging agent.

Water-based water plugging agents will preferentially enter formations with high water saturation.

(2) Single liquid water plugging agent.

Construction with single liquid water plugging agents are relatively easy.

(3) Jelly type water plugging agent.

The jelly type water plugging agents are with a certain range of selectivity, and suitable for low salinity formation with low temperature.

(4) Water glass water plugging agent.

Water glass water plugging agents have an ideal selectivity in plugging reservoirs with high temperature, high salinity and (or) high acid gas content.

能被交联的高分子主要有聚丙烯酰胺（PAM）、部分水解聚丙烯酰胺（HPAM）、羧甲基纤维素（CMC）、羟乙基纤维素（HEC）、羟丙基纤维素（HPC）、木质素磺酸钠（Na-Ls）、木质素磺酸钙（Ca-Ls）等。

交联剂多为由高价金属离子所形成的多核羟桥络离子（Cr^{3+}、Zr^{4+}、Ti^{3+}、Al^{3+}等）和醛类（甲醛、乙二醛等）。

该类堵剂很多，诸如铝冻胶、铬冻胶、锆冻胶、钛冻胶及醛冻胶等。

7.1.3.5 分散体型堵剂

分散型堵剂主要是固体分散体，用于封堵特高渗透层。例如，黏土/水泥、碳酸钙/水泥和粉煤灰/水泥等固体分散体可用在油井上封堵特高渗透层。

7.1.4 堵水剂的选择

堵水剂按下列原则优先选择：

（1）水基堵水剂。

水基堵水剂优先进入含水饱和度高的地层。

（2）单液法堵水剂。

单液法堵水剂施工方便。

（3）冻胶型堵水剂。

冻胶型堵水剂有一定的选择性，适用于温度和矿化度较低的地层。

（4）水玻璃堵水剂。

在高温高矿化度和（或）高酸性气体含量地层，水玻璃堵水剂有理想的选择性。

7.2 Profile control of water injection well

As stated earlier, water injection well profile control technique can be used to control water production of oil production wells. Besides, by adjusting water absorption profile of the reservoir that being injected water, effect of the flood volume conformance could be improved, performance of injected water sweep in medium and low permeable layers will be enhanced to a greater extent.

7.2.1 Classification of profile control agents

Water injection well profile control agent (profile control agent for short) refers to a substance injected from the injection well to adjust water absorption profile of the formation. It can be classified according to different standards:

7.2.1.1 Classified by injection process

It can be divided into single-liquid method profile control agent (such as chromic jelly) and double-liquid method profile control agent (such as water glass-calcium chloride). The single-liquid method profile control agent only needs to inject one working fluid during profile control operation into the formation while the double-liquid method profile control agent needs two (Figure 7-5). Profile control agents are usually classified according to this standard.

7.2.1.2 Classified by plugging distance

According to the plugging distance, profile control agents can be divided into percolation surface profile control agents, near wellbore area profile control agents (such as silicic acid gel) and far wellbore area profile control agents (such as colloidal dispersion jelly).

7.2 注水井调剖

如前所述，注水井调剖技术可以用来控制油井出水。此外，注水井调剖通过对注水油层吸水剖面的调整，可以提高注入水波及系数，更大程度地改善中低渗层的注入水波及效果。

7.2.1 调剖剂的分类

注水井调剖剂（简称调剖剂）是指从注水井注入地层，调整地层吸水剖面的物质，可按不同的标准对调剖剂进行分类：

7.2.1.1 按注入工艺分类

可分为单液法调剖剂（如铬冻胶）和双液法调剖剂（如水玻璃-氯化钙），其中单液法调剖剂在调剖作业时只需向地层注入一种工作液；而双液法调剖剂在调剖作业时需向地层注入两种工作液（图7-5）。调剖剂通常按这个标准分类。

7.2.1.2 按封堵距离分类

可分为渗滤面调剖剂、近井地带调剖剂（如硅酸凝胶）和远井地带调剖剂（如胶态分散体冻胶）。

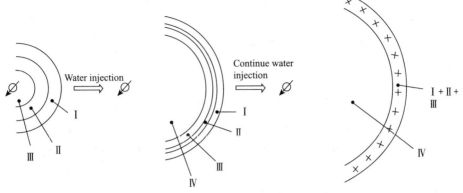

Figure7-5　Double-fluid profile control

Ⅰ—Working fluid of the first stage; Ⅱ—spacer fluid; Ⅲ—Working fluid of the second stage; Ⅳ—injection water; ∅—injection well; +—The plugging substances produced by Ⅰ and Ⅲ

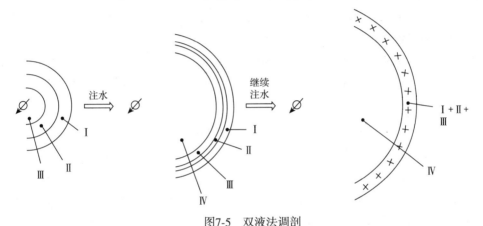

图7-5　双液法调剖

Ⅰ—第一工作液；Ⅱ—隔离液；Ⅲ—第二工作液；Ⅳ—注入水；∅—注入井；+—Ⅰ与Ⅲ相遇产生的封堵物质

7.2.1.3　Classified by usage conditions

It can be divided into high permeable layer profile control agent (such as clay/cement solidification system), low permeable layer profile control agent (such as sulfuric acid), high temperature and high salinity formation profile control agent (such as various inorganic profile control agents).

7.2.2　Single-liquid method profile control agent

7.2.2.1　Silicic acid gel

Silicic acid gel is a typical single-liquid method profile control agent. Inject sodium silicate solution into

7.2.1.3　按使用条件分类

可分为高渗透层调剖剂（如黏土/水泥固化体系）、低渗透层调剖剂（如硫酸）、高温高矿化度地层调剖剂（如各种无机调剖剂）。

7.2.2　单液法调剖剂

7.2.2.1　硅酸凝胶

硅酸凝胶是一种典型的单液法堵剂，在处理时将硅酸钠溶液

the formation, under effect of activator, it will turn to sol and then gel after a certain period of time. The gel will then plug the high permeable layer. Activators could be divided into two types:

(1) Inorganic activator.

Such as hydrochloric acid, nitric acid, sulfuric acid, sulfamic acid, ammonium carbonate, ammonium hydrogen carbonate, ammonium chloride, ammonium sulfate, sodium dihydrogen phosphate, and so on.

(2) Organic activator.

Such as formic acid, acetic acid, ammonium acetate, ethyl formate, ethyl acetate, chloroacetic acid, trichloroacetic acid, oxalic acid, citric acid, formaldehyde, phenol, catechol, resorcinol, hydroquinone, phloroglucinol, and so on.

Activator used for silicic acid gel in single-liquid method is usually hydrochloric acid, and the reaction is as follows:

$$Na_2O \cdot mSiO_2 + 2HCl \longrightarrow H_2O + mSiO_2 + 2NaCl$$

The main disadvantage of silicic acid gel is the short gelation time (generally less than 24h). What's more, the higher the formation temperature is, the shorter the gelation time will be. To prolong the gelation time, it can be activated by latent acid. Or it could be activated by thermo-sensitive activator such as lactose, xylose, and so on in 50~80℃ formation. In addition, since toughness of silicic acid gel is weak, the water glass could be thickened by HPAM and then be activated for improvement.

7.2.2.2 Inorganic acids

(1) Sulfuric acid.

Sulfuric acid can produce profile control materials by calcium and magnesium in the formation. Inject concentrated sulfuric acid or industrial waste liquid containing concentrated sulfuric acid into the well, the sulfuric acid will first react with carbonate in formation surrounding the wellbore. This will increase absorption capacity of the injection well while the produced fine calcium sulfate and magnesium sulfate crystals will follow the acid flows and deposit in certain locations (such as throat of the pore structure), forming plugs. Since there are comparatively more sulfuric acid enters the high permeable layers, more calcium sulfate and magnesium sulfate will be produced, and that is why the main plugging will occur in the high permeable layers.

Main reactions of profile control with sulfuric acid are as follows:

孔隙结构的喉部）沉积下来，形成堵塞。由于高渗透层进入硫酸多，产生的硫酸钙、硫酸镁也多，所以主要的堵塞发生在高渗透层。

用硫酸进行调剖的主要反应如下：

$$CaCO_3+H_2SO_4 \longrightarrow CaSO_4\downarrow+CO_2\uparrow+H_2O$$
$$MgCa(CO_3)_2+2H_2SO_4 \longrightarrow MgSO_4\downarrow+CaSO_4\downarrow+2CO_2\uparrow+2H_2O$$

(2) Hydrochloric acid-sulfate solution.

The system will produce profile control substances by calcium and magnesium in the formation. For example, when a hydrochloric acid-sulfate solution composed by 4.5%~12.3%HCl, 5.1%~12.5%Na_2SO_4 and 0.02%~14.5%$(NH_4)_2SO_4$ is injected into formation containing calcium carbonate, there will be precipitate produced by the following reactions. And the precipitates will work as profile control agents:

（2）盐酸—硫酸盐溶液。

该体系利用地层的钙、镁来产生调剖物质。例如，将一种配方为 4.5%~12.3%HCl，5.1%~12.5% Na_2SO_4，0.02%~14.5%（NH_4）$_2SO_4$ 的盐酸—硫酸盐溶液注入含碳酸钙的地层，则可通过下列反应产生沉淀，起调剖作用：

$$CaCO_3+2HCl \longrightarrow CaCl_2+CO_2\uparrow+H_2O$$
$$CaCl_2+SO_4^{2-} \longrightarrow CaSO_4\uparrow+2Cl^-$$

7.2.2.3 Jellies

Commonly used jelly type profile control agents include zirconium jelly, chromium jelly, aluminum jelly, phenolic resin jelly and polyethyleneimine jelly, etc. To apply plugging agent of this type, mix certain amount of polymer solution with crosslinking agent and inject the mixture into the formation. The mixture will conduct crosslinking reaction under the formation temperature and form jelly, plugging the high permeable layer. Taking zirconium jelly as an example, the following is explanation of how water plugging agent will be formed:

Zirconium jelly is produced by polymers with

7.2.2.3 冻胶类

常用的冻胶类调剖剂有锆冻胶、铬冻胶、铝冻胶、酚醛树脂冻胶、聚乙烯亚胺冻胶等。该类堵剂在应用时，是将一定量的聚合物溶液与交联剂混合后注入地层，在地层温度条件下进行交联反应，生成冻胶封堵高渗透层。下面以锆冻胶为例介绍冻胶类调剖剂的形成。

锆冻胶是用 Zr^{4+} 组成的多核羟桥络离子交联溶液中带—COO^- 的聚合物（如 HPAM）生成的。

—COO⁻ (such as HPAM). These polymers are from multi-nuclear hydroxy bridge complex ions crosslinking solution that composed of Zr^{4+}.

Zr^{4+} can be derived from $ZrOCl_2$ or $ZrCl_4$.

By the following steps, Zr^{4+} can generate zirconic multi-nuclear hydroxy bridge complex ions:

(1) Complexation:

$$Zr^{4+} + 8H_2O \longrightarrow [(H_2O)_8Zr]^{4+}$$

(2) Hydrolysis:

$$[(H_2O)_8Zr]^{4+} \longrightarrow [(H_2O)_7Zr(OH)]^{3+} + H^+$$

(3) Hydroxy bridge effect:

$$2\,[(H_2O)_7Zr(OH)]^{3+} \longrightarrow \left[(H_2O)_6Zr\underset{OH}{\overset{OH}{\diamond}}Zr(H_2O)_6\right]^{6+} + 2H_2O$$

(4) Further hydrolysis and hydroxy bridge:

Zr^{4+} 可来自 $ZrOCl_2$ 或 $ZrCl_4$。

Zr^{4+} 可通过下列步骤生成锆的多核羟桥络离子：

（1）络合：

（2）水解：

（3）羟桥作用：

（4）进一步水解和羟桥作用：

$$\left[(H_2O)_6Zr\underset{OH}{\overset{OH}{\diamond}}Zr(H_2O)_6\right]^{6+} + 2H_2O + [(H_2O)_7Zr(OH)]^{3+} \longrightarrow$$

$$\left[(H_2O)_6Zr\left\{\begin{array}{c}H_2O\ H_2O\\OH\diagup Zr\diagdown OH\\OH\diagdown\ \diagup OH\\H_2O\ H_2O\end{array}\right\}_n Zr(H_2O)_6\right]^{(2n+6)+} + nH^+ + 2nH_2O$$

Polynuclear olation complex ion of zirconium

$$\left[(H_2O)_6Zr\underset{OH}{\overset{OH}{\diamond}}Zr(H_2O)_6\right]^{6+} + 2H_2O + [(H_2O)_7Zr(OH)]^{3+} \longrightarrow$$

$$\left[(H_2O)_6Zr\left\{\begin{array}{c}H_2O\ H_2O\\OH\diagup Zr\diagdown OH\\OH\diagdown\ \diagup OH\\H_2O\ H_2O\end{array}\right\}_n Zr(H_2O)_6\right]^{(2n+6)+} + nH^+ + 2nH_2O$$

锆的多核羟桥络离子

The following is a crosslinked body produced by zirconicmulti-nuclear hydroxy bridge complex ions crosslinking zone —COO⁻polymers (such as HPAM):

下面是锆的多核羟桥络离子交联带 —COO⁻ 聚合物（如 HPAM）所产生的交联体：

Chemical water plugging and profile control technology

This crosslinked product is called zirconium jelly. For example, the zirconium jelly with 7h forming time could be made by mixing 0.75%ω (HPAM), 1.0%ω (ZrOCl$_2$) and 5.5%ω (HCl) solutions in a volume ratio of 100∶4∶3 at 60℃. The jelly could be used for high permeable layers plugging.

7.2.2.4 Swellable particles

Swellable particles are a class of polymer that will expand but not dissolve and conduct proper crosslinking in water. For example, polyacrylamide swellable particles could be produced from drying and grinding polymerized acrylamide. There shall be a small amount of crosslinking agent N,N'-methylenebisacrylamide added during polymerizing progress of the acrylamide. In water, both expansion velocity and expansion multiples of these swellable particles are high.

Swellable particles can be prepared from all properly crosslinked water soluble polymers.

There are two ways to place water swellable particles in the far wellbore area: One is to select a suitable carrier such as kerosene, ethanol or electrolyte solution (like sodium chloride or ammonium chloride solution), etc., which can inhibit expansion of swellable particles; The other is to coat the swellable particles (such as hydroxypropyl methylcellulose). Let the coated swellable particles carried into the formation by water and they will swell after the coat dissolved. Fluidized bed method could be used for coating swellable particles.

此交联体叫锆冻胶。例如，将ω（HPAM）为0.75%的溶液与ω（ZrOCl$_2$）为1.0%和ω（HCl）为5.5%的溶液按体积比100∶4∶3混合，可配得一种在60℃下成冻时间为7h的锆冻胶，用于封堵高渗透层。

7.2.2.4 水膨体

水膨体是一类适当交联遇水膨胀而不溶解的聚合物。例如，在丙烯酰胺聚合过程中加入少量交联剂N,N'-亚甲基双丙烯酰胺，聚合后干燥、磨细，就可得到聚丙烯酰胺水膨体。这种水膨体在水中的膨胀速率和膨胀倍数都高。

所有适当交联的水溶性聚合物都可制得水膨体。

为了将水膨体放置在远井地带，有两种方法：一种是选择适当的携带介质如煤油、乙醇和电解质溶液（如氯化钠溶液、氯化铵溶液）等，这些携带介质能抑制水膨体膨胀；另一种是在水膨体外表面覆膜（如覆羟丙基甲基纤维素膜），将这种覆膜的水膨体用水携带进入地层，它将在覆膜溶解至可与水相接触时才开始膨胀。用流化床法在水膨体外表面覆膜。

7.2.2.5 Jelly microspheres

Jelly microspheres are jelly dispersions with nanometer scale particle size. It can be obtained by microemulsion polymerization. For example, prepare an aqueous solution from acrylamide, any monomerwith alkenyl, N,N'-methylenebisacrylamide and ammonium persulfate, then produce oilexternal phase microemulsion with high concentration mixed surfactant (such as Span80+Tween60). After the aqueous solution is solubilized in microemulsion micelles, and monomers are copolymerized, the jelly microspheres are formed. During application, the jelly microspheres in oil external phase are needed to be reversely dispersed in water by effect of an inverse phase agent (such as OP-10). Then inject the water into formation. Since jelly microspheres are with certain expansion multiples in the formation, they could work as profile control agent through a circulation mechanism of migrating, capturing, deforming, re-migrating, re-capturing and re-deforming in preferential path. Its profile control effect will be start from where it is located and then spread further.

7.2.2.6 Lime cream

Lime cream is prepared by dispersing calcium oxide in water. Calcium oxide will react with water and forming calcium hydroxide:

$$CaO + H_2O \longrightarrow Ca(OH)_2$$

Solubility of calcium hydroxide in water is very small (at 60℃, only 0.116g calcium hydroxide will be dissolved in 100g water). It can be seen that lime cream is the suspension of calcium hydroxide in water.

$\omega(CaO)$ of lime cream is generally in range of 5%~10%. This single-liquid method profile control agent is with the following characteristics:

(1) Particle size of calcium hydroxide is quite large (about 62μm) which is particularly suitable for plugging of fractured high permeable layers. The

middle and low permeable layers will be protected since particles of calcium hydroxide cannot enter into them.

(2) Solubility of calcium hydroxide will decrease with temperature increasing (Table 7-1), thus it can be used to plug high temperature formations.

Table 7-1　Solubility of calcium hydroxide at different temperatures

Temperature /℃	Solubility/ (g/kg)	Temperature /℃	Solubility/ (g/kg)
30	1.53	70	1.06
40	1.41	80	0.94
50	1.28	90	0.85
60	1.16	100	0.77

表7-1　氢氧化钙在不同温度下的溶解度

温度/℃	溶解度/（g/kg）	温度/℃	溶解度/（g/kg）
30	1.53	70	1.06
40	1.41	80	0.94
50	1.28	90	0.85
60	1.16	100	0.77

(3) Calcium hydroxide can react with hydrochloric acid and forming water soluble calcium chloride. When it is not needed to be plugged, plugging can be removed by hydrochloric acid:

$$Ca(OH)_2 + 2HCl \longrightarrow CaCl_2 + 2H_2O$$

7.2.2.7　Clay/cement dispersion

The clay/cement dispersion is composed of suspending clay and cement in water. This dispersion is suitable for plugging ultra high permeable layers. After clay and cement entered, a filter cake will be formed in the formation (mainly at throat of the pore structure). Hydration reaction of cement will consolidate the filter cake and effectively plugs the ultra high permeable layer.

In clay/cement dispersion, ω (clay) is in range of 6%~20%, and ω (cement) is also in range of 6%~20%.

Similar with calcium hydroxide in lime cream, clay and cement will protect middle and low permeable layers since they cannot enter these layers.

能进入中、低渗透层，因此对中、低渗透层有保护作用。

（2）氢氧化钙的溶解度随温度升高而减小（表7-1），所以可用于封堵高温地层。

（3）氢氧化钙可以与盐酸反应生成可溶于水的氯化钙，在不需要封堵的时候，可用盐酸解除：

7.2.2.7　黏土/水泥分散体

黏土/水泥分散体由黏土与水泥悬浮于水中配成。此分散体适用于封堵特高渗透地层。黏土与水泥进入地层后，可在地层内（主要在孔隙结构的喉部）形成滤饼。在滤饼中，水泥的水化反应使滤饼固结，对特高渗透层产生有效封堵。

在黏土/水泥分散体中,ω(黏土)在6%~20%,ω（水泥）也在6%~20%。

类似于石灰乳中的氢氧化钙，黏土和水泥也不能进入中、低渗

If it is necessary, plugging produced by clay/cement dispersion can be removed by the conventional mud acid, i.e., acid with 12%ω (HC) and 3%ω (HF).

In addition to clay/cement dispersion, the ultra high permeable layer can also be plugged by calcium carbonate/cement dispersion and fly ash/cement dispersion.

7.2.3 Double-liquid method profile control agent

7.2.3.1 Precipitation type double-liquid method profile control agent

The precipitation type double-liquid method profile control agent refers to two liquids that can react and form precipitation. This type of profile control agent has advantages of high strength, stable shearing, good temperature resistance, good chemical stability and low cost, etc.

During profile control of the water injection wells, the commonly used first reaction solution is sodium silicate or sodium carbonate solution; The second reaction solution includes ferric chloride, calcium chloride, ferrous sulfate, and magnesium chloride. During the injection process, the two reaction solution shall be separated by a spacer liquid. They will then form precipitation in formation pores, which will physically plug the formation pores. Since both reaction solutions are aqueous and with low viscosity, they could easily enter high permeable layers selectively and plug the layers more effectively. For example, sodium silicate could react with calcium chloride and form gel (calcium silicate precipitate) to plug water flow.

The greater the viscosity ratio between the first and the second reaction solutions, the easier the fingering occurs. Since its viscosity is low, the sodium silicate

透层，所以对中、低渗透层有保护作用。

如果需要，黏土/水泥分散体产生的封堵可用常规土酸，即 ω（HC）为12%，ω（HF）为3%的酸除去。

除黏土/水泥分散体外，还可用碳酸钙/水泥分散体和粉煤灰/水泥分散体封堵特高渗透层。

7.2.3 双液法调剖剂

7.2.3.1 沉淀型双液法调剖剂

沉淀型双液法调剖剂是指两种反应液相遇后能形成沉淀的物质。沉淀型调剖剂具有强度大、对剪切稳定、耐温性好、化学稳定性好和成本低廉等优点。

在注水井调剖时，常用的第一反应液为硅酸钠或碳酸钠溶液；第二反应液有三氯化铁、氯化钙、硫酸亚铁和氯化镁等的水溶液。在注入过程中，用隔离液隔开，使其在地层孔道中形成沉淀，对被封堵地层形成物理堵塞，从而封堵地层孔道。由于这两种反应物均系水溶液，且黏度较低，与水相近，因此，能选择性地进入高渗透层产生更有效的封堵作用。如硅酸钠与氯化钙反应生成凝胶（硅酸钙沉淀）可封堵水流。

第一反应液对第二反应液的黏度比越大，指进越容易发生。黏度不高的硅酸钠溶液通常作为第一反应液，用HPAM稠化。

综上所述，作为第一反应液

solution is usually used as the first reaction solution and be thickened by HPAM.

In summary, sodium silicate is more suitable than sodium carbonate as the first reaction solution. During the selection, sodium silicate with large modulus shall be taken, and concentration is preferably in range of 20%~25%.The thickening agent shall preferably select HPAM with concentration in range of 0.4%~0.6%. As for second reaction solution, 15% concentration $CaCl_2$ and $MgCl_2 \cdot 6H_2O$ will be a good choice. Since $FeSO_4 \cdot 7H_2O$ and $FeCl_2$ are corrosive to metal equipment and pipeline, they should avoid to be used, or used after adding appropriate amount of corrosion inhibitor.

的选择，硅酸钠优于碳酸钠，而在硅酸钠中，应选模数大的，其浓度以 20%~25% 为好。所用稠化剂选用浓度 0.4%~0.6% 的 HPAM 为佳。第二反应液的选择，以浓度为 15% 的 $CaCl_2$、$MgCl_2 \cdot 6H_2O$ 为最好。由于 $FeSO_4 \cdot 7H_2O$ 和 $FeCl_2$ 对金属设备和管道有腐蚀性，故不宜使用，或加入适量的缓蚀剂后再用。

7.2.3.2 Gel type double-liquid method profile control agent

7.2.3.2 凝胶型双液法调剖剂

The gel type double-liquid method profile control agent refers to two liquids that can react and form gel. There are two main types of commonly used gel type double-liquid method profile control systems:

(1) Silicate-ammonium sulfate. Inject sodium silicate solution and ammonium sulfate solution into the formation, taking water as spacer liquid, the two solutions will react in the formation as following:

凝胶型双液法调剖剂是指两种反应液相遇后能形成凝胶的物质。常用的凝胶型双液法体系主要有以下两种：

（1）硅酸盐－硫酸铵。向地层注入硅酸钠溶液和硫酸铵溶液，用水作隔离液，两反应液在地层相遇可发生如下反应：

$$Na_2O \cdot mSiO_2+(NH_4)_2SO_4+2H_2O \longrightarrow mSiO_2+H_2O+Na_2SO_4+NH_4OH$$

Gel generated from this reaction will plug the ultra high permeable layer.

(2) Silicate-hydrochloric acid. Sodium silicate will form silicic acid under acidic conditions, its reaction formula is:

产生的凝胶可封堵高渗透层。

（2）硅酸盐－盐酸。硅酸钠在酸性条件下生成硅酸，反应式为：

$$Na_2O \cdot mSiO_2+2HCl+nH_2O \longrightarrow 2NaCl+ mSiO_2 \cdot (n+1)Na_2O$$

Silicic acid sol produced by above reaction will further coagulate and forming gel in network structure which can plug the high permeable layer with water production. Take 5%~15% content sodium silicate (modulus3~4) as the first reaction solution and 10%

生成的硅酸溶胶进一步凝聚，形成网状结构凝胶可封堵高渗透出水层。第一反应液硅酸钠（模数为 3~4）溶液的含量为 5%~15%，第二反应液为 10% 的盐酸，

hydrochloric acid as the second reaction solution, control volume ratio of the first and second reaction solutions shall be 4∶1. This formula is suitable for profile control of water wells in sandstone formation. The applicable temperature is 60~80 ℃ and water shall be used as the spacer liquid.

7.2.3.3 Jelly type double-liquid method profile control agent

The jelly type double-liquid method profile control agent refers to two liquids that can react and form jelly. Usually one reaction solution is a polymer solution, and the other reaction solution is a crosslinking agent solution. They will react and produce jelly and plug the high permeable layer (Figure7-6) in the formation. Commonly used systems of this type mainly include:

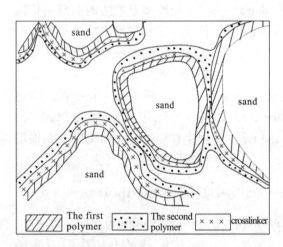
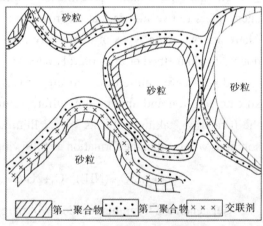

Figure7-6 Schematic diagram of profile control effect of jelly type double-liquid method profile control agent

(1) The first reaction solution is HPAN, CMC, CMHEC (carboxymethyl hydroxyethyl cellulose) or XC; The second reaction liquid is aluminum citrate. They will react and produce aluminum jelly.

(2) The first reaction solution is HPAN, CMC or XC; The second reaction solution is chromium propionate. They will react and produce chrome jelly.

(3) The first reaction solution is HPAM, CMC or XC with $Na_2S_2O_3$ (or $Na_2Cr_2O_7$); The second reaction

solution is $Na_2Cr_2O_7$ (or $Na_2S_2O_3$). During reaction, $Na_2S_2O_3$ will reduce Cr element in $Na_2Cr_2O_7$, making Cr^{6+} to Cr^{3+}:

$$CrO_7^{2-} + 3SO_3^{2-} + 8H^+ \longrightarrow 2Cr^{3+} + 3SO_4^{2-} + 4H_2O$$

Cr^{3+} will further generate multi-nuclear hydroxy bridge complex ions to crosslink polymer and forming chromium jelly.

(4) The first reaction solution is HPAM, CMC or CMHEC; The second reaction solution is aluminum acetate + zirconium lactate + zirconium oxychloride + lactic acid + triethanolamine. They will react and produce aluminum jelly + zirconium jelly.

(5) The first reaction solution is HPAM, CMC, CMHEC or XC; The second reaction solution is $Na_2Cr_2O_7$ + $NaHSO_3$ + nonionic polymer thickening agent (such as HEC). Purpose of thickening the second reaction solution is to prolong the distance between the two reaction solutions. Reaction of the two solutions will produce chromium jelly.

(6) The first reaction solution is lignosulfonate + acrylamide; The second reaction solution is persulfate. The two reaction solutions will conduct polymerization and crosslinking reaction in the formation and form jelly.

7.2.3.4 Foam type

Dissolve foaming agent in water and then alternately inject it into the formation with liquid carbon dioxide, foams will be formed (mainly at high permeable layer) and cause plugging. Foaming agents can be used are including nonionic surfactants such as polyoxyethylene alkyl phenolphenol ethers, anionic surfactants such as alkyl aryl sulfonates and cationic surfactants such as alkyl trimethyl ammonium compound salts.

Dissolve foaming agent in silicic acid sol (prepared by adding ammonium sulfate to water glass, pH 5~11), inject it into the formation, then inject natural gas or

$Na_2Cr_2O_7$）；第二反应液为 $Na_2Cr_2O_7$（或 $Na_2S_2O_3$）。相遇后 $Na_2S_2O_3$ 可将 $Na_2Cr_2O_7$ 中的 Cr^{6+} 还原为 Cr^{3+}：

Cr^{3+} 进一步生成多核羟桥络离子将聚合物交联，产生铬冻胶。

（4）第一反应液为 HPAM、CMC 或 CMHEC；第二反应液为醋酸铝+乳酸锆+氧氯化锆+乳酸+三乙醇胺，相遇后可产生铝冻胶+锆冻胶。

（5）第一反应液为 HPAM、CMC、CMHEC 或 XC；第二反应液为 $Na_2Cr_2O_7$+$NaHSO_3$+非离子聚合物稠化剂（如 HEC）。第二反应液稠化的目的是延长两反应液相遇的距离，相遇后产生铬冻胶。

（6）第一反应液为木质素磺酸盐+丙烯酰胺；第二反应液为过硫酸盐。两反应液在地层相遇后，即引发聚合和交联反应，生成冻胶。

7.2.3.4 泡沫型

将起泡剂溶于水中，然后与液体二氧化碳交替注入地层，可在地层（主要是高渗透地层）中形成泡沫，产生堵塞。所用的起泡剂包括非离子表面活性剂聚氧乙烯烷基苯酚醚、阴离子表面活性剂如烷基芳基磺酸盐和阳离子表面活性剂如烷基三甲基季铵盐。

将起泡剂溶于硅酸溶胶（由硫酸铵加入水玻璃中配成，pH 值为 5~11）中注入地层，然后注入

nitrogen gas, there will be foams with liquid dispersion medium formed. Then silicic acid sol will gelate, and foams will be formed. These foams are with gel as dispersion medium. Foaming agent used in this process shall be quaternary ammonium salt surfactant.

7.2.3.5 Flocculation type

If clay suspension and HPAM solution are divided into several slugs by spacer liquid and are alternately injected into the formation, they will form flocculation when they meet. And the flocculation will effectively plug the ultra high permeable layers.

7.2.4 Selection of profile control agent

7.2.4.1 For high permeable layers

Options available: Zirconium jelly, chromium jelly, phenolic resin jelly, swellable particles, lime cream, clay/cement dispersion, precipitation type double-liquid method profile control agents, foam type double-liquid method profile control agents and flocculation type double-liquid method profile control agents, etc.

7.2.4.2 For low permeable layers

Options available: Sulfuric acid, ferrous sulfate, jelly microspheres, jelly type double-liquid method profile control agents and precipitation type double-liquid method profile control agents, etc.

7.2.4.3 For high temperature and high salinity formations

Shall mainly select inorganic profile control agents such as sulfuric acid, ferrous sulfate, lime cream, clay/cement dispersion or precipitation type double-liquid method profile control agents, etc.

7.2.4.4 For near wellbore areas

Options available: Silicic acid gel, zirconium jelly, chromium jelly, swellable particles, lime cream and clay/cement dispersion, etc.

天然气或氮气，则可在地层中先产生以液体为分散介质的泡沫，随后硅酸溶胶胶凝，就可产生以凝胶为分散介质的泡沫。所用的起泡剂为季铵盐表面活性剂。

7.2.3.5 絮凝体型

若将黏土悬浮体与HPAM溶液分成几个段塞，中间以隔离液隔开，交替注入地层，它们在地层中相遇会形成絮凝体，这种絮凝体能有效地封堵特高渗透层。

7.2.4 调剖剂的选择

7.2.4.1 高渗透层

可选择锆冻胶、铬冻胶、酚醛树脂冻胶、水膨体、石灰乳、黏土/水泥分散体、沉淀型双液法调剖剂、泡沫型双液法调剖剂和絮凝体型双液法调剖剂等。

7.2.4.2 低渗透层

可选择硫酸、硫酸亚铁、冻胶微球、冻胶型双液法调剖剂、沉淀型双液法调剖剂等。

7.2.4.3 高温高矿化度地层

主要使用无机调剖剂如硫酸、硫酸亚铁、石灰乳、黏土/水泥分散体、沉淀型双液法调剖剂等。

7.2.4.4 近井地带

可选择硅酸凝胶、锆冻胶、铬冻胶、水膨体、石灰乳、黏土/水泥分散体等。

7.2.4.5 For far wellbore areas

Options available: Colloidal dispersion jelly (jelly made from low mass concentration polymer and low mass concentration crosslinking agent), jelly microspheres, jelly type double-liquid method profile control agents and precipitation type double-liquid method profile control agents, etc.

7.2.4.5 远井地带

可选择胶态分散体冻胶（由低质量浓度的聚合物和低质量浓度的交联剂配制的冻胶）、冻胶微球、冻胶型双液法调剖剂、沉淀型双液法调剖剂等。

第8章 油井化学清防蜡技术
Oil well chemistry paraffin (wax) inhibition and removal technology

According to the data, paraffin (wax) content of crude oil produced in the most countries all over the world exceeds 2%, of which are more than 10% in most of oilfields in China, and even reach to 40%-50% (Table 8-1). Therefore, during process of high paraffin (wax) crude oil production, paraffin (wax) deposit in oil well is a problem often encountered in oil recovery.

有关资料表明：世界各国生产的原油含蜡量大多数超过2%，其中我国大部分油田含蜡量超过10%，有的甚至高达40%~50%(表8-1)。在开采高含蜡原油时，油井结蜡是采油过程中经常遇到的一个问题。

Table 8-1　Paraffin (wax) content of major oil fields in China

Oilfield	Daqing	Jilin	Liaohe	Jidong	Dagang	Huabei	Shengli	Zhongyuan	Nanyang	Hanjiang
Paraffin (wax) content/%	26.2	22.3	16.8	20.5	14.1	21.2	20.8	21.4	30.9	14.6
Oilfield	Jiangsu	Sichuan	Changqing	Qinghai	Yumen	Tuha	Eastern Xinjiang	Bohai	Nanhai	Shenbei
Paraffin (wax) content/%	14.6	17.1	10.2	20.0	10.0	12.0	16.0	10.0	20.5	40.9

表8-1　国内主要油田含蜡情况

油田	大庆	吉林	辽河	冀东	大港	华北	胜利	中原	南阳	江汉
含蜡量 /%	26.2	22.3	16.8	20.5	14.1	21.2	20.8	21.4	30.9	14.6
油田	江苏	四川	长庆	青海	玉门	吐哈	新疆东部	渤海	南海	沈北
含蜡量 /%	14.6	17.1	10.2	20.0	10.0	12.0	16.0	10.0	20.5	40.9

Paraffin (wax) deposit in oil well will affect normal production, which will reduce oil production and increase load, and even cause problems such as production shutdown, paraffin (wax) obstruction and rod parting. Therefore, oil well paraffin (wax) inhibition and removal is one of the most important measures in oil well management.

油井结蜡会影响油井的正常生产，使得油井减产、负荷增大，严重时甚至会造成油井停产、蜡卡、抽油杆断脱等问题。因此，油井清、防蜡是油井管理中极为重要的措施之一。

8.1 Chemical structure features of paraffin (wax)

8.1.1 Definition and structure of paraffin (wax)

Petroleum is a complex mixture with various components. It is mainly in form of hydrocarbons, which accounting for more than 75% of all petroleum components. Phase state of each component hydrocarbon will change with its conditions (temperature and pressure), and it shows liquid, gas-liquid or gas-liquid-solid. Strictly speaking, paraffin (wax) in crude oil refers to n-alkane with higher carbon number. Alkane of $C_{16}H_{34}$-$C_{64}H_{130}$ is usually called paraffin, in which n-alkane of C_{18-35} is soft wax, and isoparaffin of C_{35-64} is hard wax.

Pure paraffin is white, slightly transparent crystal. Its density is 880-905kg/m3 and melting point is 49-60℃; However, paraffin formed during the actual oil recovery process is usually black or brown mixture of semi-solid or solid substance with high carbon normal alkyl (asphaltene, colloid, inorganic scale, mud sand, rust and oil-water emulsion, etc.). , i.e. commonly known as wax.

Paraffin (wax) formed in different crude oil production process has different compositions and properties. The typical chemical structure of paraffin (wax) is shown in Figure 8-1(a). Broadly speaking, isoparaffins with long carbon chains as well as cycloalkanes or aromatic hydrocarbons with long chain alkyl groups also belong to category of paraffin (wax), their structures are shown in Figures 8-1(b), (c), and (d). It can be seen that paraffin (wax) formed in production process can be divided into two categories, namely paraffin and microcrystalline wax (or mierocrystalline wax). The n-alkane wax is called paraffin., which can be formed in large lumps by needle-like crystals and

8.1 蜡的化学结构特征组成

8.1.1 蜡的定义与结构

石油是一种成分复杂的混合物，主要以烃（碳氢化合物）的形式存在，占石油成分的75%以上；各组分烃的相态随着所处的状态（温度和压力）不同而发生变化，呈现出液相、气液两相或气液固三相。严格地说，原油中的蜡是指碳数比较高的正构烷烃，通常把$C_{16}H_{34}$-$C_{64}H_{130}$的烷烃称为蜡，其中C_{18-35}的正构烷烃为软蜡，C_{35-64}的异构烷烃为硬蜡。

纯蜡是白色、略带透明的结晶体，密度为880~905kg/m³，熔点为49~60℃；但实际采油过程中结出的蜡通常是与高碳正构烷基混合在一起的半固体或固体物质（沥青质、胶质、无机垢、泥砂、铁锈以及油水乳化物等），呈现黑色或棕色，即俗称的蜡。

不同原油，不同条件下结出的蜡，其组成和性质都有较大的差异，蜡的典型化学结构式如图8-1（a）所示。但是，广义地讲，高碳链的异构烷烃和带有长链烷基的环烷烃或芳香烃也属于蜡的范畴，结构如图8-1（b）~图8-1（d）所示。由此可见，生产过程中结出的蜡可以分为两大类，即石蜡和微晶蜡（或称地蜡）。正构烷烃蜡称为石蜡，能够形成大晶块蜡，为针状结晶，是造成蜡沉积而导致油井堵塞的

is the main cause of paraffin (wax) deposition and blockage of oil wells. Alkane isomers, long straight chain cycloalkane and aromatic hydrocarbon are the main content of microcrystalline wax. Its relative molecular mass of the microcrystalline wax is large. It is mainly formed at tank bottom and in oil sludge, which will obviously affect the formation and growth of large lumps wax. In general, paraffin (wax) with carbon number above 20 will become threat to oil well production.

主要原因。支链烷烃、长直链环烷烃和芳烃主要形成微晶蜡，其相对分子质量较大，主要存在于罐底和油泥中，当然也会明显影响大晶块蜡结晶的形成和增长。一般来说，蜡的碳数高于20都会成为油井生产的威胁。

8.1.2 Characteristics of paraffin (wax)

Both paraffin and microcrystalline wax are characterized by different range of carbon numbers, and number of n-alkane, isoparaffins, cycloalkane (Table 8-2). Among them, composition of paraffin is mainly n-alkane, microcrystalline wax is mainly cycloalkane.

8.1.2 蜡的特征

石蜡和微晶蜡的特征主要是碳数范围、正构烷烃数量、异构烷烃数量、环烷烃数量不同（表8-2）。其中，石蜡以正构烷烃为主，而微晶蜡以环烷烃为主。

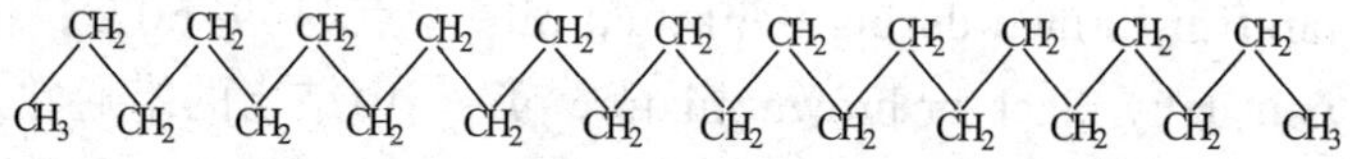

(a) N-alkanes

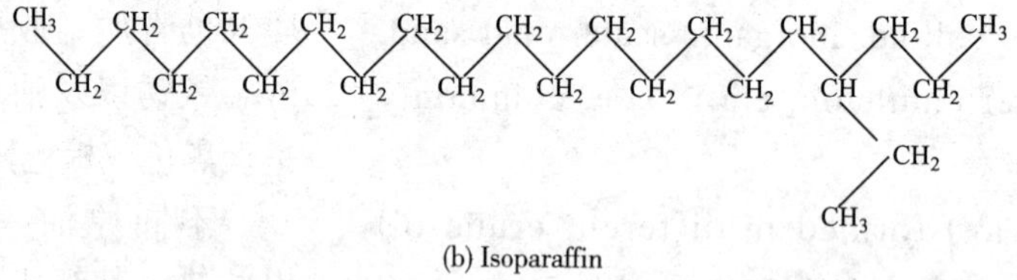

(b) Isoparaffin

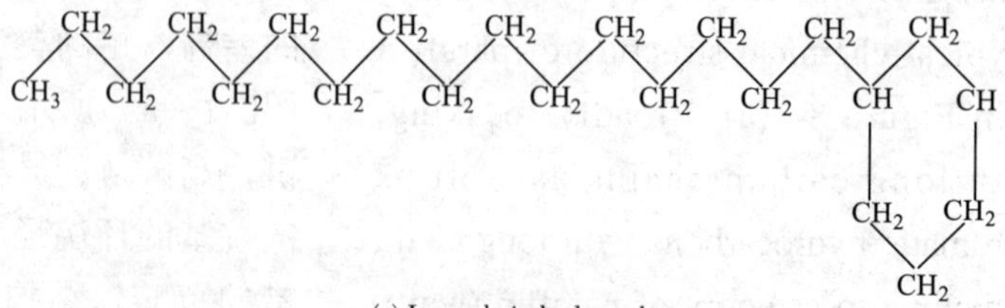

(c) Long chain hydrocarbon

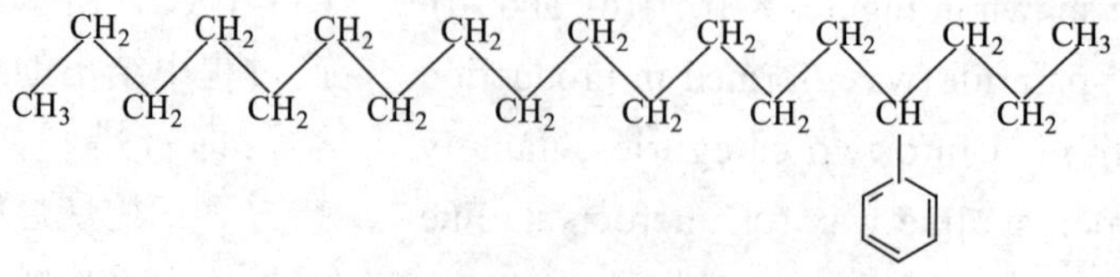

(d) Long chain hydrocarbon

Figure 8-1 Typical chemical structure of paraffin (wax)

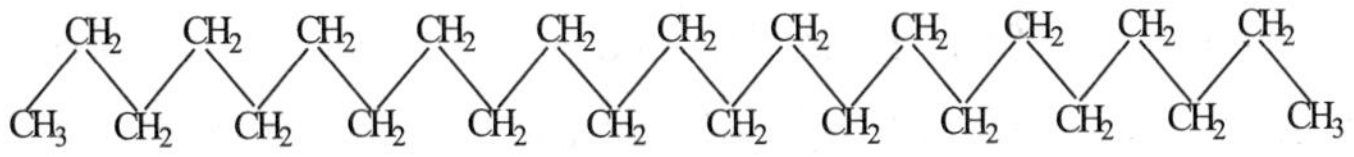

(a) 正构烷烃

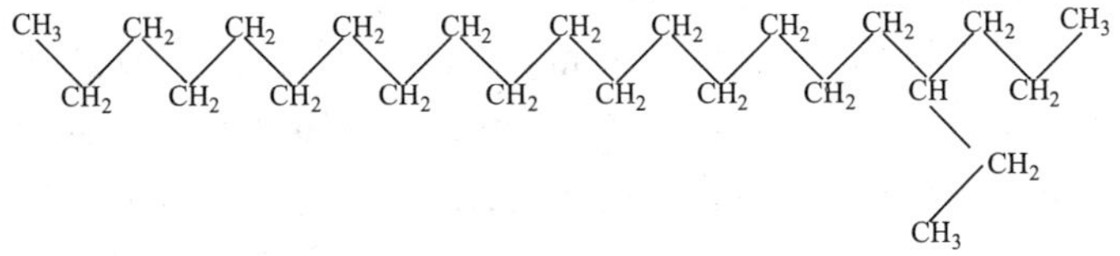

(b) 异构烷烃

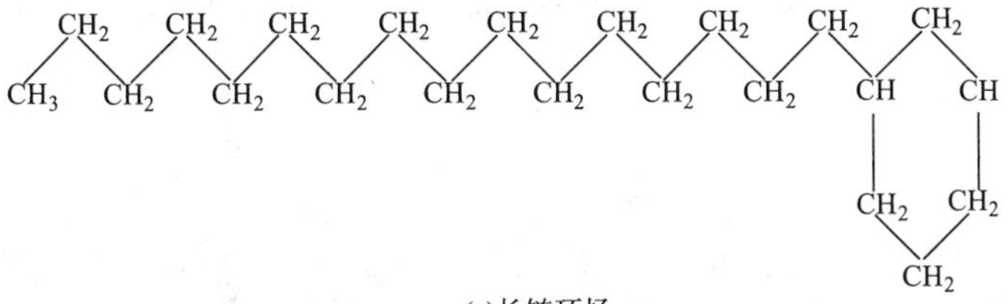

(c) 长链环烃

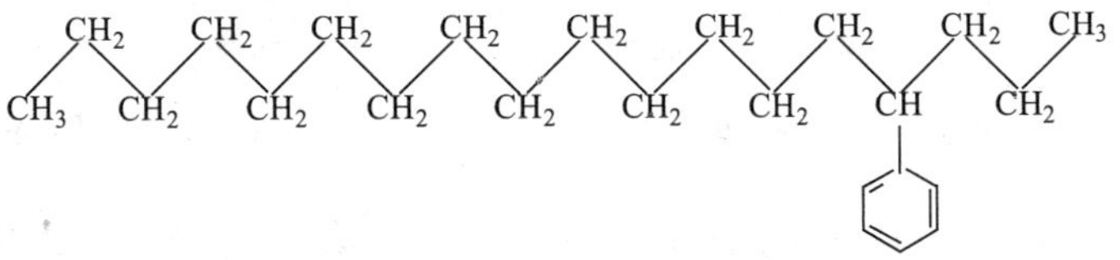

(d) 长链环烃

图8-1 石蜡的典型化学式

Table 8-2 Composition of paraffin and microcrystalline wax

Item	Paraffin	Microcrystalline wax
N-alkane/%	80-90	0-15
Isoparaffin/%	2-15	15-30
Cycloalkane/%	2-8	65-75
Melting point range/℃	50-65	60-90
Range of average relative molecular mass	350-430	500-800
Typical carbon number range	16-36	30-60
Crystallinity range/%	80-90	50-65

表8-2 石蜡及微晶蜡的组成

项 目	石 蜡	微晶蜡
正构烷烃/%	80~90	0~15
异构烷烃/%	2~15	15~30
环烷烃/%	2~8	65~75
熔点范围/℃	50~65	60~90
平均相对分子量范围	350~430	500~800
典型碳数范围	16~36	30~60
结晶度范围/%	80~90	50~65

Crystal form of paraffin (wax) is often altered by influence of the crystalline medium. In most cases, paraffin (wax) forms orthorhombic lattice [Figure.

蜡的晶型常常受蜡的结晶介质的影响而改变，在多数情况下，蜡形成斜方晶格［图8-2

8-2(a)], but it is possible to form hexagonal lattice by changing conditions [Figure. 8-2(b)]. If cooling rate is slow and there are some impurities (such as colloid, asphalt or other addition agents), it may also form transitional crystal structure. Orthorhombic lattice structure is star-shaped (needle-like) or plate layer (flake-like). This structure is most likely to form large wax crystal lumps. The main crystal forms of paraffin are shown in Figure 8-3.

（a）], 但改变条件也可能形成六方晶格［图8-2（b）］。如果冷却速度比较慢，并且存在一些杂质（如胶质、沥青或其他添加剂），也会形成过渡型结晶结构。斜方晶结构为星状（针状）或板状层（片状），这种结构最容易形成大块蜡晶团，石蜡的主要晶型如图8-3所示。

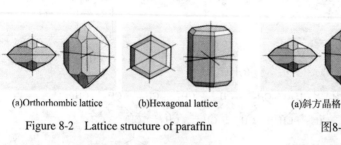

(a) Orthorhombic lattice   (b) Hexagonal lattice

Figure 8-2 Lattice structure of paraffin

(a) 斜方晶格   (b) 六方晶格

图8-2 石蜡的晶格结构

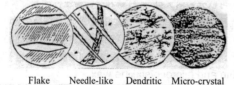

Flake   Needle-like   Dendritic   Micro-crystal

Figure 8-3 Main crystal forms of paraffin

片状   针状   树枝状   微晶状

图8-3 石蜡的主要晶型

Paraffin (wax) contained in crude oil of some oilfields of China has different proportions of n-alkane carbon in total paraffin (wax). Generally speaking, when the proportion is in normal distribution and peak carbon number is about 25, paraffin (wax) inhibition and removal is relatively easy.

国内部分油田原油中所含的蜡，其正构烃碳数占总含蜡量的比例各有不同，从总体上看都呈正态分布，碳数高峰值约在25，清防蜡比较容易。

8.1.3 Process and mechanism of paraffin (wax) deposition

8.1.3.1 Process of paraffin (wax) deposition

Paraffin (wax) deposition in oil wells can be divided into three stages (Figure. 8-4):

(1) Paraffin (wax) precipitation: When temperature drops below paraffin (wax) precipitation point, paraffin (wax) will precipitate from crude oil in crystalline

8.1.3 结蜡过程及机理

8.1.3.1 结蜡过程

油井结蜡可分为三个阶段（图8-4）：

（1）析蜡阶段：当温度降到析蜡点（蜡的晶体形成阶段温度称为析蜡点）以下时，蜡以结晶形式从原油中析出。

form. Paraffin (wax) precipitation point refers to the temperature of paraffin or wax crystal formed.

(2) Crystal grows: Temperature continues to decrease, and paraffin (wax) precipitated by crystallization grows.

(3) Deposition: Wax crystals are deposited on surface of pipe and continue to grow until block the pipe.

（2）聚集阶段：温度继续降低，结晶析出的蜡聚集长大。

（3）沉积阶段：蜡晶沉积在管壁表面上，蜡晶继续聚集长大直至堵死油管。

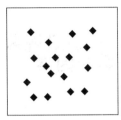  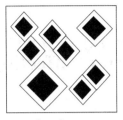 

Paraffin (wax) precipitation　　　Crystal grows　　　Deposition

Figure 8-4　wax deposition growth stage

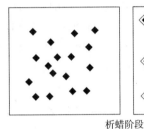 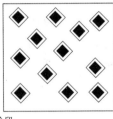 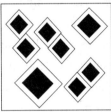 

析蜡阶段　　　聚集阶段　　　沉积阶段

图8-4　结蜡生长阶段

8.1.3.2　Mechanism of paraffin (wax) deposition

Before oil field production, paraffin (wax) is generally in liquid phase and completely dissolved in the crude oil under higher temperature and pressure formations. During the production process, crude oil passes from formation, throw well bottom, hole to wellhead, temperature and pressure gradually decrease. When pressure is reduced to a certain extent, dissolved paraffin (wax) and colloidal light components are gradually lost, which can destroy its dissolve equilibrium condition in crude oil, i.e. escaping of light components leads to decrease in paraffin (wax) solubility in crude oil; When temperature decrease

8.1.3.2　结蜡机理

油田在开发之前，原油在地层中处于高温高压条件下，一般以液相存在，完全溶解在原油中。在开采过程中，原油从地层中经过井底、井筒到井口，温度和压力随之逐渐下降，当压力降低到一定程度时，溶解的石蜡和胶质的轻质组分逐渐损失，破坏了石蜡溶解在原油中的平衡条件，即轻组分的逸出导致原油对蜡的溶解能力下降；温度的降低使蜡的溶解度进一步降低，致使

further, paraffin (wax) solubility will continuously reduce, causing precipitation, growth, aggregation and deposition of paraffin (wax) on metal surface of the oil well facility, that is, paraffin (wax) deposition. In late stage of oilfield production, the changes of oil geology and process conditions will lead to change of paraffin (wax) deposition mechanism, and its range will expand. Once paraffin (wax) is formed, the deposition is mainly based on membrane adsorption and droplet adsorption.

8.2 Paraffin (wax) deposition in oil wells and factors affecting the deposition

8.2.1 Paraffin (wax) deposition in oil wells

Crude oil properties of different oilfields are quite different, wand its deposition of paraffin (wax) are also different. In order to take the best oil well paraffin (wax) inhibition and removal measures, the deposition process of oil well paraffin (wax) must be studied. Generally, deposition process of oil well paraffin (wax) is with the following features:

(1) The higher the paraffin (wax) contents in crude oil, the more serious the paraffin (wax) deposition will show in oil wells. In low water cut stage, paraffin (wax) deposition of oil wells is quite severe and it is needed to be removed 2-3 times a day. In middle and high water cut stages, speed of deposition is reduced and paraffin (wax) will be removed once every 2-3 days or even once in ten days.

(2) Under the same temperature conditions, paraffin (wax) deposition in light oil is more serious than in heavy crude oil.

(3) In early production stage, paraffin (wax) deposition is more serious than in late stage.

(4) High productivity or high oil temperature at the wellhead are not with serious paraffin (wax) deposition; otherwise, paraffin (wax) deposition is severe.

石蜡以晶体析出、长大、聚集并沉积在油井设施的金属表面，即出现结蜡现象。油田开发后期，由于采油地质、工艺条件的变化，导致油井的结蜡机理也发生变化，结蜡范围扩大。结蜡一旦形成，蜡沉积主要以薄膜吸附和液滴吸附为主。

8.2 油井结蜡现象和影响结蜡的因素

8.2.1 油井结蜡现象

不同油田，原油性质有较大差异，油井结蜡规律也不同，为了制定油井清防蜡措施，必须研究油井结蜡现象，一般具有下列现象：

（1）原油含蜡量愈高，油井结蜡愈严重。低含水阶段油井结蜡严重，一天清蜡2~3次，中高含水阶段结蜡有所减轻，2~3天清蜡一次甚至十几天清蜡一次。

（2）在相同温度条件下，稀油比稠油结蜡严重。

（3）开采初期较后期结蜡严重。

（4）高产井及井口出油温度高的井结蜡不严重，或不结蜡；反之结蜡严重。

（5）油井工作制度改变，结蜡点深度也改变，缩小油嘴，结蜡点上移，反之亦然。

（6）表面粗糙的油管比表面

(5) Depth of paraffin (wax) precipitation point will be changed with different oil well working systems. It will move up when size of choke reduced, and vice versa.

(6) Oil pipe with rough surface is easier to form paraffin (wax) deposition than those with smooth surface; Pipes will easily get paraffin (wax) deposition again if previous paraffin (wax) removal is not made completely.

(7) Sand wells are easily to be deposited by paraffin (wax).

(8) For flowing wells, the most severe paraffin (wax) deposition is not formed at wellhead or bottom hole but at a certain depth of the well bore. Figure 8-5 shows paraffin (wax) deposition profile of a well.

光滑的油管容易结蜡；油管清蜡不彻底的易结蜡。

（7）出砂井易结蜡。

（8）自喷井结蜡严重的地方既不在井口也不在井底，而是在井的一定深度上。图8-5为某井的结蜡剖面。

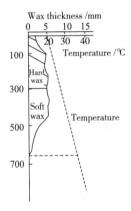

Figure 8-5 Paraffin (wax) deposition profile of a well

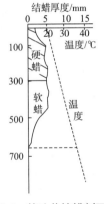

图8-5 某油井结蜡剖面图

8.2.2 Analysis on factors of paraffin (wax) deposition

According to observation of oil wells paraffin (wax) deposition and study of deposition process, the main factors affecting paraffin (wax) deposition are crude oil composition (content of light components, wax, colloid and asphalt), oil well production conditions (temperature, pressure, gas-oil ratio and production), impurities in crude oil (mechanical impurity and water, etc.), smoothness and surface property of pipe (Table 8-3). Among them, crude oil composition is the internal

8.2.2 结蜡因素分析

通过对油井结蜡现象的观察及结蜡过程的研究，影响结蜡的主要因素是原油的组成（轻质成分、蜡、胶质和沥青的含量）、油井的开采条件（温度、压力、气油比和产量）、原油中的杂质（机械和水等）、管壁的光滑程度及表面性质（表8-3）。其中，原油组成是影响结蜡的内在因

factor of paraffin (wax) deposition, while temperature and pressure are external conditions.

Table 8-3 Statistics on factors affecting paraffin (wax) deposition

Factors	Influence law
Crude oil composition	1. Light components: The higher of light components content, the lower of paraffin (wax) crystallization temperature. Paraffin (wax) will not be easy to precipitate in this case. 2. Content of paraffin (wax): The higher of paraffin (wax) content, the higher of paraffin (wax) crystallization temperature, which will lead to serious paraffin (wax) deposition in oil wells. 3. Colloid and asphalt: Content of colloid and asphaltene indicates lower possibility of paraffin (wax) deposition; but when paraffin (wax) deposition on pipe surface contains colloid and asphaltene, hard wax will be formed, which is not easy to be washed away by oil flows
Oil well production conditions	1. Temperature: dissolution amount of paraffin (wax) reduces with decreasing of temperature, and tendency of oil well paraffin (wax) deposition will increase. 2. Pressure and dissolved gas: When pressure is higher than formation saturation pressure, the initial crystallization temperature of paraffin (wax) decreases with the pressure reduction. Below the saturation pressure, the dissolved gas will release and reduces solubility of paraffin (wax). 3. Production: Large production means high oil flow rate, which will take paraffin (wax) crystal away easily, hence reduce possibility of paraffin (wax) deposition on pipe wall
Impurities in crude oil	1. Mechanical impurities: It has little effect on initial crystallization temperature. But mechanical impurities in crude oil may become the crystalline core of the paraffin (wax) precipitation and accelerate deposition process. 2. Water cut: High water cut will slow paraffin (wax) deposition process and prevent paraffin (wax) deposition on pipe surface
Smoothness and surface property of pipe	When pipe surface is rough, paraffin (wax) crystals are tending to adhere on it and form deposition; the stronger the hydrophilicity of the pipe surface, the less likely to form paraffin (wax) deposition.

表8-3 结蜡影响因素统计表

影响因素	影响规律
原油组成	1.轻质成分：轻质成分含愈多，蜡结晶温度越低，蜡不易析出。 2.含蜡量：含蜡量愈高，蜡的结晶温度就愈高，油井结蜡也就越严重。 3.胶质和沥青：胶质、沥青质的存在，不易结蜡；但当沉积在管壁上的蜡中含有胶质、沥青质时将形成硬蜡，不易被油流冲走
开采条件	1.温度：蜡溶解量随着温度降低而减少，增大油井结蜡趋势。 2.压力和溶解气：压力高于地层饱和压力下，蜡的初始结晶温度随压力的降低而降低；低于饱和压力下，溶解气分离，降低蜡的溶解能力。 3.产量：产量大即流速高，冲刷作用强，易带走蜡晶，不易沉积在管壁上
原油中杂质	1.机械杂质：对蜡的初始结晶温度影响不大，但原油中的机械杂质将成为石蜡析出的结晶核心，加速结蜡的过程。 2.含水率：含水升高，减缓结蜡过程，不利于蜡沉积到管壁上
管壁光滑程度及表面性质	管壁粗糙时，蜡晶体容易黏附在上面形成蜡；管壁亲水性越强越不容易结蜡

8.3 Oil well chemistry paraffin (wax) inhibition and removal technology

There are many methods for oil well paraffin (wax) inhibition and removal, including mechanical paraffin

8.3 油井化学清防蜡技术

油井清防蜡方法有很多，包括机械清蜡、热力清蜡、磁防

(wax) removal, thermal paraffin (wax) removal, magnetic paraffin (wax) removal, chemical and other oil well paraffin (wax) inhibition and removal measures. Among them, chemical paraffin (wax) inhibition and removal technology is with many advantages including simple process, convenient on-site application, high efficiency, removal and control at same time, which will not affect normal oil well production. It already becomes a widely used in oilfield application. According to its mechanism, paraffin inhibitors used in this technology can be divided into three types: polycyclic aromatic hydrocarbon type, surfactant type and polymeric type; According to solvent, the paraffin removers can be classified into oil-soluble type, water-soluble type, and oil-in-water emulsion type.

8.3.1 Classification and mechanism of paraffin inhibitors

Chemical agents that inhibit the precipitation, growth, aggregation, and/or deposition of paraffin (wax) crystals in crude oil are known as paraffin inhibitor. There are three types of commonly used paraffin inhibitors:

(1)Polycyclic aromatic hydrocarbon type paraffin inhibitor.

Polycyclic aromatic hydrocarbon is an aromatic hydrocarbon, in which two or more benzene rings share two adjacent carbon atoms (Figure 8-6); The polycyclic aromatic hydrocarbons used for paraffin inhibitor are mainly derived from fractions of coal tar, and all of them are mixed polycyclic aromatic hydrocarbons with their derivatives (Figures 8-7).

蜡、化学清防蜡等油井清防蜡措施，其中化学清防蜡技术具有工艺简单、现场应用方便、清防蜡效率高、清防并重，并且不影响油井正常生产等优点，是目前油田应用比较广泛的一种清防蜡技术。按照其作用机理可分稠环芳香烃型防蜡剂、表面活性剂型防蜡剂、高分子型防蜡剂三种类型；清蜡剂根据其溶剂可分为油溶型、水溶型、水包油乳液型三种类型。

8.3.1 防蜡剂分类及其作用机理

能抑制原油中蜡晶析出、长大、聚集和（或）在固体表面上沉积的化学剂称为防蜡剂。常用的防蜡剂有三种类型：

（1）稠环芳香烃型防蜡剂。

稠环芳香烃是指两个或两个以上的苯环分别共用两个相邻的碳原子而成的芳香烃（图8-6）；用于防蜡的稠环芳香烃类主要来自煤焦油中的馏分，都是混合稠环芳烃，也包括其衍生物（图8-7）。

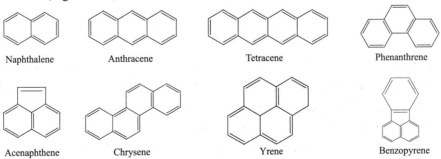

Figure 8-6 Polycyclic aromatic hydrocarbons

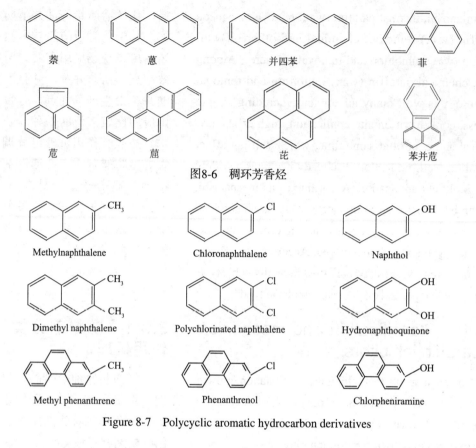

图8-6 稠环芳香烃

Figure 8-7 Polycyclic aromatic hydrocarbon derivatives

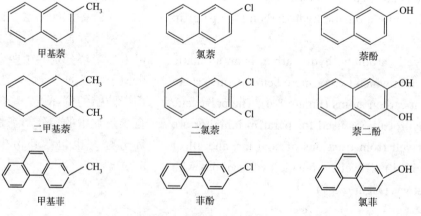

图8-7 稠环芳香烃衍生物

In crude oil, solubility of polycyclic aromatic hydrocarbon type paraffin inhibitor is lower than paraffin, the latter is enabling to be dissolved in paraffin inhibitor and injected into bottom hole and produced together with crude oil. During the production process, as temperature and pressure decrease, the polycyclic

稠环芳香烃型防蜡剂在原油中的溶解度低于石蜡，将其溶于溶剂中注入井底并随原油一起采出。在采出过程中，随着温度和压力的降低，稠环芳香烃首先析出，给石蜡的析出提供了大量晶

aromatic hydrocarbons will be precipitate first, which could provide a large number of crystalline cores for paraffin precipitation. This causes paraffin precipitates on crystalline cores of polycyclic aromatic hydrocarbon. These cores will distort arrangement of paraffin crystals and prevent their growth, thus further help paraffin (wax) removal. The polycyclic aromatic hydrocarbons can also be added into rod shape or granular weighting agent. Then the weighting agent could be placed at bottom hole, so that it can be slowly dissolved and its effect will be prolonged.

(2) Surfactant type paraffin inhibitor.

Surfactant type paraffin inhibitors can be classified into two types: oil-soluble (Figure. 8-8) and water-soluble (Figure. 8-9). Mechanism of these two types is quite different.

核，使石蜡在稠环芳香烃晶核上析出，影响了蜡晶的排列，使晶核扭曲变形，不利于蜡晶的继续长大，起到防蜡作用。也可将稠环芳香烃掺入加重剂，制成棒状或颗粒状固体投入井底，使其缓慢溶解，延长使用效果。

（2）表面活性剂型防蜡剂。

表面活性剂型防蜡剂可分为油溶性（图8-8）和水溶性（图8-9）两大类，二者的作用原理不同。

$RArSO_3M$ $M: 1/2Ca, Na, K, NH_4$

Alkylbenzene sulfonate

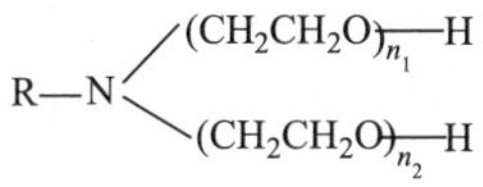

$R-N\begin{cases}(CH_2CH_2O)_{n_1}-H\\(CH_2CH_2O)_{n_2}-H\end{cases}$ $n_1+n_2=2\sim 4$ $R:C_{16}\sim C_{22}$

Polyoxyethylene fatty amin

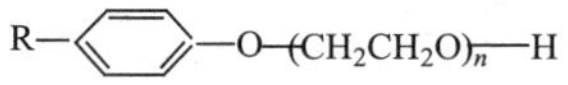

$R-\langle\rangle-O-(CH_2CH_2O)_n-H$ $n=3\sim 4$ $R:C_9, C_{12}$

Alkyl phenol polyoxyethylene ether

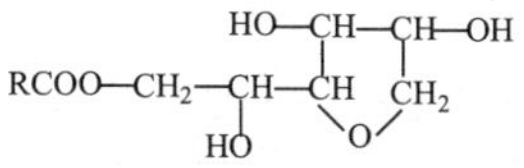

$RCOO-CH_2-\underset{HO}{\overset{HO-CH-CH-OH}{CH}}-\underset{O}{CH_2}$ Span-xx

Sorbitan monopalmitate

Figure 8-8 Oil-soluble surfactant type paraffin inhibitors

$RArSO_3M$ $M: 1/2Ca, Na, K, NH_4$

烷基苯磺酸盐

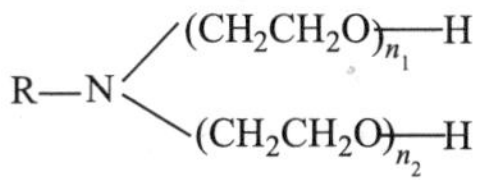

$R-N\begin{cases}(CH_2CH_2O)_{n_1}-H\\(CH_2CH_2O)_{n_2}-H\end{cases}$ $n_1+n_2=2\sim 4$ $R:C_{16}\sim C_{22}$

聚氧乙烯脂肪胺

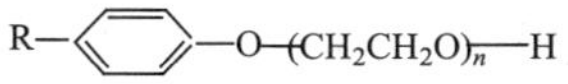

$R-\langle\rangle-O-(CH_2CH_2O)_n-H$ $n=3\sim 4$ $R:C_9, C_{12}$

烷基酚聚氧乙烯醚

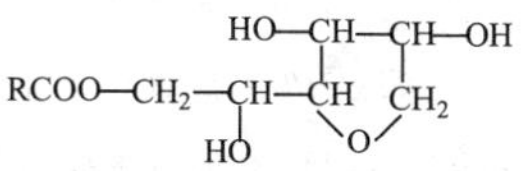

$RCOO-CH_2-\underset{HO}{\overset{HO-CH-CH-OH}{CH}}-\underset{O}{CH_2}$ Span-xx

山梨糖醇酐单羧酸酯

图8-8 油溶性表面活性剂型防蜡剂

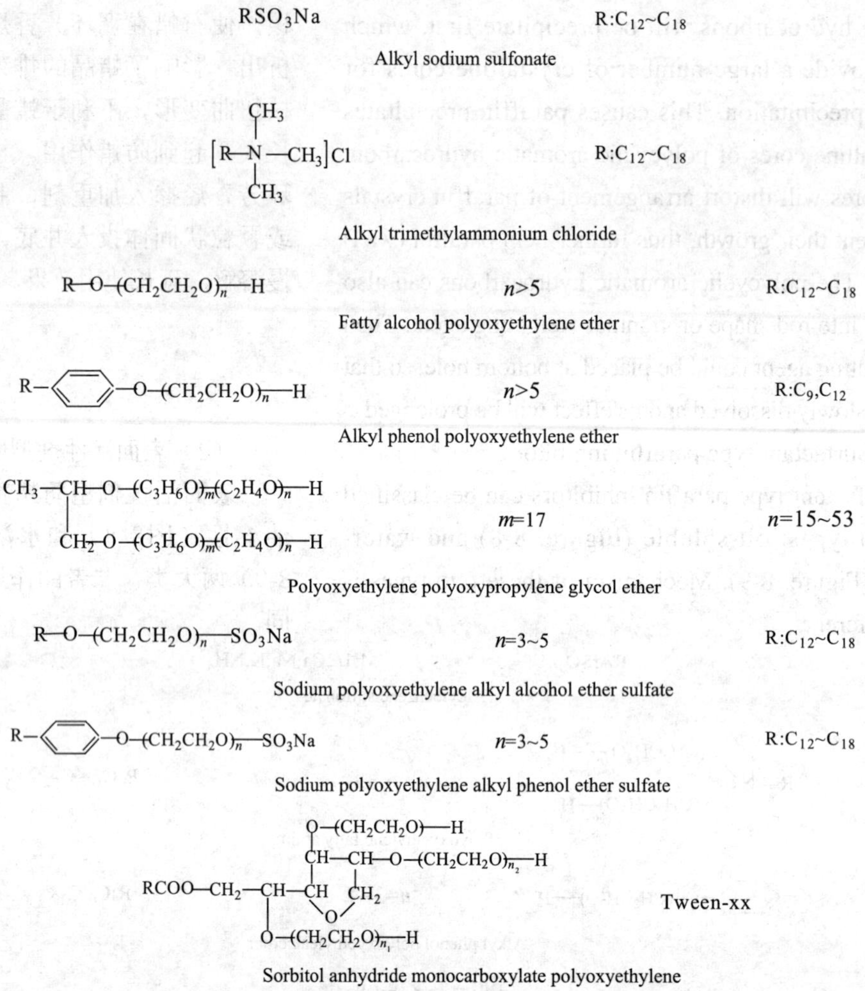

RSO$_3$Na   R:C$_{12}$~C$_{18}$

Alkyl sodium sulfonate

[R—N(CH$_3$)$_3$]Cl   R:C$_{12}$~C$_{18}$

Alkyl trimethylammonium chloride

R—O—(CH$_2$CH$_2$O)$_n$—H $n>5$ R:C$_{12}$~C$_{18}$

Fatty alcohol polyoxyethylene ether

R—C$_6$H$_4$—O—(CH$_2$CH$_2$O)$_n$—H $n>5$ R:C$_9$,C$_{12}$

Alkyl phenol polyoxyethylene ether

CH$_3$—CH—O—(C$_3$H$_6$O)$_m$(C$_2$H$_4$O)$_n$—H
CH$_2$—O—(C$_3$H$_6$O)$_m$(C$_2$H$_4$O)$_n$—H $m=17$ $n=15~53$

Polyoxyethylene polyoxypropylene glycol ether

R—O—(CH$_2$CH$_2$O)$_n$—SO$_3$Na $n=3~5$ R:C$_{12}$~C$_{18}$

Sodium polyoxyethylene alkyl alcohol ether sulfate

R—C$_6$H$_4$—O—(CH$_2$CH$_2$O)$_n$—SO$_3$Na $n=3~5$ R:C$_{12}$~C$_{18}$

Sodium polyoxyethylene alkyl phenol ether sulfate

RCOO—CH$_2$—CH—CH—CH—O—(CH$_2$CH$_2$O)$_{n_2}$—H (Tween-xx)
(Sorbitol anhydride with polyoxyethylene chains n, n_1, n_2)

Sorbitol anhydride monocarboxylate polyoxyethylene

Figure 8-9   Water-soluble surfactant type paraffin inhibitors

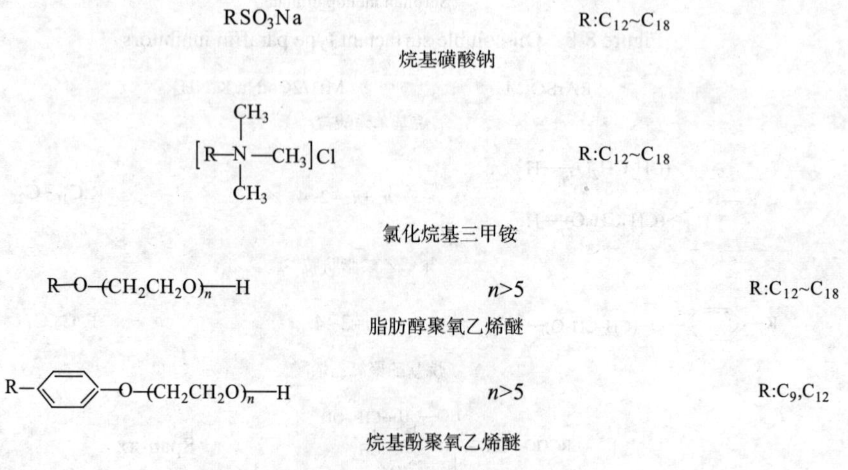

RSO$_3$Na   R:C$_{12}$~C$_{18}$

烷基磺酸钠

[R—N(CH$_3$)$_3$]Cl   R:C$_{12}$~C$_{18}$

氯化烷基三甲铵

R—O—(CH$_2$CH$_2$O)$_n$—H $n>5$ R:C$_{12}$~C$_{18}$

脂肪醇聚氧乙烯醚

R—C$_6$H$_4$—O—(CH$_2$CH$_2$O)$_n$—H $n>5$ R:C$_9$,C$_{12}$

烷基酚聚氧乙烯醚

图8-9   水溶性表面活性剂型防蜡剂

$$\begin{array}{l}CH_3-CH-O-(C_3H_6O)_m-(C_2H_4O)_n-H\\ CH_2-O-(C_3H_6O)_m-(C_2H_4O)_n-H\end{array} \qquad m=17 \qquad n=15\sim53$$

聚氧乙烯聚氧丙烯丙二醇醚

$$R-O-(CH_2CH_2O)_n-SO_3Na \qquad n=3\sim5 \qquad R:C_{12}\sim C_{18}$$

聚氧乙烯烷基醇醚硫酸脂钠盐

$$R-\text{C}_6H_4-O-(CH_2CH_2O)_n-SO_3Na \qquad n=3\sim5 \qquad R:C_{12}\sim C_{18}$$

聚氧乙烯烷基苯酚醚硫酸脂钠盐

$$\begin{array}{l} O-(CH_2CH_2O)-H\\ CH-CH-O-(CH_2CH_2O)_{n_2}-H\\ RCOO-CH_2-CHCH_2\\ O-(CH_2CH_2O)_{n_1}-H\end{array} \qquad \text{Tween-xx}$$

山梨糖醇酐单羧酸酯聚氧乙烯醚

图8-9 水溶性表面活性剂型防蜡剂（续）

Oil-soluble surfactant inhibitor prevents further growth of paraffin (wax) crystals by changing their surface properties (from non-polar to polar surfaces) [Figure. 8-10(a)]; Water-soluble surfactant inhibitor prevents paraffin (wax) deposition by changing surface property of paraffin (wax), such as pipes, sucker rods and equipment surfaces, normally changing property of these surfaces from non-polar to polar [Figure 8-10(b)] because non-polar surface is easy to be adhered.

油溶性表面活性剂是通过改变蜡晶表面性质（由非极性表面变为极性表面），从而抑制了蜡晶的进一步长大［图8-10（a）］；水溶性表面活性剂是通过改变结蜡表面（如油管、抽油杆和设备表面）性质，使非极性的结蜡表面变成极性表面，从而防止蜡的沉积［图8-10（b）］。

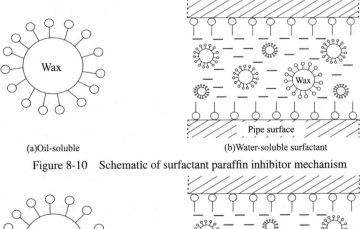

Figure 8-10 Schematic of surfactant paraffin inhibitor mechanism

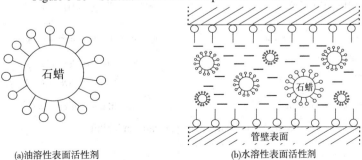

图8-10 表面活性剂型防蜡剂机理示意图

(3) Polymeric type paraffin inhibitor.

Polymeric type paraffin inhibitor is oil-soluble comblike structure polymer (Figure 8-11). Its molecular structure is composed by two parts: polar chain and non-polar chain; wherein the polar chain element will distort crystal form of paraffin (wax) crystals. This will prevent the crystals grow into network structure, and the non-polar chain element will participates crystallization of paraffin (wax) (Figure 8-12). In addition, long molecular chains of these polymers can form a network structure throughout crude oil. In this case, the small crystalline cores formed will be controlled in dispersed state, i.e. they cannot aggregate or grow each other, let alone deposit on surfaces of oil pipe or sucker rod, then oil flows will take them away easily.

（3）高分子型防蜡剂。

高分子型防蜡剂是油溶性的梳状聚合物（图8-11），分子结构中具有极性链节和非极性链节两部分；其中极性链节使蜡晶的晶型产生扭曲，不利于蜡晶继续长大形成网络结构，非极性链节参与蜡结晶（图8-12）。此外，这类聚合物的分子链较长，可在油中形成遍及整个原油的网络结构，使形成的小晶核处于分散状态，不能相互聚集长大，也不易在油管或抽油杆表面上沉积，而易被油流带走。

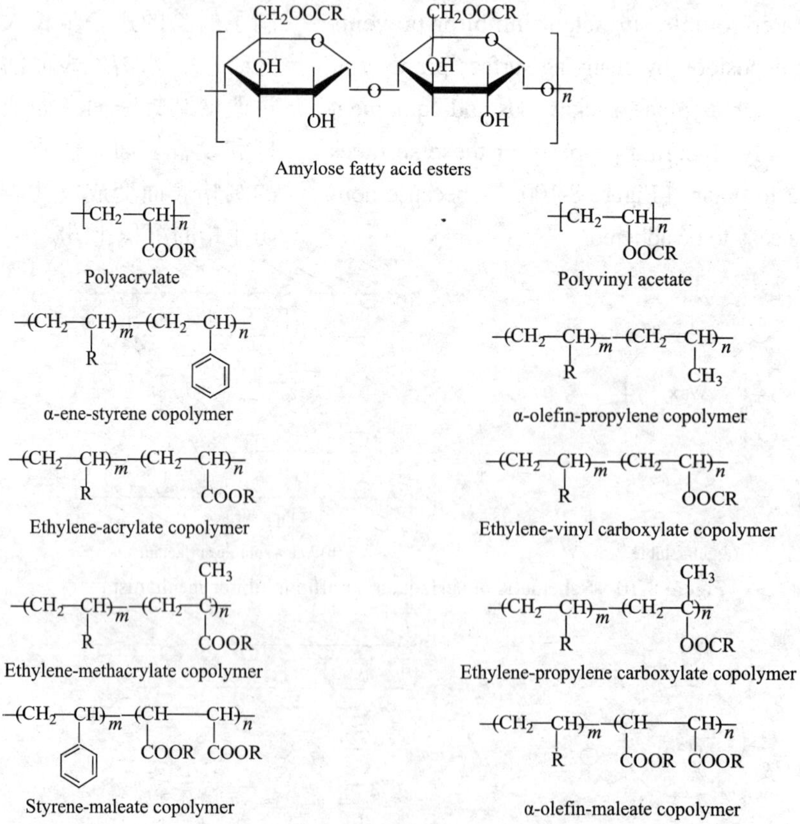

Figure 8-11 Polymeric type paraffin inhibitor

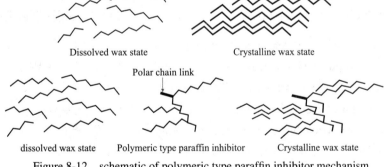

图8-11 高分子型防蜡剂

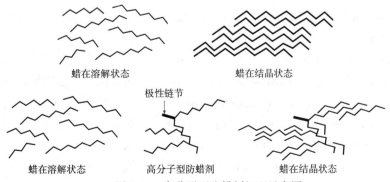

Figure 8-12　schematic of polymeric type paraffin inhibitor mechanism

图 8-12　高分子型防蜡剂机理示意图

Side chain length of polymeric type paraffin inhibitor is directly related to removal effect. The most favorable condition for paraffin (wax) precipitation is when average number of carbon atoms in side chain of the inhibitor is close to peak carbon number of paraffin (wax) in crude oil. Under this condition, the best paraffin (wax) removal effect will be achieved.

高分子防蜡剂侧链的长短与防蜡效果直接有关，当侧链平均碳原子数与原油中蜡的峰值碳数相近时，最有利蜡的析出，可获得最佳防蜡效果。

8.3.2 Classification and mechanism of paraffin remover

8.3.2 清蜡剂分类及其作用机理

Chemical paraffin remover is an agent that achieves the purpose of paraffin removal by chemical reaction. By dissolve or disperse the deposited paraffin (wax), the paraffin remover is able to completely dissolve paraffin (wax)or suspend small particles in crude oil. Then paraffin (wax) will flow out with the crude oil. The whole process of paraffin remover involves infiltration, dissolution and dispersion. Chemical paraffin remover has the characteristics of thorough paraffin removal, low cost, simple process, good effect, long cycle and wide application range (flowing wells and pumping wells).

化学清蜡剂是借助于化学药剂的作用达到清蜡的目的，其作用过程是将已沉积的蜡溶解或分散开，使其在原油中处于溶解或小颗粒悬浮状态而随油井液流流出，涉及渗透、溶解和分散等过程。此方法清蜡彻底、成本低、工艺简单、效果好、周期长、适用范围广（自喷井和抽油井）。

8.3.2.1 Oil-soluble paraffin remover

8.3.2.1 油溶型清蜡剂

Oil-soluble paraffin remover is a strong paraffin (wax) solvent. Its main mechanism is organic solvent dissolution to wax, colloid and asphaltene. The main oil-soluble paraffin remover mainly includes:

① Aromatic hydrocarbons: Benzene, toluene, xylene, trimethylbenzene, ethylbenzene, cumene, mixed aromatic hydrocarbons.

② Distillate oil: Light hydrocarbons, gasoline, kerosene, diesel, etc.

③ Other solvents: Carbon disulfide, carbon tetrachloride, chloroform, tetrachloroethylene, etc.

The main disadvantages of oil-soluble paraffin removers are toxic, low flash point, flammable, explosive, and not safe enough during usage. Among them, carbon disulfide and carbon tetrachloride are

油溶型清蜡剂是溶蜡能力很强的溶剂，主要作用机理是有机溶剂对蜡、胶质、沥青质的溶解作用。主要有：

①芳烃：苯、甲苯、二甲苯、三甲苯、乙苯、异丙苯、混合芳烃。

②馏分油：轻烃、汽油、煤油、柴油等。

③其他溶剂：二硫化碳、四氯化碳、三氯甲烷、四氯乙烯等。

油基清蜡剂的主要缺点是有毒、闪点低、易燃易爆，使用时不够安全。其中，二硫化碳、四氯化碳等是油田早期使用的清

used in the early years. Although their paraffin (wax) remove effect is excellent, they have been banned due to toxicity, corrosiveness and catalyst poisoning caused by themselves in crude oil processing. Plant oil obtained by extracting or distilling from stems and leaves of some woody plants (such as pine and eucalyptus) and herbs (such as peppermint and citronella) can be used as oil-solution paraffin remover. The main component of this oil-soluble paraffin remover is terpene (Figure. 8-13), which is highly valued because of its low toxicity, low flammability and degradability.

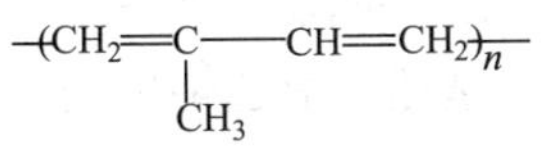

$$-(CH_2-\underset{\underset{CH_3}{|}}{C}-CH=CH_2)_n-$$

Figure 8-13 terpene

In order to further improve paraffin (wax) removal effect, the oil-soluble paraffin remover can be compounded. Table 8-4 is formula of a compounding oil-solution paraffin remover.

蜡剂，其清蜡效果优异，但由于其毒性以及在原油加工中造成的腐蚀性和催化剂中毒等问题，已经被禁止使用。一些有木本植物（如松树、樟树）和草本植物（如薄荷、香茅）的茎、叶等抽提或蒸馏等方法得到的植物油能溶液蜡，可用作油基清蜡剂，主要成分是萜烯（图8-13），该类清蜡剂低毒、低燃、可降解而为人们所重视。

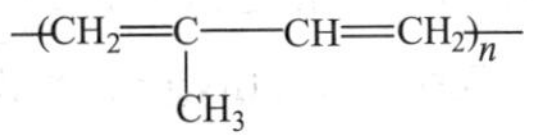

$$-(CH_2-\underset{\underset{CH_3}{|}}{C}-CH=CH_2)_n-$$

图8-13 萜烯

为了进一步提高油基清蜡剂的清蜡效果，可复配使用，表8-4是一种复配油基清蜡剂的配方。

Table 8-4 A formulation of oil-solution paraffin remover

Composition	w (composition) /%
Kerosene	45-85
Benzene	5-45
Ethylene glycol monobutyl ether	0.5-6
Isopropanol	1-15

表8-4 油基清蜡剂的一种配方

成 分	w（成分）/%
煤油	45~85
苯	5~45
乙二醇丁醚	0.5~6
异丙醇	1~15

8.3.2.2 Water-soluble paraffin remover

Water-soluble paraffin remover is composed of water, surfactant, mutual solvent and/or alkaline by a certain ratio. It is suitable for paraffin (wax) removal in oil wells with high water cut, and its typical formulation is shown in Figure 8-14. This kind of paraffin remover

8.3.2.2 水溶型清蜡剂

水基清蜡剂是由水、表面活性剂、互溶剂和（或）碱按一定比例组成的清蜡剂，适合于含水量较高的油井清蜡，其典型配方如图8-14所示。该类清蜡剂通过

will be entered gaps between paraffin (wax) blocks or gaps between paraffin (wax) blocks and sidewall by penetration and dispersion, reduces their adhesion, which could make them fall off and take away with produced liquid. Different components in water-soluble paraffin remover have different mechanism on paraffin (wax) remove as following indicates:

Surfactant is used to change wettability of paraffin (wax)surface that, which can make paraffin (wax) easy to peel off and disperse. Commonly used surfactants are alkylsulfonate, alkylbenzene sulfonate, fatty alcohol polyoxyethylene ether, alkylphenol ethoxylates, fatty alcohol polyoxyethylene sulfate sodium, alkylphenol polyoxyethylene sulfate sodium, tween, etc.

Mutual solvent is used to increase mutual solubility of water and oil. Commonly used mutual solvents are:

(1) Alcohols: Isopropanol, n-propanol, polyethylene glycol, glycerol, etc.

渗透、分散，沿蜡块间或蜡块与井壁间的缝隙渗入，降低蜡块间或蜡块与井壁间的黏附力，从井壁上脱落，随采出液流出油井。水基清蜡剂中不同组分具有不同清蜡作用机理，具体如下所述：

表面活性剂的作用是改变结蜡表面的润湿性，使其易于剥落分散。常用的表面活性剂有烷基磺酸盐、烷基苯磺酸盐、脂肪醇聚氧乙烯醚、烷基酚聚氧乙烯醚、脂肪醇聚氧乙烯硫酸钠、烷基酚聚氧乙烯硫酸钠、吐温等。

互溶剂的作用是增加水和油的相互溶解度。常用的互溶剂有：

（1）醇类：异丙醇、正丙醇、乙二醇、丙三醇等。

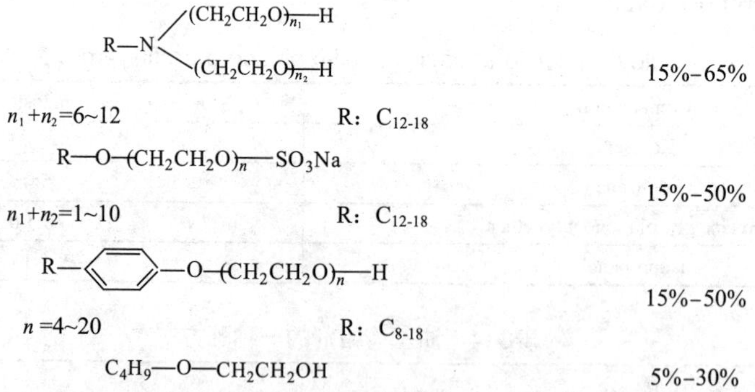

$R-N\begin{cases}(CH_2CH_2O)_{n_1}-H\\(CH_2CH_2O)_{n_2}-H\end{cases}$

$n_1+n_2=6\sim12$ R: C_{12-18} 15%–65%

$R-O-(CH_2CH_2O)_n-SO_3Na$

$n_1+n_2=1\sim10$ R: C_{12-18} 15%–50%

$R-\langle\bigcirc\rangle-O-(CH_2CH_2O)_n-H$

$n=4\sim20$ R: C_{8-18} 15%–50%

$C_4H_9-O-CH_2CH_2OH$ 5%–30%

Figure 8-14   A formulation of water-solution paraffin remover

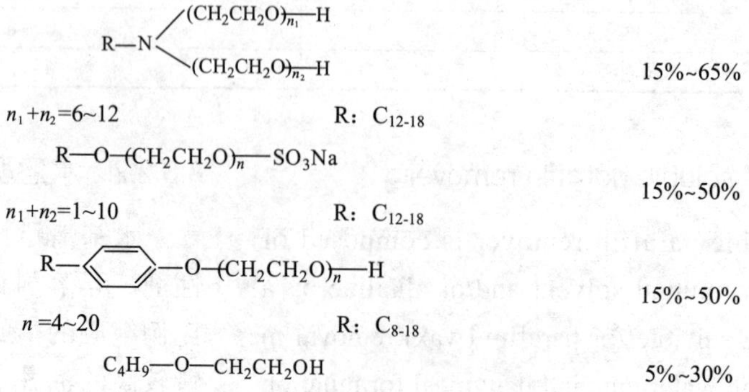

$R-N\begin{cases}(CH_2CH_2O)_{n_1}-H\\(CH_2CH_2O)_{n_2}-H\end{cases}$

$n_1+n_2=6\sim12$ R: C_{12-18} 15%~65%

$R-O-(CH_2CH_2O)_n-SO_3Na$

$n_1+n_2=1\sim10$ R: C_{12-18} 15%~50%

$R-\langle\bigcirc\rangle-O-(CH_2CH_2O)_n-H$

$n=4\sim20$ R: C_{8-18} 15%~50%

$C_4H_9-O-CH_2CH_2OH$ 5%~30%

图8-14   水基清蜡剂的一种配方

(2) Ethers: Butyl ether, amyl ether, hexyl ether, heptyl ether, octyl ether, etc.

(3) Alcohol ether: Ethylene glycol monobutyl ether, butanediol ethyl ether, carbitol, glycerin ether, etc.

Alkaline will react with polar substances such as asphaltenes in paraffin (wax). Products of the reaction are easily dispersed in water, thus paraffin (wax) can be removed from surfaces by water-soluble paraffin remover. The available alkaline not only includes real alkaline such as sodium hydroxide and potassium hydroxide, but also includes some salts that show alkaline in water like sodium silicate, sodium orthosilicate, sodium phosphate, sodium pyrophosphate, sodium hexametaphosphate, etc.

Water-soluble paraffin remover is characterized by no sulfur and chlorine, thus will be non-corrosive to all equipments; it is with high flash point and could be used safely. Its specific gravity is large (generally greater than 1.0), when it is used for high water cut crude oil, it could be added from the casing and will easily sink into the bottom hole. It also with good stability so it could be easily transported and stored.

8.3.2.3 Oil-in-water emulsion paraffin remover

Oil-in-water emulsion paraffin remover is generally composed of an oil-soluble paraffin remover and a water-soluble paraffin remover, i.e. paraffin remover is emulsified into water. This kind of paraffin remover is sent from annulus to below paraffin deposition sections, then it will be mixed with the oil, this is to say, the oil-soluble paraffin remover and the water-soluble paraffin remover will be separated and works, producing good paraffin (wax) removal effect.

The functions of oil-in-water emulsion paraffin remover can complement each other and relieve environmental pollution.

The main mechanism of oil-in-water emulsion paraffin remover is to release organic solvent and oil-soluble surfactants by demulsification under temperature of the bottom hole. Both of the organic solvent and oil-soluble surfactants are with good solubility to paraffin (wax), hence the goal of paraffin (wax) removal will be achieved.

8.3.3 New chemical paraffin (wax) inhibition and removal technology and the trend of development

In the future, the chemical paraffin remover will be developed with environmentally friendly, efficient, multifunctional and low selectivity. And researches on compounding promoters will be carried out to improve effect of paraffin remover. Its development trend is mainly to develop new and multi-effect polymer paraffin inhibitor, which will improve the paraffin inhibit effect. The main trend of chemical paraffin (wax) inhibition and removal technology in the future will be comprehensively utilizing advantages of each other, and fully take their complementary advantages. Therefore, research on paraffin (wax) deposition, inhibition and removal in oil field can be carried out in the following aspects:

(1) Conduct further researches on microcosmic mechanism of paraffin inhibition and removal to provide theoretical basis of new paraffin inhibitors and removers development.

(2) Develop surfactants with superior performance and low cost to improve paraffin removal efficiency and reduce costs.

(3) Make full use of natural polymer resources to develop new polymer paraffin inhibitors with efficient, non-toxic and inexpensive.

(4) Carry out researches on paraffin inhibition and removal process and combine different methods. For example, combination of thermal paraffin removal technology and emulsion paraffin remover will surely receive better removal effect.

蜡具有良好溶解性能的有机溶剂和油溶性表面活性剂，从而起到清蜡效果。

8.3.3　油井化学清防蜡新技术及发展趋势

未来化学清蜡剂的发展方向是开发环保、高效、多功能、低选择性的产品；并开展复配助剂的研究，提高蜡溶剂效果。化学防蜡剂的发展趋势主要是开发新型、多效高分子型防蜡剂，提高其防蜡效果。综合利用化学防蜡剂与化学清蜡剂各自的优点，充分发挥两者优势互补的特性，将成为今后化学清防蜡技术的发展趋势。为此，对油田蜡沉积和清防蜡的研究可以进一步开展以下几方面工作：

（1）进一步开展蜡沉积和清防蜡的微观机理研究，为开发新型清防蜡剂提供理论依据。

（2）开发性能优越、价格低廉的表面活性剂，提高清蜡效率，降低使用成本。

（3）充分利用天然高分子资源，开发高效、无毒、价廉的新型聚合物防蜡剂。

（4）开展清防蜡工艺的研究，将不同的防蜡方法有机结合，例如将热清蜡技术和乳液型清蜡剂结合使用，必将收到更好的清蜡效果。

In short, the development of environmental friendly, low cost, low selectivity and multi-effect (such as paraffin inhibit, remove, pour point depression, viscosity reduction, demulsification, etc.) paraffin inhibitor and remover to meet oilfield application requirements in China, which will be the direction of future development.

总之，开发研制环保、低成本，低选择性，一剂多效（兼有防蜡、清蜡、降凝、降黏、破乳等作用）清防蜡剂以满足我国不同含蜡油田的开采需要，将成为未来清防蜡剂的发展方向。

第9章 化学防砂
Chemical sand control

In China, unconsolidated sandstone reservoirs are widely distributed and in large reserves. Sand production in oil and gas wells is the main contradiction in exploitation of such reservoirs. Sand production will lead to production shutdown, extra equipment maintenance or other problems of oil and gas wells, which will increase crude oil production costs and oilfield management difficulties. Sand control is one of the indispensable technological measures for development of oil and gas reservoirs with sand production. Sand control methods can be divided into mechanical sand control and chemical sand control, in which the latter is with unique advantages for controlling sand production in unconsolidated sandstone oil wells.

9.1 Causes and hazards of sand production in oil and gas wells

9.1.1 Causes of sand production in oil and gas wells

Whether a formation produces sand or not depends on degree of particle cementation, i.e. formation strength. Under normal circumstances, formation stress exceeds its strength will cause sand production. Causes for sand production in oil and gas wells will influence choice of method and formula of sand control greatly. Sand production in oil and gas wells can be attributed into two parts: geological and produced factors. Table 9-1 lists the details.

我国疏松砂岩油藏分布范围广、储量大，油气井出砂是这类油藏开采的主要矛盾。出砂会导致油气井停产作业、设备维修等，增加原油生产成本和油田管理难度。防砂是开发易出砂油气藏必不可少的工艺措施之一。防砂方法可分为机械防砂和化学防砂，其中化学防砂对于治理疏松砂岩油井出砂具有独特优势。

9.1 油气井出砂的原因及危害

9.1.1 油气井出砂的原因

地层是否出砂取决于颗粒的胶结程度即地层强度。一般情况下，地层应力超过地层强度就可能出砂。油气井出砂原因对于防砂及防砂剂配方的选择有很大的影响，油气井出砂的原因可以归结为地质和开采两种，见表9-1。

Table 9–1 Statistics on causes of sand production in oil and gas wells

Causes of sand production		Specific details
Geological factors	Uncemented formation	There is no cement between the particles, and formation sand can flow under certain conditions; shortly after production, sand is produced continuously, and its content is relatively stable in wellhead
	Partially cemented formation	Relatively low formation cement content, formation sand is partially cemented, hence the cementation condition is poor and strength is low; after production, fluctuation of sand content is large
	Friable sand formation	Relatively high formation cement content and strong cementation force in formation sand; formation strength is quite high, but because the brittleness of cement is stronger than sand particles, this kind of formation is easy to break; the sand production changes periodically
Produced factors	Excessive large production pressure	When other conditions remain unchanged, scouring force is increasing with growth of production pressure difference, which will make sand production easier
	Water breakthrough in reservoir	The reservoir cement is mainly composed of clay, when injection water soaks the formation, the clay will expand and loosen, which will reduce cementation strength, and make the particle displace. Then degree of formation sand production will become more severe
	Speed of production and operation	Frequent operation, improper measures or operation with improper speed will easily lead to sand production
	Poor management	Frequent open and shut wells will destroy stable sand bridges and cause sand production in the formation

表9–1 油气井出砂的原因统计表

出砂原因		具体表现形式
地质因素	未胶结地层	颗粒之间无胶结物，地层砂在一定的条件下可以流动；投产后立即出砂并连续不断，井口含砂量相对稳定
	部分胶结地层	地层胶结物含量较少，地层砂部分被胶结，胶结差，强度低；投产后出砂含量波动变化大
	脆性砂地层	地层胶结物含量较多，地层砂间胶结力较强，地层强度较好，但因胶结物的脆性比砂粒强，故这种地层易破碎；出砂规律呈周期性变化
开采原因	生产压过大	其他条件相同时，生产压差愈大，冲刷力愈大，越容易出砂
	油层见水	油层胶结物以黏土为主，注入水浸泡地层，使黏土遇水膨胀变松散，降低胶结强度，进而发生颗粒位移，加剧地层出砂程度
	作业及开采速度	频繁作业或作业过程措施不当以及不恰当的开采速度，容易导致出砂加剧
	管理不善	频繁开关井，使稳定的砂桥破坏，造成地层出砂

9.1.2 Mechanism and hazards of sand production in oil and gas wells

Under production conditions, formation stability is related to the stress field of formation matrix. Formation matrix adapts to state of the stress field in a complicated way. The sustained stress of formation matrix is actually a joint force by overburden pressure, pore pressure, flowing pressure gradient at near wellbore zone, interfacial tension and the friction formed by

9.1.2 油气井出砂的机理与危害

生产条件下地层稳定性与地层基质所受应力场作用有关。地层基质以复杂的方法适应应力场状态，所受应力是上覆地层压力、孔隙压力、近井地带地层流体流动压力梯度、界面张力、流体通过基质颗粒间空隙流动时与颗粒

fluid flowing through the interstices between matrix particles. Ground stress adapts to formation stability is determined by intrinsic strength of formation medium and the formation productivity coefficient under certain conditions. During production process, stability of various factors will be broke, and sand arches and bridges will be formed around the wellbore as shown in Figure 9-1.

摩擦而形成的摩阻。地应力适应地层稳定性的方式是在一定条件下由地层介质本征强度和地层产能系数所决定的，在生产过程中因各种因素而破坏失稳后便形成了井眼周围的稳定砂拱和砂桥，如图9-1所示。

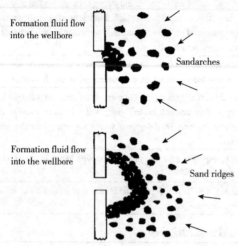

Figure 9-1　Schematic of sand bridge and sand arch

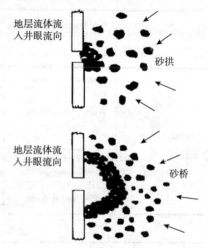

图9-1　砂桥和砂拱示意图

Sand production in oil and gas wells is one of the important problems for unconsolidated sandstone reservoirs. It causes hazards from four aspects as shown in Table 9-2.

油气井出砂是疏松砂岩油气藏面临的重要问题之一，其危害有四个方面，见表9-2。

Table 9-2　Hazard statistics of sand production in oil and gas wells

Hazards	Specific details
Oil and gas well production reduction or shutdown	Sand production in oil and gas wells is easy to cause producing formation covered by sand, tubing sand block and oil storage tank sand accumulation, thus production will be forced shutdown
Equipment and pipeline abrasion	Formation sands carried in oil-gas flow are highly abrasive, which will corrode equipments and pipelines and increase investment costs
Casing damage	Long-term serious sand production will cause imbalance of internal and external forces, which will lead to sudden formation collapse and resulting in casing damage
Original structural damage or permeability reduction	With sand production in oil and gas wells, formation sand migration will be intensified. In this case, structure of far wellbore zone will be more loosen and formation sand deposits in near wellbore zone will be increased, resulting in a significant permeability decrease

表9-2　油气井出砂的危害统计表

危害类型	表现形式
油气井减产或停产	油气井出砂，极易造成砂埋产层，油管砂堵及地面管汇和储油罐积砂，从而被迫停产
设备及管线磨蚀	油气流中携带的地层砂粒磨蚀性强，可以腐蚀设备和管线，增大投资成本
套管损坏	长期严重的出砂造成内外受力不平衡，发生突发性地层坍塌，导致套管损坏
原始构造破坏或渗透率下降	油气井出砂后，地层砂运移加剧，远井地带则变得结构疏松加剧，近井地带地层砂沉积较多，造成渗透率显著下降

9.2 Classification of chemical sand control

Chemical sand control refers to a oil wells sand production reduce technique by injecting a certain amount of chemical agents near the wellbore to cement formation sand, or by fixing cementing agent with solid particles near bottom hole and forming an artificial well wall with certain compressive strength and permeability. It is as shown in Figure 9-2.

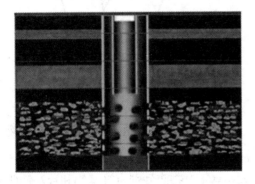

Figure 9-2　Schematic of chemical sand control

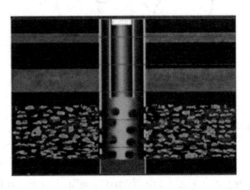

图9-2　化学防砂示意图

Advantages of chemical sand control are simple operation, no residue in well after construction and sand control effect is obvious; it can be used for sand control of abnormal high pressure reservoirs. Its disadvantage is that it will cause certain damage to formation permeability, especially when sand control construction is repeated; the injectant may aging, which makes its validity period limited; besides, success rate of chemical is not as good as mechanical sand control, and it is not suitable for sand control of open hole wells. By far, chemical sand control technology has been rapidly developed, and the sand control agents used are classified into chemical bridging and chemical cementing according to their chemical principles.

9.2.1　Chemical bridging sand control agent

Chemical bridging sand control is to bridge the

9.2 化学防砂分类

化学防砂是指向井眼附近挤入一定数量的化学剂以胶结地层砂，或将携带有固体颗粒的胶结剂至井底附近，形成具有一定抗压强度和渗透率能力的人工井壁，从而减轻油井出砂的技术，如图9-2所示。

化学防砂的优点是操作简单，施工后井内无遗留物；可用于异常高压油层的防砂；防砂效果明显。缺点是：对地层渗透率有一定伤害，特别是重复施工时；注入剂存在老化现象，使其有效期有限；成功率不如机械防砂；化学防砂不适用于裸眼井防砂。截至目前，化学防砂技术已经得到了快速发展，所用防砂剂按照化学原理分为化学桥接和化学胶结两类。

9.2.1　化学桥接防砂剂

化学桥接防砂法是由桥接剂

loose sand grains at their contact points by a bridging agent (Figure 9-3). A bridging agent is a chemical agent capable of bridging loose sand grains at their contact points. Its main content is inorganic cationic polymer and organic cationic polymer.

将松散砂粒在它们接触点处桥接起来，达到防砂的目的（图9-3）。桥接剂是指能将松散砂粒在接触点处桥接起来的化学剂，主要为无机阳离子型聚合物和有机阳离子型聚合物。

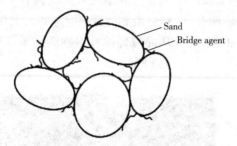

Figure 9-3 Schematic of bridge formation between sand particles

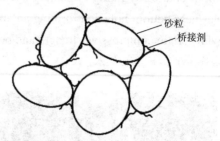

图9-3 桥接剂在砂粒间产生桥接示意图

9.2.1.1 Inorganic cationic polymer

The most typical inorganic cationic polymer sand control agents are hydroxy aluminum and hydroxy zirconium. This is because multi-nuclear hydroxy bridge complex ions composed by high valent inorganic cations could bridge the loose sand grains with negatively charged surface. In this case sand production from the formation is reduced, as shown in Figure 9-4.

9.2.1.1 无机阳离子型聚合物

最典型的无机阳离子型聚合物防砂剂有羟基铝和羟基锆两类。由于高价无机阳离子的多核羟桥络离子，可将表面带负电的松散砂粒桥接起来，减少砂从地层产出，如图9-4所示。

$$\left[(H_2O)_4Al\underset{OH}{\overset{H_2O}{\underset{H_2O}{\overset{OH}{\fbox{Al}}}}}\underset{OH}{\overset{OH}{}}_n Al(H_2O)_5\right]^{(n+4)+}$$

$$\left[(H_2O)_6Zr\underset{OH}{\overset{H_2O\ H_2O}{\underset{H_2O\ H_2O}{\overset{OH}{\fbox{Zr}}}}}\underset{OH}{\overset{OH}{}}_n Zr(H_2O)_6\right]^{(2n+6)+}$$

Figure 9-4 Multinuclear hydroxy bridge ion

图9-4 多核羟桥络离子

9.2.1.2 Organic cationic polymer

Organic cationic polymer with quaternary ammonium salt structure on its branch (Figure 9-5) will control sand production by bridging its cationic chain link element and loose sand grains with negatively

9.2.1.2 有机阳离子型聚合物

支链上有季铵盐结构的有机阳离子型聚合物（图9-5），通过其阳离子链节将表面带负电的松散砂粒桥接起来，起到防砂作用。

charged surface. —$CONH_2$ in modified polyacrylamide can also form hydrogen bond with hydroxyl groups on the sandstone surface and plays the role of bridging.

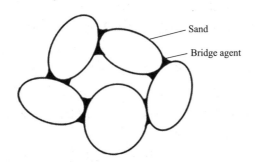

Figure 9-5　Acrylamide with (2-acrylamido-2-methyl) propyltrimethylammonium chloride copolymer

9.2.2　Chemical cementing sand control agent

Chemical cementing sand control is to bond loose sand grains at their contact points by cementing agent (Figure 9-6). Cementing agent refers to a chemical agent capable of cementing loose sand grains at their contact points. It includes inorganic type, organic type and compound type cementing agent.

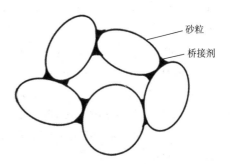

Figure 9-6　Schematic of the bridge between sand particles

9.2.2.1　Inorganic type cementing agent

Inorganic cementing agent will form contact cement, basal cement or cement of others by reacting with reservoir fluids or solids, or by depositing on surface of gravels. Inorganic sand control agents include silicic acid, calcium silicate, cements, calcium hydroxide, silicon tetrachloride, fluoboric acid sand control agent, mortar, etc., which are mostly used in artificial well walls (Figure 9-7). At present, the most frequently used agents are mainly cements, including cement mortar, diesel emulsified cement slurry, water carrying dry lime sand and so on.

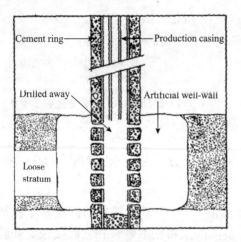

Figure 9-7　Schematic of the artificial well wall

(1) Cement mortar artificial well wall : Taking cement as cementing agent and quartz sand as proppant in proportion, mix them with appropriate amount of water, and carried the mixture to the well by oil. Squeezed it at outside of casing and let it accumulated on section with sand production. It will then form an artificial well wall with certain strength and permeability after solidification, preventing sand production of the reservoir. It is suitable for sand control of oil wells with sand production, low pressure oil wells, shallow wells (well depth about 1000m) and thin oil reservoir wells (less than 20m).

The main working principle of other artificial well wall methods are as above described, but application will be slightly different according to reservoir conditions. For example, the artificial sand wall formed by water carrying dry lime sand method is suitable for late stage low pressure oil and water wells, oil and water wells already with sand production, multilayered reservoirs, high water cut production wells and oil and water wells whose sand control section is in range of 50m. The artificial well wall of diesel emulsified cement slurry is suitable for shallow wells, wells with formation sand production less than 500L/m, oil or water wells with reservoir sections in range of 15m as well as early sand control of oil and water wells. Resin walnut shell artificial well wall is suitable for oil wells with small amount sand production, whole well sand control with perforation interval less than 20m and early sand control of water wells; Resin mortar artificial well wall is suitable for sand control of oil and water well patterns with high absorption capacity and late stage oil and water wells with reservoir sections in range of 20m.

Moreover, working principle of inorganic cementing agents such as silicic acid, calcium silicate, calcium hydroxide and silicon tetrachloride is similar to that of cement mortar.

(2) Fluoboric acid sand control agent: It is a high efficiency compound sand control agent with dual functions of "plugging removal and sand control". Taking 5%–8% HBF_4 as its main content, the sand control agent is with a variety of addition agents. Its mechanism is that HBF_4 reacts with water and form $BF_3(OH)^-$, $BF_2(OH)_2^-$ and $BF(OH)_3^-$ slowly. At the same time, stepwise hydrolyses and form HF. HF will then dissolve small pieces of clay and stabilize clay particles, which could prevents particle migration and reduces sand production of oil well.

9.2.2.2 Organic type cementing agent

Organic cementing agent will synthesize through or under the ground. It will bond reservoir gravels and form contact cements, and is widely used in artificial cementing method and artificial well wall method. Type of organic sand control agent includes resin type, jelly type, polyurethane type, coke type, polyvinyl chloride type, oxidized organic substance, silicon polymer sand control agent and so on. At present, the most frequent used type in applications are resins, such as phenolic resin, epoxy resin, urea-formaldehyde resin, furan resin, etc.

(1) Phenolic resin consolidation sand layer: Use

phenol and formaldehyde as main materials and made phenol-aldehyde solution according to certain proportion. The solution will be then heated and decocted into resin solution. Squeeze the pre-prepared resin solution into sandstone reservoirs, use diesel for pore adding and then hydrochloric acid as curing agent. The mixture will react and solidify at reservoir temperature, cementing the loose sandstones and prevent sand production in oil and water wells as shown in Figure 9-8.

苯酚和甲醛为主料，碱性物质为催化剂，按比例混合制成酚醛溶液，酚醛溶液经加热熬制成树脂溶液。将预制备的树脂溶液挤入砂岩油层，以柴油增孔，再挤入盐酸作固化剂，在油层温度下反应固化，将疏松砂岩胶结以防止油水井出砂，如图9-8所示。

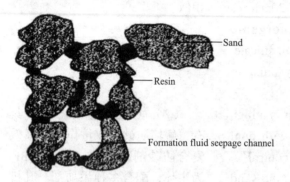

Figure 9-8 Schematic diagram of resin-consolidated formation sand

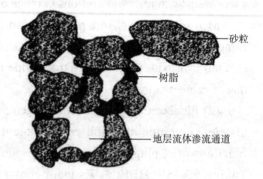

图9-8 树脂固结地层砂示意图

To make up the shortcoming that phenolic resin is not suitable for consolidation of low temperature reservoirs, phenolic resins have been modified in recent years, such as epoxy resin and silicone modification. These modification methods have the effect of lowering the intrinsic consolidation temperature, extending effective period and increasing cementation strength. After modification, the initial resin degradation temperature is increased by 60–80℃.

(2) Phenol-aldehyde solution underground synthetic sand control: Stir phenol and formaldehyde in proportion, add catalyst and use diesel as pore adding agent. After being squeezed into sand production section, the phenol-aldehyde solution will gradually form resin under reservoir temperature and deposit on surface of sand grains. It will solidify and cement sand grains in reservoir firmly. As the continuous phase, diesel will not participate in reaction and is filling the

针对酚醛树脂不适用于低温油藏固结问题，近年来对酚醛树脂的改性，如环氧树脂、有机硅改性，降低自身固结温度、增加有效期和胶结强度，改性后树脂降解起始温度提高60~80℃。

（2）酚醛溶液地下合成防砂：将加有催化剂的苯酚与甲醛，按比例配料搅拌均匀，并以柴油为增孔剂。酚醛溶液挤入出砂层后，在油层温度下逐渐形成树脂并沉积于砂粒表面，固化后将油层砂胶结牢固，而柴油不参加反应为连续相充满孔隙，使胶结后的砂岩保持良好的渗透性，从而起到提高砂岩的胶结强度，防止油气层出砂的方法。

pores, so that the cemented sandstone maintains good permeability. Therefore, this method could improve strength of sandstone cementation and prevent reservoir sand production.

Except applied in different reservoir conditions, the main principle of other resin sand control agents is basically similar with that of phenolic resin. For example, urea-formaldehyde resin is suitable under conditions that temperature 80-140℃ and water cut 60%–98.2%; while high temperature foam resin could be used at a temperature as high as 350℃.

9.2.2.3 Compound type cementing agent

The most frequent used compound type cementing agent currently is mainly composed of resin and inorganic material. The inorganic material is normally walnut shell, mortar, quartz sand, etc. It is mostly used in artificial well walls, and the working mechanism is similar to that of resin (Figure 9-8). Inorganic materials are added to increase compressive strength and permeability of the artificial well walls.

Compound type sand control is with advantages of high sand blocking precision, good sand control effect, high construction success rate and long sand control effective period. It is also suitable for conditions of high fluid producing intensity and large production pressure difference. It is applicable to sand control that with conditions of:

(1) Fine sandstone and fine sand with serious heterogeneity;

(2) Middle and late stages of oil and gas wells with particularly severe sand production;

(3) Oil and gas wells with large liquid production, high liquid production intensity and large production pressure difference.

其他树脂类防砂剂的主要原理和酚醛树脂的原理基本相似，但适应油藏条件不一致，如脲醛树脂适用温度80~140℃，含水量60%~98.2%；高温泡沫树脂防可耐350℃高温。

9.2.2.3 复合型胶结剂

目前，复合型胶结剂主要采用树脂与无机物复合，其中无机物有核桃壳、砂浆、石英砂等复合形成防砂剂，多用于人工井壁，防砂机理与树脂类似（图9-8），其中无机物的添加是增加人工井壁的抗压强度和渗透能力。

复合防砂具有挡砂精度高，防砂效果好，同时能够承受较高采液强度和较大生产压差，施工成功率高、防砂有效期长等优点。复合化学防砂技术适用于：

（1）非均质性严重的细砂岩及粉细砂防砂；

（2）出砂特别严重的油气井的中后期防砂；

（3）产液量大，采液强度高以及较大生产压差的油气水井防砂。

第10章 Chemical flooding 化学驱

Oil recovery is generally low. During water flooding development, the oil recovery can only reach 30%-40% normally. Instead of being recovered, most of crude oil will remain in reservoir. The reason for low oil recovery is heterogeneity of reservoir, which will cause displacement medium to protrude into oil well along high permeable layer but not affecting reservoir with lower permeability. There is a concept called sweep efficiency that shall be explained here. Sweep efficiency refers to ratio that volume of reservoir affected by oil displacement agent to oil content of the entire reservoir. However, due to wettability and capillary force of the reservoir surface, oil affected by oil displacement agent in reservoir will not be completely produced. Thus there is a concept of displacement efficiency. Displacement efficiency refers to ratio of oil amount produced from reservoir affected by displacement medium to entire oil reserves of the reservoir. According to concept of sweep efficiency and displacement efficiency, it can be known that:

Oil recovery=sweep efficiency×displacement efficiency

It can be seen from the above formula that there are two ways to improve oil recovery: one is to increase sweep efficiency; the other is to improve displacement efficiency.

The primary method of increasing sweep efficiency is to change mobility of the displacement medium and

原油采收率低，水驱开发时原油采收率一般仅能达到30%~40%，大部分原油仍滞留在储层中无法采出。原油采收率低的原因是油层的非均质性，使驱替介质沿高渗透层突入油井而波及不到渗透率较低的油层。这里涉及一个波及系数的概念。波及系数是指驱油剂波及到的油层容积与整个油层含油容积的比值。然而，驱油剂波及到的油层，由于油层表面润湿性和毛管力的作用，油也无法全部采出，因而又有一个洗油效率的概念。洗油效率是指驱替介质波及到的油层所采出的油量与这部分油层储量的比值。根据波及系数和洗油效率的概念，可得：

原油采收率 = 波及系数 × 洗油效率

从上式可以看出，提高原油采收率有两个途径：一个是提高波及系数；另一个是提高洗油效率。

提高波及系数的主要方法是改变驱替介质和（或）油的流度。

(or) oil. Mobility is a measure of whether a fluid can pass through porous medium. Its definition formula is:

$$\lambda = k/\mu$$

In above formula, λ-Mobility of the fluid;
k-Effective permeability of fluid in porous media;
μ-Viscosity of the fluid.

Since mobility of displacement medium is much higher than that of oil, during oil displacement, the displacement medium will be easily protruded into oil well along the high permeable layer. To increase sweep efficiency of the displacement medium, mobility of displacement medium shall be reduced, and (or) mobility of oil shall be increased.

The main method of improving displacement efficiency is to change rock surface wettability and reduce the adverse effects of capillary resistance.

Chemical flooding is a displacement method in which a chemical agent (surfactant, polymer or alkali) is added to displacement medium. It can improve oil recovery by increasing sweep efficiency and (or) displacement efficiency, and is taken as an effective way of oil recovery improvement.

10.1 Polymer flooding

10.1.1 Definition of polymer flooding

Polymer flooding refers to the oil displacement method using polymer solution as oil displacement agent. As a kind of chemical flooding, polymer flooding is also known as polymer solution flooding, polymer enhanced water flooding, viscous water flooding or viscosity enhancing water flooding.

Polymer flooding is an independent oil displacement method, but it can also be used as an auxiliary flooding method. When it is used as an auxiliary flooding method, it could be applied with other methods to enhance oil recovery (such as surfactant flooding or alkaline flooding). In this process, polymer slug is used for mobility control. It could help to protect slugs of the previous oil recovery enhancing methods going smoothly through the formation, hence improve oil recovery effect of the main oil recovery enhancing methods.

Figure 10-1 is diagram of polymer flooding slug. Since polymers are with effect of salt sensitivity, fresh water slugs shall be injected before and after the polymer solution.

塞用于流度控制，以保护它前面的其他提高采收率方法的段塞平稳地通过地层，充分发挥其他提高采收率方法的驱油作用。

图 10-1 为聚合物驱的段塞图。考虑到聚合物的盐敏作用，所以在聚合物溶液的前后注入淡水段塞。

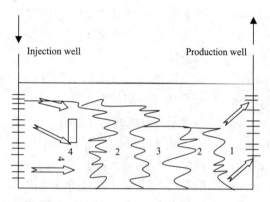

Figure 10-1　Diagram of polymer flooding slug
1-Remaining oil; 2-Fresh water; 3-Polymer solution; 4-Water

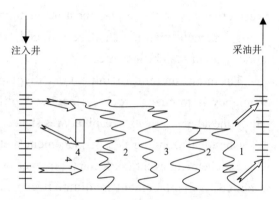

图10-1　聚合物驱段塞图
1—剩余油；2—淡水；3—聚合物溶液；4—水

10.1.2　Mechanism of oil recovery enhancement by polymer flooding

The mechanism of oil recovery enhancement by polymer flooding can be seen from relative permeability curves of water flooding and polymer flooding (Figure 10-2).

Polymer flooding will improve oil recovery by reducing water-oil mobility ratio and increasing sweep efficiency of the displacement liquid.

Therefore, polymer flooding mainly enhances oil recovery by the following mechanisms:

10.1.2.1　Mechanism of viscosity enhancement

Polymer can increase sweep efficiency by enhancing water viscosity and reducing the water-oil

10.1.2　聚合物驱提高采收率的机理

聚合物驱提高采收率的机理可以从水驱和聚合物驱的相对渗透率曲线看出（图 10-2）。

聚合物驱是通过降低水油流度比，提高驱动液的波及系数，从而提高原油采收率的。

因此，聚合物驱主要通过下列机理提高采收率：

10.1.2.1　增黏机理

聚合物可以通过增加水的黏度，降低水油流度比，从而提高

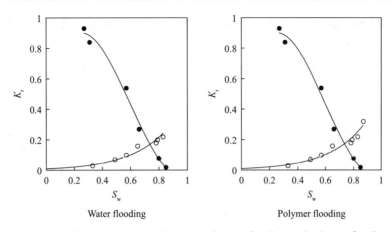

Figure 10-2 Relative permeability curve of water flooding and polymer flooding

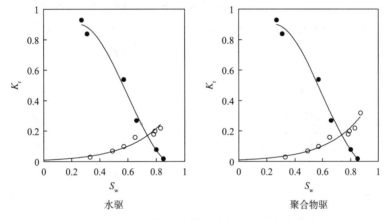

图10-2 水驱与聚合物驱的相对渗透率曲线

mobility ratio. Polymers increasing water viscosity is mainly by:

(1) Structures formed by polymer molecules entangle with each other in water.

(2) Hydrophilic groups in polymer chain elements that solvated in water.

(3) If the polymer is ionic, it will dissociate in water and produce a number of chain elements with charges of same type. This will make conformation formed by polymer molecules looser, thus with better ability of viscosity enhancement.

10.1.2.2 Mechanism of permeability reduction

Polymers can increase sweep efficiency by reducing effective permeability of water and making water-oil mobility ratio lower. The reason polymer can

波及系数。聚合物之所以能增加水的黏度，主要由于：

（1）水中聚合物分子互相纠缠形成结构。

（2）聚合物链节中亲水基团在水中溶剂化。

（3）若为离子型聚合物，则其在水中解离，产生许多带电符号相同的链节，使聚合物分子在水中形成的无视线团更松散，因而有更好的增黏能力。

10.1.2.2 降低渗透率机理

聚合物可通过减小水的有效渗透率，降低水油流度比，从而提高波及系数。聚合物之所以能

reduce effective permeability of water is mainly relying on polymer retention in pore structure of rocks. There are two forms for polymer retention in pore structure of rocks:

(1) Adsorption.

Adsorption refers to the phenomenon that polymer molecules are concentrated on surface of the rock by dispersion forces, hydrogen bonding or other forces.

(2) Capture.

Although conformation formed by polymer molecules in water is with a smaller radius than throat radius, they can be retained outside the throat by bridging (Figure 10-3). This retention is called capture.

减小水的有效渗透率，主要由于它可在岩石孔隙结构中产生滞留。聚合物在岩石孔隙结构中有两种滞留形式：

（1）吸附。

吸附是指聚合物分子通过色散力、氢键或其他作用力在岩石表面所产生的浓集。

（2）捕集。

聚合物分子在水中所形成的无视线团的半径小于喉道半径，但是它们可通过架桥而滞留在喉道外（图10-3）。这种滞留叫捕集。

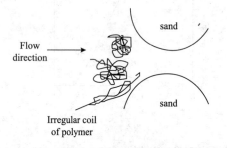

Figure 10-3　Capture of polymer molecules outside the throat

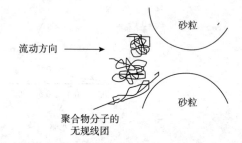

图10-3　聚合物分子在喉道外的捕集

Polymer flooding mainly improves sweep efficiency through these two mechanisms, which will finally resulting oil recovery increase.

Figure 10-4 shows comparison of polymer flooding and water flooding. It can be seen from Figure 10-4 that polymer flooding is with higher horizontal sweep efficiency than water flooding. Figure 10-5 shows that polymer flooding is with higher vertical sweep efficiency than water flooding.

Since mobility of polymer solution in low permeable layer is less than that in high permeable layer, during the following water flooding, water will cross flow between the high and low permeable layers, increasing sweep efficiency of the water flooding.

聚合物驱主要通过这两个机理提高波及系数，从而提高原油采收率。

图10-4是聚合物驱与水驱的对比。从图10-4可以看出聚合物驱比水驱有更高的平面波及效率。图10-5说明聚合物驱比水驱有更高的纵向波及效率。

由于聚合物溶液在低渗透层的流度小于在高渗透层的流度，所以在随后的水驱中，水可以在高渗透层与低渗透层之间窜流，提高水驱的波及系数。

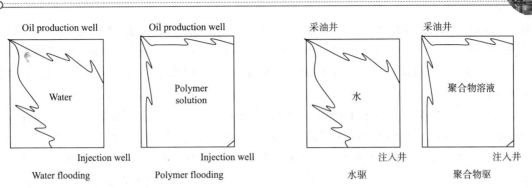

Figure 10-4　Horizontal sweep efficiency of water flooding and polymer flooding

图10-4　水驱与聚合物驱的平面波及效率

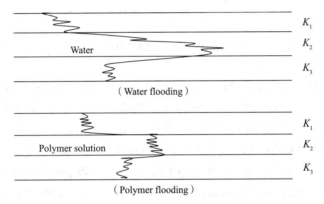

Figure 10-5　Vertical sweep efficiency of water flooding and polymer flooding
（$K_2 > K_3 > K_1$）

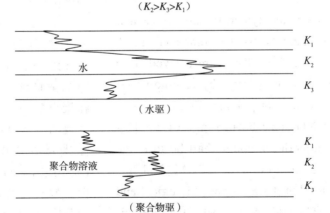

图10-5　水驱与聚合物驱的纵向波及效率
（$K_2 > K_3 > K_1$）

10.1.3　Polymer that could be used for polymer flooding

Polymer used for flooding should meet the following requirements: good water solubility, strong

10.1.3　聚合物驱用的聚合物

聚合物驱用的聚合物应满足下列条件，即水溶性好、稠化

thickening ability, heat stability, shear stability, chemical factors stability, biological effect stability, low retention, low cost and shall be with extensive supply sources, etc.

According to above requirements, there are two polymer types that commonly used in oil fields: HPAM (partially hydrolyzed polyacrylamide) and XG (Xanthan gum).

Molecular formula of HPAM is:

$$—(CH_2—\underset{CONH_3}{CH})_m—(CH_2—\underset{COOM}{CH})_n—$$

In above formula, M could be Na, K or NH_4. In polymer flooding, relative molecular mass of HPAM shall be in range of 1×10^{-3}-15×10^{-3}, its degree of hydrolysis (the percentage of carboxyl containing chain elements in polymer chain elements) shall between 1%-45%, its mass concentration shall be in range of 250-2000mg/L. And it could be used under reservoir temperature above 90℃.

XG is a group of biopolymer. Relative molecular mass of most of them are 2×10^6, but some could reach as high as 13×10^6-15×10^6. They are derived from fermentation of the carbohydrates by Xanthomonas campestris bacteria. During polymer flooding, mass concentration of XG shall be in range of 1×10^2-2×10^3mg/L and injection amount shall be in range of 0.01-0.25V_p. Compared with HPAM, XG has good shear stability, but it shows poor biological and thermal stability. XG could not be used if the reservoir temperature is higher than 70℃.

With deepening of the oilfield exploration and development, reservoir conditions becomes more and more severe, both temperature and salinity are continually increasing. These conditions require the polymers used in flooding could withstand higher temperature and salinity. By adding salt or temperature

能力强、对热稳定、对剪切稳定、对化学因素稳定、对生物作用稳定、滞留量低、来源广、便宜等。

基于上述条件，目前油田常用的两类聚合物是HPAM（部分水解聚丙烯酰胺）和XG（黄胞胶）。

HPAM的分子式为：

式中，M为Na、K或NH_4。在聚合物驱中，使用HPAM的相对分子质量在1×10^{-3}~15×10^{-3}范围，水解度（含羧基的链节在聚合物链节中所占的百分数）在1%~45%范围，它的质量浓度在250~2000mg/L范围内，油藏使用温度可达90℃以上。

XG是生物聚合物，相对分子质量为2×10^6，有的高达13×10^6~15×10^6，由黄单胞杆菌属细菌将碳水化合物发酵制得。在聚合物驱中，使用XG的质量浓度在1×10^2~2×10^3mg/L范围，注入量在0.01~0.25V_p范围。与HPAM相比，XG的剪切稳定性好，但生物稳定性和热稳定性较差，油藏使用温度低于70℃。

随着油田勘探开发的不断深入，油藏条件越来越苛刻，温度和矿化度不断升高，这也就要求聚合物驱所用的聚合物能够耐受更高的温度和矿化度。通过在聚合物分子主链上引入抗盐、耐温的功能性单体（2-丙烯酰胺基甲

resistant functional monomers (2-acrylamido methyl propane sulfonic acid, N-vinyl pyrrolidone or sodium styrene sulfonate, etc.) on main chain of polymer molecular, corresponding function of the polymer will be significantly improved. Based on this principle, some new functional polymers have also been developed, including branched polymers, amphoteric polymers, molecular composite polymers, comb-shaped polymers, hydrophobic associative polymers and so on. It is worth mentioning that performance of the polymers can be further improved by hydrophobic modification, especially on viscosity increasing as well as salt and shear resistance. This principle works on most of polymers including hyperbranched and amphoteric polymer.

(1) Copolymer containing side groups with high molecular weight or rigid side groups.

Adding rigid groups on main chain of polymer will enhance rigidity of the polymer molecular chain, and inhibit hydrolysis of amide group to a certain extent, thereby achieving purpose of temperature and salt resistance improvement of the polymer. 2-acrylamido 2-methylpropanesulfonic acid (AMPS), N-vinyl pyrrolidone (NVP) and sodium styrene sulfonate (SSS) are three relatively common used monomers that provide rigid pendant groups. Temperature resistance of the polymer can be remarkably improved by copolymerizing them with acrylamide and acrylic acid.

(2) Amphoteric copolymer.

As a functional polymer, amphoteric copolymer refers to a copolymer in which anionic group (carboxylic group, sulfonic group) and cationic group (quaternary ammonium salt) are simultaneously added on molecular chain of the polymer. Since number of positive and negative charges on molecular chain is equal in aqueous solution of the amphoteric polymer, the electrostatic force between groups is attracting each other, which will cause shrink of molecular chain, resulting decrease

基丙磺酸、N-乙烯基吡咯烷酮、对苯乙烯磺酸钠等），利用其功能性基团可以显著地提高聚合物的耐温、抗盐性能。在此基础上，其他一些新型功能性聚合物也被开发出来，包括支化聚合物、两性聚合物、分子复合型聚合物、梳形聚合物、疏水缔合聚合物等。值得一提的是，无论是超支化聚合物还是两性聚合物等，均可以通过疏水改性使聚合物的性能得到进一步的提高，尤其是增黏性、抗盐性以及抗剪切性等。

（1）含大侧基或刚性侧基的共聚物。

在聚合物主链上引入刚性基团，可以增强聚合物分子链的刚性，可以在一定程度上抑制酰胺基的水解，从而达到提高聚合物耐温抗盐性的目的。2-丙烯酰胺基2-甲基丙磺酸（AMPS）、N-乙烯基吡咯烷酮（NVP）、对苯乙烯磺酸钠（SSS）是三种比较常用的可以提供刚性侧基的单体，与丙烯酰胺、丙烯酸共聚后，可以显著改善聚合物的耐温性。

（2）两性共聚物。

两性共聚物作为一种功能性聚合物，是指在聚合物分子链上同时引入阴离子基团（羧酸基、磺酸基）和阳离子基团（季铵盐）的共聚物。在两性聚合物的水溶液中，由于分子链上的正负电荷数目相等，基团间的静电作用力表现为相互吸引，分子链收缩，导致聚合物在清水中的黏度降低；

of polymer viscosity in clean water. However in saline solution, mutual attraction between the groups in polymer is weakened or shielded, thus molecular chain will stretch and hydrodynamic volume of polymer molecule will increase, resulting increase of the solution viscosity.

(3) Molecular composite polymer.

Molecular composite polymer refers to aggregate formed by interaction of two polymers by hydrogen bonding, hydrophobic or electrostatic interaction. Polymer acids, polymer alkali and polymer salts are main components of some common composite polymers. For example, compound anionic and cationic polymer in a certain ratio by mutual synergy of Coulomb forces between molecular chains, a homogeneous composite type supramolecular structure will be formed. In this polymer system, the molecular chain is stretched and the hydrodynamic volume is increased. Comparing with single component polymer, viscosity of composite polymer solution is significantly increased. In addition, the composite polymers could form dynamic network structures, which will shield small molecule electrolytes to some extent and improve properties of viscosity increase and salt tolerance in oil displacement agent.

(4) Comb-shaped polymer.

Combining effect of volume and electrical repulsion makes performance of comb-shaped polymers significantly different with traditional linear polymers, mainly in the following aspects: Rigidity of the main chain is enhanced; molecular chain is not easy to be curled and entangled; in aqueous solution, structure of macromolecular chain is regular, its conformation is stretched, hydrodynamic volume is large, and its shape is similar to a comb. Since with these structural features, comb-shaped polymer solution is able to show good performance on adhesion, as well as excellent salt, shear and aging resistance.

而在盐水中，聚合物中基团间的相互吸引力会被削弱或者屏蔽，分子链变得舒展，聚合物分子流体力学体积增大，溶液的黏度增加。

（3）分子复合型聚合物。

复合型聚合物是指两种聚合物通过氢键、疏水或静电等相互作用形成的聚集体。聚合物酸、聚合物碱和聚合物盐是常见的一些复合型聚合物的主要成分。例如，阴离子型聚合物与阳离子型聚合物按照一定比例复配，二者通过分子链间库仑力的相互协同作用，可形成均相复合型超分子结构。该聚合物体系中，聚合物的分子链伸展、流体力学体积增大。相比单一组分聚合物，复合型聚合物溶液的黏度显著增加。此外，复合型聚合物之间可以形成一些动态网络结构，对屏蔽小分子电解质起到了一定的作用，进而可以改善驱油剂的增黏、抗盐性能。

（4）梳形聚合物。

与传统线性聚合物相比，体积排斥和电性排斥的共同作用，使梳形聚合物的性能表现出明显的不同：主链刚性增强，分子链不易卷曲、缠结，大分子链在水溶液中，结构规整、构象伸展、流体力学体积大，并排列成类似梳子的形状。这些结构特点使梳型聚合物溶液表现出良好的增黏、抗盐、抗剪切和抗老化性能。

(5) Hydrophobically associating polymer.

Due to their special association in aqueous solutions, hydrophobically associating polymers show properties that different from conventional polymers. In aqueous solution of the polymer, hydrophobic groups will aggregate with each other due to hydrophobic interaction and form a spatial network structure. This structure will increase the hydrodynamic volume and let the solution show good viscosity-enhancing properties. In salt solution, due to increase of the solution polarity, the hydrophobic association is enhanced. What's more, since with good salt resistance, it shows obvious salt thickening effect within a certain degree of salinity. Besides, the polymer is with a relatively low molecular weight (generally around several million), and the molecular chain is not easy to break during shearing, showing excellent shear resistance. These special properties enable the water soluble hydrophobically associative polymer be used as oil displacing agent for tertiary oil recovery.

10.2 Alkaline flooding

10.2.1 Definition of alkaline flooding

Alkaline flooding refers to oil displacement method that using alkali solution as oil displacing agent.

Although cost of chemical agents for alkaline flooding is low and its operation is relatively simple, it is with complicated oil displacement mechanism. That is why the scale and extent of alkaline flooding field test is much smaller than that of polymer flooding.

Alkaline flooding is suitable for early stage of secondary oil recovery. During this period, the oil saturation is high and amount of surface active substance produced by reaction of acid and alkali is also large. In addition, since k_{rw} and λ_w are low in this period, both sweep efficiency and the recovery factor is high.

In alkaline flooding, polymer is commonly used to control mobility as shown in Figure 10-6. Since Ca^{2+} and Mg^{2+} in formation water can react with alkali and causing extra alkali consumption, a slug of fresh water shall be injected before the alkali solution. Polymer solution slug injected after alkaline solution is used as a mobility control slug. It allows alkali solution pass through the formation smoothly.

碱驱中常用聚合物控制流度，如图 10-6 所示。地层水中的 Ca^{2+}、Mg^{2+} 可与碱反应，增加碱耗，因此在碱溶液段塞之前注了一段塞的淡水。碱溶液之后注入的聚合物溶液段塞是作为流度控制段塞使用的，它可使碱溶液平稳地通过地层。

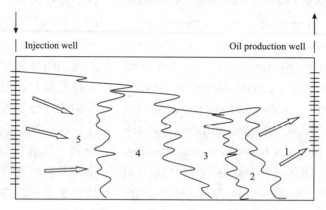

Figure 10-6　Slug diagram of alkaline flooding
1-Residual oil; 2-Fresh water; 3-Alkaline solution; 4-Polymer solution; 5-Water

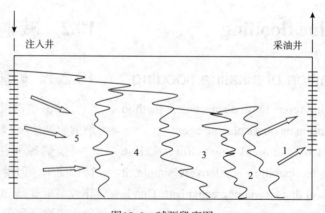

图10-6　碱驱段塞图
1—剩余油；2—淡水；3—碱溶液；4—聚合物溶液；5—水

10.2.2　Mechanism of oil recovery enhancement by alkaline flooding

Mechanism of alkaline flooding is complicated. The following mechanisms have been proposed to explain the reasons why alkaline flooding could improve

10.2.2　碱驱提高采收率机理

碱驱机理复杂，至今已提出如下几种机理解释碱驱之提高原油采收率的原因：

oil recovery:

10.2.2.1 Mechanism of low interfacial tension

Under low alkali mass fraction (less than 1×10^{-2}) and an optimal salt content, the surface active substance formed by reaction of alkaline and acidic components of the crude oil (fatty acid, naphthenic acid, colloidal acid and asphaltenic acid, etc.) can reduce oil-water interfacial tension to 10^{-2}mN/m or less.

10.2.2.1 低界面张力机理

低的碱质量分数（小于 1×10^{-2}）和一个最佳的盐含量下，碱与原油中酸性成分（脂肪酸、环烷酸、胶质酸、沥青质酸等）反应生成的表面活性物质，可使油水界面张力降至 10^{-2}mN/m 以下。

10.2.2.2 Emulsification-carrying mechanism

Under condition of low alkali mass fractions and salt content, the surfactant formed by reaction of alkali and petroleum acid will emulsify the remaining oil. The alkaline solution will then carry them go through the formation. According to this mechanism, alkaline flooding shall be with the following characteristics:

(1) Can generate emulsion in forms of very small oil beads;

(2) Can improve oil displacement efficiency of the alkaline flooding by the emulsion;

(3) Oil production will not be increased before breakthrough of alkaline solution;

(4) Coalescence properties of the oil beads will influence the process greatly.

10.2.2.2 乳化-携带机理

在低的碱质量分数和低的含盐量下，由碱与石油酸反应生成的表面活性剂可使地层中的剩余油乳化，并被碱水携带着通过地层。根据此机理，碱驱应有如下特点：

（1）可以形成极小油珠的乳状液；

（2）通过乳状液提高碱驱的洗油效率；

（3）碱水突破前采油量不可能增加；

（4）油珠的聚并性质对过程有较大的影响。

10.2.2.3 Emulsification-capturing mechanism

Under condition of low alkali mass fractions and salt content, oil is emulsified in alkaline solution phase due to the low interfacial tension. Since oil bead is with large radius, it will be captured when it moves forward. Capture of the oil bead will increase water flow resistance, i.e. reduce water mobility. This means mobility ratio is improved, sweep efficiency is increased and oil recovery is enhanced. According to this mechanism, alkaline flooding shall be with the following characteristics:

(1) Oil can form emulsion in alkaline solution phase;

10.2.2.3 乳化-捕集机理

在低的碱质量分数和低的含盐量下，由于低界面张力使油乳化在碱水相中，但油珠半径较大，因此当它向前移动时，就被捕集，增加了水的流动阻力，即降低了水的流度，从而改善了流度比，增加了波及系数，提高了采收率。根据此机理，碱驱应有如下特点：

（1）油可在碱水相中形成乳状液；

(2) Dispersed oil beads will be captured in smaller channels, which will improve Sweep efficiency of alkaline flooding;

(3) Amount of oil production may increase before breakthrough of alkaline solution;

(4) The coalescence properties of oil beads are beneficial to oil displacement process.

10.2.2.4 Mechanism of oil wet reversal to water wet

Under condition of high alkali mass fractions and low salt content, alkali can be desorbed by changing water solubility of the oil-soluble surfactant (the surfactant is adsorbed on rock surface). This will restore original hydrophilicity of the rock surface, making oil wet rock surface reversed to water wet, and thus improving the displacement efficiency. At the same time, the relative permeability of oil and water can be changed to form a favorable mobility ratio and improve the sweep efficiency.

10.2.2.5 Mechanism of water wet reversal to oil wet

Under condition of high alkali mass fractions and low salt content, surfactant formed by reaction of alkali and petroleum acid will be mainly distributed to the oil phase and adsorbed on rock surface. This will cause rock surface property change from water wet to oil wet. In this way, the discontinuous remaining oil phase will become continuous oil phase, providing channels for crude oil flow. At the same time, in a continuous oil phase, the low interfacial tension may result generation of water-in-oil emulsions. Water droplets in these emulsions will plug the flow channels. In porous medium that being plugged, these water droplets will produce high pressure gradient. The high pressure gradient could overcome capillary resistance reduced by the low interfacial tension, allowing oil to drain from the continuous oil phase between emulsified water

（2）分散的油珠会被捕集在较小孔道，改善了碱驱的波及系数；

（3）碱水突破前采油量可以增加；

（4）油珠的聚并性质对驱油过程有利。

10.2.2.4 由油湿反转为水湿机理

在高的碱质量分数和低的含盐量下，碱可以通过改变吸附在岩石表面的油溶性表面活性剂在水中的溶解度而解吸，恢复岩石表面原来的亲水性，使岩石表面由油湿反转为水湿，提高洗油效率。同时也可使油水相对渗透率发生变化，形成有利的流度比，提高波及系数。

10.2.2.5 由水湿反转为油湿机理

在高的碱质量分数和高的含盐量下，碱与石油酸反应生成的表面活性剂主要分配到油相并吸附到岩石表面上来，使岩石表面从水湿转变为油湿。这样，非连续的剩余油变成连续的油相，为原油流动提供通道。与此同时，在连续的油相中，低界面张力可能导致油包水乳状液的形成，这些乳状液中的水珠将起到堵塞流通孔道的作用，并在有水珠堵塞的孔隙介质中产生高的压力梯度。高的压力梯度能克服被低界面张力所降低的毛管阻力，使油从乳化水珠与砂粒之间的连续油相这条通道排泄出去，留下高含水率

droplets and sand grains. High water content emulsion will be then left (in which saturation of remaining oil could be as low as 5%) and saturation of remaining oil in formation will be reduced.

10.2.2.6 Mechanism of spontaneous emulsification and aggregation

With the optimal alkali mass fraction, crude oil can be emulsified into alkaline solution spontaneously. This spontaneous emulsification phenomenon is caused by aggregation and diffusion of surfactant. The surfactant is produced by reaction of petroleum acid in the oil and alkali in the alkaline solution, which will first be concentrated at the interface and then diffused into alkaline solution.

Carboxylic acid is the primary component of petroleum acid in the oil, which can react with alkali (such as sodium hydroxide) and produce carboxylic sodium. Aggregation of carboxylic sodium in water is determined by its mass fraction. Figure 10-7 is a phase diagram of sodium dodecanoate-water system. It can be seen from the diagram that if the temperature is kept constant, increasing mass fraction of water in dense system of sodium dodecanoate, the following changes will be conducted on sodium dodecanoate micelles in the system:

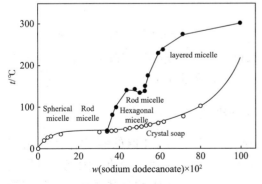

Figure 10-7 Phase diagram of sodium dodecanoate-water system

Layered micelles → Hexagonal rod-like micelles → Rod-like micelles → Spherical micelles

10.2.2.7 Mechanism of solubilizing rigid membrane

During tertiary oil recovery, the oil is in a dispersed state, and the asphaltene can form a rigid membrane on oil-water interface. This membrane makes it difficult for oil bead to pass through pore throat structure, thus water cannot displace remaining oil effectively. Injection of alkaline solution will increase water solubility of the asphaltene, reduce its rigidity and improve flowing ability of the remaining oil.

10.2.3 Alkalis that could be used for alkaline flooding

In addition to substances with structure of alkali (such as NaOH, KOH and NH$_4$OH), alkalis that could be used for alkaline flooding also includes some salts (such as Na$_2$CO$_3$, Na$_2$SiO$_3$ and Na$_4$SiO$_4$, etc.). This is because these salts can be hydrolyzed in water and produce OH$^-$:

$$CO_3^{2-} + H_2O \longrightarrow OH^- + HCO_3^-$$

$$HCO_3^- + H_2O \longrightarrow OH^- + H_2CO_3$$

$$SiO_3^{2-} + H_2O \longrightarrow OH^- + HSiO_3^-$$

$$HSiO_3^- + H_2O \longrightarrow OH^- + H_2SiO_3$$

$$SiO_4^{4-} + H_2O \longrightarrow OH^- + HSiO_4^{3-}$$

$$HSiO_4^{3-} + H_2O \longrightarrow OH^- + H_2SiO_4^{2-}$$

$$H_2SiO_4^{2-} + H_2O \longrightarrow OH^- + H_3SiO_4^-$$

$$H_3SiO_4^- + H_2O \longrightarrow OH^- + H_4SiO_4$$

They are all called latent alkali.

By the following reaction, sodium carbonate and sodium bicarbonate can buffer the pH value:

$$CO_3^{2-} + H_2O \rightleftharpoons OH^- + HCO_3^-$$

They are a pair of buffer substances. Therefore, sodium carbonate and sodium bicarbonate can be compounded to produce buffering latent alkali system.

10.2.2.7 增溶刚性膜机理

在三次采油时，油处在分散状态，沥青质可在油水界面上形成一层刚性膜。这种膜的存在，使油珠通过孔喉机构时不易变形通过，使水不能有效排驱剩余油。碱水的注入，增加了沥青质的水溶性，使它刚性减小，提高了剩余油的流动能力。

10.2.3 碱驱用碱

碱驱中的碱除一般具备碱结构的物质（如 NaOH、KOH、NH$_4$OH）外，还包括盐（如 Na$_2$CO$_3$、Na$_2$SiO$_3$、Na$_4$SiO$_4$ 等）。由于这些盐均可在水中水解产生 OH$^-$：

所以它们都可称为潜在碱。

碳酸钠与碳酸氢钠可通过下面反应：

起 pH 值的缓冲作用，它们是一对缓冲物质。因此，可用碳酸钠与碳酸氢钠复配，产生有缓冲作用的潜在碱体系。

10.2.4 Crude oil suitable for alkaline flooding

Since petroleum acid that could be used to produce surfactant is the necessary condition of alkaline flooding, crude oil shall be with sufficient acid value to conduct alkaline flooding (acid value refers to quantity of potassium hydroxide. The specific quantity shall be able to ensure the process from 1g crude oil being neutralized to jump of pH value. Its unit is mg/g). Generally, acid value of crude oil shall be larger than 0.2 mg/g. It can be seen from the statistical results (Figure 10-8) that the larger the acid value of crude oil, the higher the density is, as well as viscosity and water-oil mobility ratio. Therefore, requirements of alkaline flooding for acid value and viscosity of the crude oil are inconsistent.

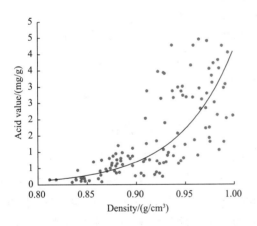

Figure 10-8   Statistical relationship between acid value and density of crude oil

Acid value in a certain range is a necessary condition but not a sufficient condition for alkaline flooding. There is no consistent relationship between enhancement extent of oil recovery by alkaline flooding and acid value of the crude oil. Crude oil with acid value less than 0.2mg/g is not suitable to conduct alkaline flooding. However, crude oil with high acid value may not necessarily show better flooding effect than those with lower acid value. This is because product from reaction between acidic component of the oil and alkali may not necessarily be interfacial active.

10.3 Surfactant flooding

10.3.1 Definition of surfactant flooding

Surfactant flooding refers to oil displacement method that using surfactant system as oil displacing agent.

Surfactant systems for oil displacement include dilute surfactant systems (such as active water or micellar solutions) and concentrated surfactant systems (such as upper phase microemulsion, middle-phase microemulsion and lower phase microemulsion). Since foams and emulsions involved in foam flooding and emulsion flooding are for stable surfactant flooding, these two types of flooding are also included in surfactant flooding.

The following oil displacement methods are included by surfactant flooding:

10.3.1.1 Active water flooding

The oil displacement method using active water as displacement medium is called active water flooding. Active water belongs to dilute surfactant system. The surfactant concentration in it is less than the critical micelle concentration. There is no need to add any promoter in active water other than surfactants.

10.3.1.2 Micelle solution flooding

Micelle solution also belongs to dilute surfactant system. The surfactant concentration in it is greater than the critical micelle concentration, but its mass fraction will not exceed 2×10^{-2}.

Since surfactant concentration is greater than the critical micelle concentration, there are micelles formed. Under condition of surfactant concentration exceeds critical micelle concentration, the micelles are

10.3 表面活性剂驱

10.3.1 表面活性剂驱定义

以表面活性剂体系作为驱油剂的驱油法叫表面活性剂驱。

驱油用的表面活性剂体系有稀表面活性剂体系（如活性水、胶束溶液）和浓表面活性剂体系（如上相微乳、中相微乳和下相微乳）。由于泡沫驱、乳状液驱是用表面活性剂稳定驱油用的泡沫和乳状液，所以也包括在表面活性剂驱之中。

表面活性剂驱包括下列的驱油法：

10.3.1.1 活性水驱

以活性水作为驱替介质的驱油方法叫活性水驱。活性水属稀表面活性剂体系，其中的表面活性剂浓度小于临界胶束浓度。活性水中，除表面活性剂外，不再加入其他助剂。

10.3.1.2 胶束溶液驱

胶束溶液也属于稀表面活性剂体系，其中表面活性剂浓度大于临界胶束浓度，但其质量分数不超过 2×10^{-2}。

由于表面活性剂浓度大于临界胶束浓度，所以存在胶束。在表面活性剂浓度超过临界胶束浓度不多的条件下，胶束大多呈球

mostly spherical. However, if surfactant concentration is 10 times higher than critical micelle concentration, shapes of the micelles are generally non-spherical and varied, such as rod-like, hexagonal rod-like or layered, etc. In addition, other than surfactants, in order to reduce interfacial tension between micelle solution and the oil, alcohols and salts are often added in micelle solution.

10.3.1.3 Microemulsion flooding

Micro emulsion belongs to concentrated surfactant system (mass fraction of surfactant exceeds 2×10^{-2}). It consists of five components: water, oil, surfactant, co-surfactant (such as alcohol) and salts. There are two basic types and one transition type concluded into micro emulsion. The former includes water external phase micro emulsion and oil external phase micro emulsion, and the latter is middle-phase micro emulsion. If condition changes, the system could transform from water external phase to oil external phase, or conversely from oil external phase to water external phase. During this process, it will go through a transitional stage called middle-phase micro emulsion (Figure 10-9).

10.3.1.4 Foam flooding

Foam flooding is a method of oil recovery enhancement that using foams as oil displacing agent. Its main components are water, gas and foaming agent.

形，但当表面活性剂浓度高于10倍于临界胶束浓度时，胶束一般是非球形的，如形成棒状胶束、棒状胶束六角束、层状胶束等。此外，为了降低胶束溶液与油之间的界面张力，在胶束溶液中，除表面活性剂外，通常还需加入醇和盐。

10.3.1.3 微乳驱

微乳属浓表面活性剂体系（表面活性剂的质量分数超过2×10^{-2}），由水、油、表面活性剂、助表面活性剂（如醇）和盐五种组分组成。它有两种基本类型和一种过渡类型。前者为水外相微乳和油外相微乳，后者为中相微乳。由于条件变化，当体系由水外相微乳转变为油外相微乳时，或相反，由油外相微乳转变为水外相微乳时，经过中相微乳这一过渡阶段（图10-9）。

10.3.1.4 泡沫驱

泡沫驱是以泡沫作为驱油剂的一种提高采收率方法，其主要成分是水、气和起泡剂。矿场使

Figure 10-9　Transformation of micro emulsion type

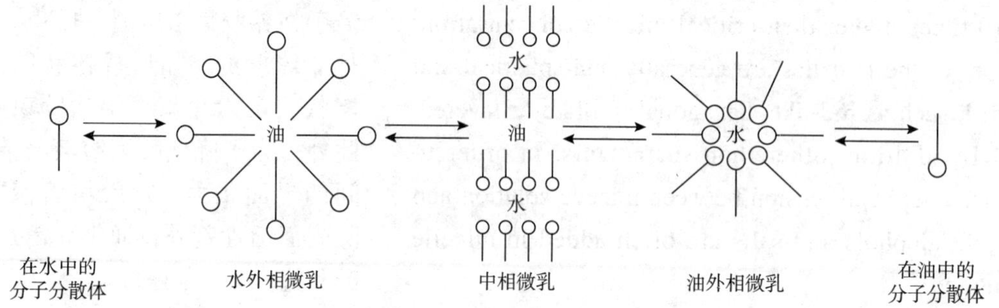

图10-9 微乳类型的转变

To apply foam flooding on field application, foaming agent solution and gas shall be alternatively injected into the reservoir, or inject them simultaneously into the reservoir through oil pipe and casing. Forming mechanism of foams in the reservoir mainly includes: extruding, liquid-membrane separation and liquid-membrane lag. Forming mechanism of foams is shown in Figure 10-10.

用时，交替向储层中注入起泡剂溶液和气体，也可分别通过油管和套管将二者同时注入储层。储层中泡沫的生成机理主要包括：卡断、液膜分离和液膜滞后，泡沫生成机理如图10-10所示。

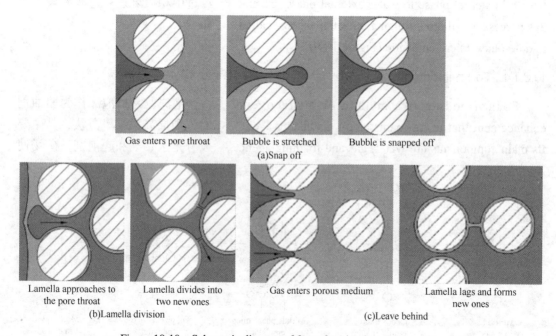

Figure 10-10 Schematic diagram of foam forming in porous media

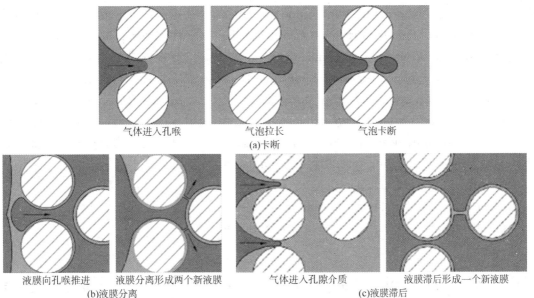

图10-10 多孔介质中泡沫生成机理示意图

Water used to prepare foams can be fresh water or saline solution. Gas options for foam forming include nitrogen, carbon dioxide, natural gas, refinery gas or flue gas. And the foaming agent is mainly surfactants, such as alkylsulfonate, alkylbenzene sulfonate, polyoxyethylene alkyl alcohol ether-15, polyoxyethylene alkyl phenol ether-10, polyoxyethylene alkyl alcohol ether sulfate, polyoxyethylene alkyl alcohol ether carboxylate and so on.

In addition to all mechanisms of surfactant flooding, foam flooding is also with two mechanisms for oil recovery enhancement.

(1) Increasing sweep efficiency of displacement medium by superposition of the Jamin effect;

(2) Bubbles in oil displacement can be deformed according to shape of the channel. This will effectively displace oil in pores and improve the oil displacement efficiency.

10.3.1.5　Emulsion flooding

Emulsion flooding refers to the method that injecting oil-in-water emulsion formed by emulsifier

配制泡沫的水可用淡水，也可用盐水。配制泡沫的气体可用氮气、二氧化碳、天然气、炼厂气或烟道气。配制泡沫用的起泡剂主要是表面活性剂，如烷基磺酸盐、烷基苯磺酸盐、聚氧乙烯烷基醇醚–15，聚氧乙烯烷基苯酚醚–10、聚氧乙烯烷基醇醚硫酸酯盐、聚氧乙烯烷基醇醚羧酸盐等。

除具有表面活性剂驱的所有机理外，泡沫驱还有两个提高采收率机理。

（1）通过贾敏效应的叠加，提高驱替介质的波及系数；

（2）驱油泡沫中的气泡，可依孔道的形状而变形，能有效地将波及到孔隙中的油驱出，提高洗油效率。

10.3.1.5　乳状液驱

乳状液驱油是指将乳化剂（表面活性剂、碱、助剂）与原油在

(surfactant, alkali or promoter) and crude oil under high-speed agitation into the formation to displace the remaining oil after water flooding. Its main mechanisms for oil recovery enhancement are: Shunting effect generated from plugging the large pore throat by emulsion will increase the sweep efficiency, so that residual oil formed by detouring flow will be reduced; by squeezing and scraping laterally, the emulsion will improve oil displacement efficiency and reduce saturation of residual oil at edge. In addition, emulsion is also with important application value in profile control and water shutoff.

10.3.2　Mechanism of oil recovery enhancement by surfactant flooding

Mechanism of oil recovery enhancement by surfactant flooding is roughly the same as that of alkaline flooding. The former is effected by direct addition of surfactant, while the latter is effected by surfactant that being formed by acidic component of the crude oil and an alkali.

Mechanisms for recovery factor enhancement of various surfactant systems are as following.

10.3.2.1　Mechanism of active water flooding

Active water flooding mainly enhances oil recovery by the following mechanisms:

(1) Mechanism of low interfacial tension.

By adsorbing at oil-water interface, the surfactant could lower oil-water interfacial tension. As the oil-water interfacial tension decreases, its adhesion work will also decrease, which means oil will be easily washed off from the formation surface. In this way, oil displacement efficiency is improved.

(2) Mechanism of wettability alteration.

Surfactants used for oil displacement are with better hydrophilicity than their lipophilicity. By adsorbing on the formation surface, it can reverse the

高速搅拌下形成的水包油型乳状液注入地层，驱替水驱后残余油的驱替方式。其提高采收率的机理主要有：利用乳状液堵塞大孔喉后产生的分流作用提高波及系数，使绕流形成的残余油减少；通过乳状液侧向挤油、刮油改善洗油效率，降低边缘残余油饱和度。此外，乳状液在调剖堵水方面也有重要的应用价值。

10.3.2　表面活性剂驱提高采收率机理

表面活性剂驱提高采收率的原因与碱驱大体相同，因前者是通过外加表面活性剂起作用，后者则是通过碱与原油酸性成分就地生成表面活性剂起作用。

各种表面活性剂体系提高采收率作用机理如下所述。

10.3.2.1　活性水驱机理

活性水驱主要通过下列机理提高采收率：

（1）低界面张力机理。

表面活性剂吸附在油水界面，可降低油水界面张力。随油水界面张力降低，黏附功减小，即油易从地层表面洗下来，提高了洗油效率。

（2）润湿反转机理。

驱油用表面活性剂的亲水性均大于亲油性，它在地层表面吸附，可使亲油的地层表面（由天然表面活性物质通过吸附或沉积

formation surface from lipophilic to hydrophilic (the lipophilic formation surface is formed by adsorption or deposition of natural surface active substance). Increased contact angle of the oil on rock surface (Figure 10-11) reduces the adhesion work and improves oil displacement efficiency.

形成）反转为亲水，油对岩石表面的润湿角增加（图 10-11），减小了黏附功，也提高了洗油效率。

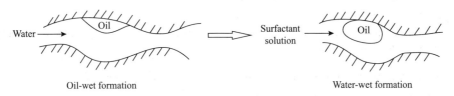

Figure 10-11　Surfactant reverses wettability of the formation surface

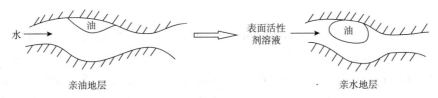

图10-11　表面活性剂使地层表面润湿反转

(3) Mechanism of emulsification.

HLB value of the surfactants used for flooding is generally in range of 7-18. It adsorbs at the oil-water interface and can stabilize the oil-in-water emulsion. In this case, after being emulsified, the crude oil will not be easily adhered back on surface of the formation in process of moving forward, which will improve the oil displacement efficiency. Moreover, the emulsified oil will produce superimposed Jamin effect in high permeable layer, allowing water be propelled more uniformly in the formation, increasing the sweep efficiency.

(4) Mechanism of density improvement of the surface charges.

When the surfactant used in oil displacement is an anionic surfactant, it can be adsorbed on oil beads and the formation surface, which will increase density of the surface charges (Figure 10-12). Increase of the electrostatic repulsion between oil beads and formation surface makes the oil beads easy to be carried by

（3）乳化机理。

驱油用表面活性剂的 HLB 值一般在 7~18，其吸附在油水界面上，可稳定水包油乳状液。原油乳化后再向前移动中不易重新黏附回地层表面，可提高洗油效率。而且，乳化的油在高渗透层产生叠加的贾敏效应，使水较均匀地在地层推进，提高了波及系数。

（4）提高表面电荷密度机理。

当驱油所用表面活性剂为阴离子型表面活性剂时，其在油珠和地层表面上吸附，可提高表面电荷密度（图10-12），增加油珠与地层表面之间的静电斥力，使油珠易于被驱替介质携带，从而

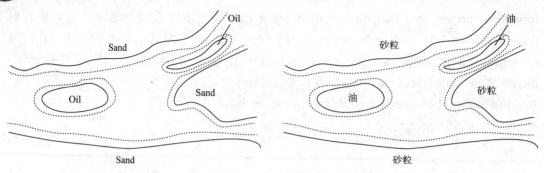

Figure 10-12 Effect of increasing surface charge density during oil displacement

图10-12 驱油过程中提高表面电荷密度的作用

displacement medium, thereby improving the oil displacement efficiency.

(5) Mechanism of coalescence and oil zones formation.

After more and more oil is washed from the formation surface, the crude oil will collides with each other during forward movement. When energy of the collision is sufficient to overcome repulsive energy between them caused by electrostatic repulsion, the oil beads may coalesce. Coalescence of the oil beads will form oil zones (Figure 10-13). As the oil zone moves forward, dispersed oil beads will be continuously collected. In this case the oil zone will be expanded (Figure 10-14) continuously, and finally be recovered from the well.

提高了洗油效率。

（5）聚并形成油带机理。

当从地层表面洗下来的油越来越多时，原油在向前移动过程中会发生相互碰撞。当碰撞的能量足以克服它们之间由静电斥力产生的相斥能量时，油滴间可发生聚并。油滴的聚并可形成油带（图10-13）。油带向前流动时，不断将遇到的分散的油滴聚并过来，使油带不断扩大（图10-14），最后从油井采出。

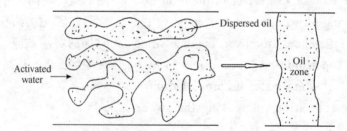

Figure 10-13 Displaced oil is coalesced to oil zone

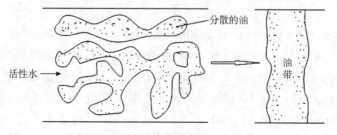

图10-13 被驱替的油滴聚并为油带

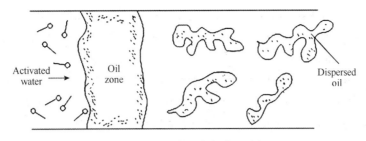

Figure 10-14 Oil zone expands in forward movement

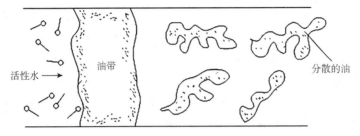

图10-14 油带在向前移动中不断扩大

10.3.2.2 Mechanism of micelle solution flooding

Comparing with active water, there are two distinctive features of micelle solution: One is that since surfactant concentration is larger than critical micelle concentration, micelles can be found in the solution; the other is there are co-surfactants like alcohols and salts added beside surfactants in the micelle solution. Mechanism of micelle solution flooding is same as active water flooding. This is because micelles can solubilize oil and improve oil displacement efficiency of the micelle solution. In addition, promoters like alcohols and salts will adjust polarity of oil and water phase, making lipophilicity and hydrophilicity of the surfactant in sufficient balance. Therefore, it can adsorb at oil-water interface to the maximum extent, resulting in ultra-low interfacial tension and strengthen the low interfacial tension mechanism of micelle solution flooding.

In micelle solution flooding, mass fraction of surfactant shall be less than 0.02, and there are very few micelles in the solution. Therefore, the solubilization mechanism is not important, but the ultra-low interfacial tension between micelle solution and oil is important because it makes a qualitative leap in oil displacement efficiency enhancement.

The significance of ultra-low interfacial tension in micelle solution flooding on oil displacement improving can be reflected by capillary number. Under condition of water flooding, the capillary number is generally 10^{-6}. It is impossible to increase capillary number from 10^{-6} to 10^{-2} only by increase viscosity and flow rate of the drive liquid, but this goal could be achieved by reducing interfacial tension between drive liquid and the oil.

In order to increase capillary number from 10^{-6} to 10^{-2} under water flooding conditions, the interfacial tension between oil displacement medium and oil needs to be reduced by four magnitude orders. During water flooding, interfacial tension between oil and water is in magnitude order range of 10mN/m, so interfacial tension between oil displacement medium and the oil needs to be reduced to 10^{-3}mN/m (i.e. less than 10^{-2}mN/m). Then the requirement of increased the capillary number by four magnitude orders could be met.

Figure 10-15 is a typical diagram of relation between oil-water interfacial tension and mass fraction of surfactant. It can be seen from the figure that in

超低界面张力对提高胶束溶液洗油能力的意义可通过毛管数说明。水驱条件下,毛管数一般为10^{-6}。要将毛管数由10^{-6}提高到10^{-2},仅靠提高驱动液黏度和驱动液流速是达不到的,但可通过减小驱动液与油之间的界面张力达到。

为使毛管数由水驱条件下的10^{-6}提高到10^{-2},需要将驱油介质与油之间的界面张力降低四个数量级。水驱时,油水的界面张力在10mN/m的数量级范围,因此需将驱油介质与油之间的界面张力降到10^{-3}mN/m(即低于10^{-2}mN/m),即可达到毛管数提高四个数量级的要求。

图10-15为一个典型的油水界面张力与表面活性剂质量分数的关系图。由图可以看到,在一个加有醇和盐的胶束溶液中,当ω(石油磺酸盐)$=0.1\times10^{-2}$时,界面张力可低至2.6×10^{-4}mN/m,从而使这种胶束溶液在驱油时的

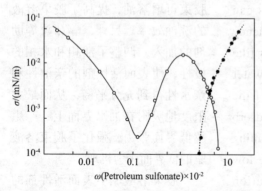

Figure 10-15 A typical diagram of the relation between oil-water interfacial tension and surfactant mass fraction

Condition: Surfactant: TRS10-410 (a petroleum sulfonate); Alcohol: IBA (isobutanol); Oil phase: Dodecane; Water phase: TRS10-410+IBA+NaCl ω(NaCl)=0.015 m(TRS10-410)/m(IBA)=5/3

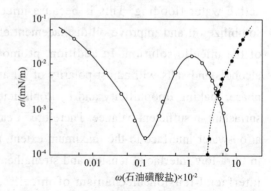

图10-15 一个典型的油水界面张力与表面活性剂质量分数关系图

条件:表面活性剂:TRS10-410(一种石油磺酸盐);醇:IBA(异丁醇);油相:十二烷;水相:TRS10-410+IBA+NaCl ω(NaCl)=0.015 m(TRS10-410)/m(IBA)=5/3

a micelle solution added alcohol and salt, when ω (petroleum sulfonate) $= 0.1 \times 10^{-2}$, the interfacial tension can be as low as 2.6×10^{-4} mN/m. Therefore, during oil displacement capillary number of the micelle solution will exceed 10^{-2}, thus with excellent oil displacement effect.

10.3.2.3 Mechanism of micro emulsion flooding

The micro emulsion flooding is with all mechanisms of micelle solution flooding, namely:

(1) Mechanism of low interfacial tension;
(2) Mechanism of wettability alteration;
(3) Mechanism of emulsification;
(4) Mechanism of solubilization;
(5) Mechanism of density improvement of the surface charges;
(6) Mechanism of coalescence and oil zones forming.

Since micro emulsion belongs to concentrated surfactant system, micro emulsion flooding is more prominent than micelle solution flooding in mechanisms of solubilization and surface charges density improvement.

10.3.3 Surfactants that could be used for surfactant flooding

Surfactant that could be used for surfactant flooding should meet the following requirements: low interfacial tension, low adsorption capacity, high solubilization parameters, salt tolerance, temperature resistance, extensive supply sources, low cost. It shall be compatible with formation fluids also.

There are four types of surfactants that being used mainly:

10.3.3.1 Sulfonate surfactant

Petroleum sulfonate is one of the important class in sulfonate surfactants. It can be prepared according to the following process: sulfonating petroleum with high

毛管数超过 10^{-2}，因而有优异的洗油能力。

10.3.2.3 微乳驱机理

微乳驱有胶束溶液驱的全部机理，即：

（1）低界面张力机理；
（2）润湿反转机理；
（3）乳化机理；
（4）增溶机理；
（5）提高表面电荷密度机理；
（6）聚并形成油带机理。

由于微乳属浓表面活性剂体系，所以微乳驱在增溶机理和提高表面电荷密度机理上比胶束溶液驱更突出。

10.3.3 表面活性剂驱用的表面活性剂

表面活性剂驱用的表面活性剂要满足下列条件：低界面张力、低吸附量、高增溶参数、与地层流体配伍、耐盐、耐温、来源广、成本低等。

主要用的表面活性剂有四类：

10.3.3.1 磺酸盐型表面活性剂

石油磺酸盐是一类重要的磺酸盐型表面活性剂，它可用磺化剂（如三氧化硫）将芳香烃含量

aromatic content or petroleum fractions by sulfonating agent (such as sulfur trioxide) and then neutralize the sulfonation product with alkali:

$$RArH + SO_3 \longrightarrow RArSO_3H$$
$$RArSO_3H + NaOH \longrightarrow RArSO_3Na + H_2O$$

Synthetic sulfonate is also an important class in sulfonate surfactants. It could be prepared by synthetic methods with corresponding hydrocarbons (e.g. alkanes, alkylbenzenes, alkyltoluenes or alkylxylenes, etc.). Alkylsulfonates, alkyl aryl sulfonates, α-olefin sulfonates and the like belong to sulfonates of this type.

Among all synthetic sulfonates, the α-olefin sulfonates show particularly good performance on salt-tolerance and high-valent metal ions tolerance.

10.3.3.2　Carboxylate surfactant

Petroleum carboxylate is an important class of carboxylate surfactants. It can be obtained from saponifying petroleum carboxylic acid with alkali. The petroleum carboxylic acid could be prepared through gas phase oxidation of petroleum fractions. However, performance of salt-tolerant and high-valent metal ions resistant of carboxylate surfactant is not as good as sulfonate surfactant.

10.3.3.3　Nonionic surfactant

Nonionic surfactant mainly includes the polyoxyethylene nonionic surfactants. It can be obtained from polyoxyethylation by alkanol, alkylphenols or sorbitan fatty acid ester. The frequently used types are:
(1) Peregal surfactant;
(2) OP surfactant;
(3) Tween surfactant.

高的石油或石油馏分磺化再用碱中和制成：

合成磺酸盐是另一类重要的磺酸盐型表面活性剂。可由相应的烃类（如烷烃、烷基苯、烷基甲苯、烷基二甲苯等）用相应的合成方法制得。烷基磺酸盐、烷基芳基磺酸盐、α-烯烃磺酸盐等属这一类磺酸盐。

在合成磺酸盐中，α-烯烃磺酸盐特别耐盐、耐高价金属离子。

10.3.3.2　羧酸盐型表面活性剂

石油羧酸盐是一类重要的羧酸盐型表面活性剂，它由石油馏分通过气相氧化得石油羧酸，再用碱皂化石油羧酸得石油羧酸盐。但是，羧酸盐型表面活性剂的耐盐、耐高价金属离子的能力远不如磺酸型表面活性剂。

10.3.3.3　非离子型表面活性剂

主要有聚氧乙烯基的非离子型表面活性剂。它由烷基醇、烷基苯酚或山梨糖醇酐脂肪酸酯聚氧乙烯化制得。常用的有：
（1）平平加型表面活性剂；
（2）OP 型表面活性剂；
（3）Tween 型表面活性剂。

10.4 Combination flooding

10.4.1 Definition of combination flooding

Combination flooding is a flooding that combines two or more types of oil displacement components. Oil displacement components mentioned in this section are the main agents (polymer, alkali and surfactant) in chemical flooding. Combination flooding could be composed by various different ways. For example, alkali + polymer is called intensified alkaline flooding or alkali enhanced polymer flooding, surfactant + polymer is called thickened surfactant flooding or surfactant enhanced polymer flooding, alkali + surfactant is called alkali-enhanced surfactant flooding or surfactant-enhanced alkaline flooding, and alkali (A) + surfactant (S) + polymer (P) is called ASP flooding. A pseudo-ternary phase diagram can be used to illustrate the various combinations in a chemical flooding (Figure 10-16).

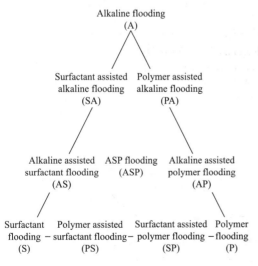

Figure 10-16 Combination of various displacement methods in chemical flooding

10.4 复合驱

10.4.1 复合驱定义

复合驱是指两种或两种以上驱油成分组合起来的驱动。本节所讲的驱油成分是指化学驱中的主剂（聚合物、碱、表面活性剂），可按不同的方式组成各种复合驱，如碱＋聚合物叫稠化碱驱或碱强化聚合物驱；表面活性剂＋聚合物叫稠化表面活性剂驱或表面活性剂强化聚合物驱；碱＋表面活性剂叫碱强化表面活性剂驱或表面活性剂强化碱驱；碱（A）＋表面活性剂（S）＋聚合物（P）叫 ASP 三元复合驱。可用准三组分相图表示化学驱中各种驱动的组合（图10-16）。

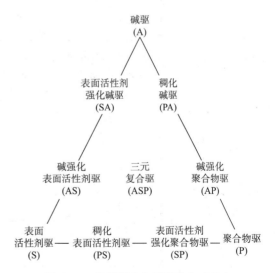

图10-16 化学驱中各种驱替方式组合

Figure 10-17 is segment diagram of an ASP flooding. In order to reduce the effects of exchangeable cations such as Ca^{2+}, Mg^{2+}, the formation shall be pre-washed by saline solution (sodium chloride). Sacrificial solution injected before the ASP system is intended to reduce consumption of various chemical agents in the ASP system. The optional sacrificial agents includes alkali substances (sodium carbonate), polycarboxylic acids (oxalic acid), oligomers (polyethylene glycol) and modified lignosulfonates (sulfonated lignin sulfonate), which will protect the ASP system by mechanisms such as reaction with Ca^{2+}, Mg^{2+} or competitive adsorption. Polymer solution injected behind the ASP system is used as control agent of mobility, which will ensure the ASP system passing through formation smoothly.

图 10-17 是 ASP 三元复合驱的段塞图。为了减小地层 Ca^{2+}、Mg^{2+}等可交换阳离子的影响，用盐水（氯化钠）预冲洗地层。注在 ASP 体系前的牺牲剂溶液是为了减少 ASP 体系中各种化学剂的消耗。可用的牺牲剂如碱性物质（碳酸钠）、多元羧酸（草酸）、低聚物（聚乙二醇）和改性木质素磺酸盐（磺甲基化木质素磺酸盐），它们通过与 Ca^{2+}、Mg^{2+}反应或竞争吸附等机理保护 ASP 体系。注在 ASP 体系后面的聚合物溶液时流度控制剂，它可使 ASP 体系平稳地通过地层。

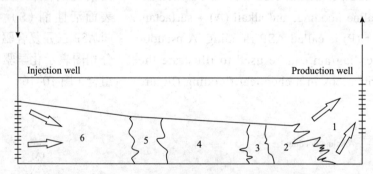

Figure 10-17　Segment diagram of ASP flooding
1-Residual oil; 2-Saline solution; 3-Sacrificial agent solution; 4-ASP system;
5-Polymer solution; 6-Water

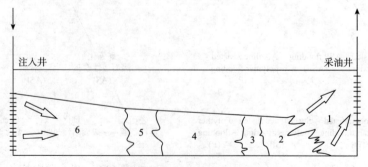

图10-17　ASP三元复合驱段塞图
1—剩余油；2—盐水；3—牺牲剂溶液；4—ASP体系；5—聚合物溶液；6—水

10.4.2 Advantages of combination flooding

Under normal conditions, combination flooding is with higher recovery factors than single flooding. Figure 10-18 shows the comparison on recovery factors after being conducted alkali / polymer flooding, alkaline flooding, polymer flooding, polymer flooding after alkaline flooding, and alkaline flooding after polymer flooding. It can be seen from the figure that recovery factor of the remaining oil of the alkali/ polymer flooding is 5 times than that of alkaline flooding and 3 times than that of polymer flooding.

10.4.2 复合驱的优点

复合驱通常比单一的驱动有更高的采收率。图 10-18 是碱+聚合物驱与单纯碱驱、单纯聚合物驱、先碱驱后聚合物驱和先聚合物驱后碱驱的比较。从图中可以看到，碱+聚合物驱的剩余油采收率是碱驱的 5 倍，是聚合物驱的 3 倍。

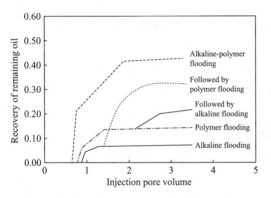

Figure 10-18 Comparison of flooding methods
(Crude oil viscosity is 180mPa·s, polymer is PAM, alkali is Na_4SiO_4)

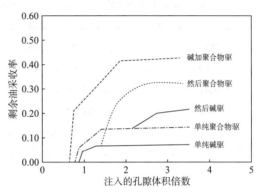

图10-18 驱动方式的对比
（原油黏度为180mPa·s，聚合物为PAM，碱为Na_4SiO_4）

Since polymer, surfactant and alkali in the combination flooding will produce synergistic effect and optimize their effect in flooding, the combination flooding will show better performance on oil displacement.

10.4.2.1 Effect of polymer

(1) Improve mobility ratio of surfactant and (or) alkali solution to oil.

(2) By thickening the oil displacement medium, it could slow diffusion speed of surfactant and alkali down, thereby reducing their consumption.

(3) Could react with calcium and magnesium ions

复合驱之所以有更好的驱油效果，主要是由于复合驱中的聚合物、表面活性剂和碱之间有协同效应，它们在协同效应中起各自的作用。

10.4.2.1 聚合物的作用

（1）改善了表面活性剂和（或）碱溶液对油的流度比。

（2）对驱油介质的稠化，可减小表面活性剂和碱的扩散速度，从而减小它们的药耗。

（3）可与钙、镁离子反应，

to protect surfactant, making it difficult to form low surface active calcium and magnesium salts.

(4) Could improve stability of the oil-in-water emulsion formed by alkali and surfactant, thereby increasing sweep efficiency (according to the emulsification -capture mechanism) and (or) displacement capacity (according to the emulsification -carrying mechanism) greatly.

10.4.2.2 Effect of surfactant

(1) Could reduce interfacial tension between polymer solution and oil, thus increase the displacement capacity.

(2) Could emulsify oil and increasing viscosity of the oil displacement medium. The more oil being emulsificd, the higher the viscosity of the emulsion will be.

(3) If surfactant forms complexation structure with polymer, the surfactant could increase viscosity-enhancing capacity of the polymer.

(4) Could supplement deficiency of the surfactant which was produced by reaction of alkali and petroleum acid.

10.4.2.3 Effect of alkali

(1) Could improve thickening capacity of polymer (HPAM).

(2) Could react with petroleum acid and produce surfactant. The surfactant will then emulsify oil and increase viscosity of oil displacement medium, thereby improving mobility control capacity of the polymer.

(3) Surfactant produced by reaction of alkali and petroleum acid will generate synergistic effect with the synthetic surfactants.

(4) Could react with calcium or magnesium ions or exchange ions with clay. In this way it could be used as sacrificial agent and protect polymers and surfactants.

(5) Could increase amount of negative charges at sandstone surface. By doing this, it will reduce adsorption of polymer and surfactant on sandstone surface.

保护了表面活性剂，使它不易形成低表面活性的钙、镁盐。

（4）提高了碱和表面活性剂所形成的水包油乳状液的稳定性，使波及系数（按乳化－捕集机理）和（或）洗油能力（按乳化－携带机理）有较大的提高。

10.4.2.2 表面活性剂的作用

（1）可以降低聚合物溶液与油的界面张力，提高它的洗油能力。

（2）可使油乳化，提高了驱油介质的黏度。乳化的油越多，乳状液的黏度越高。

（3）若表面活性剂与聚合物形成络合结构，则表面活性剂可提高聚合物的增黏能力。

（4）可补充碱与石油酸反应产生表面活性剂的不足。

10.4.2.3 碱的作用

（1）可提高聚合物（HPAM）的稠化能力。

（2）与石油酸反应产生的表面活性剂，可将油乳化，提高了驱油介质黏度，因而加强了聚合物控制流度的能力。

（3）与石油酸反应产生的表面活性剂与合成的表面活性剂有协同效应。

（4）可与钙、镁离子反应或与黏土进行离子交换，起牺牲剂作用，保护了聚合物与表面活性剂。

（5）可提高砂岩表面的负电性，减少砂岩表面对聚合物和表

(6) Alkali could improve biological stability of biopolymers.

Due to interaction of each component, the flooding efficiency of combination flooding is higher than other flooding types. It is also with advantages of small chemical agent consumption and low cost.

面活性剂的吸附量。

（6）碱可提高生物聚合物的生物稳定性。

由于各成分的相互作用，因此复合驱的驱油效率高，化学剂消耗少，成本降低。

第11章 原油破乳与起泡原油的消泡
Crude oil demulsification and defoaming of foamed crude oil

11.1 Crude oil emulsion

11.1.1 Generation of crude oil emulsion

(1) The intrinsic factor of crude oil emulsion generation is that there are water and sufficient natural emulsifier contained in crude oil.

(2) Corrosion inhibitors, bactericides, wetting agents, and various production enhancement chemical agents that are often used in petroleum production are all emulsifiers that could promote generation of the emulsions.

(3) Various enhanced oil recovery methods could help to generate stable emulsions.

(4) Some changes in mine shaft and ground gathering system will promote generation and stability of the emulsion. These changes include sudden pressure drop, associated gas emission, pressure increase on oil and water caused by pumping, the pigging and the oil-gas mixing delivery, etc.

11.1.2 Classification of crude oil emulsion

There are mainly two types of emulsions composed of crude oil and water: water-in-oil type and oil-in-water type. In addition, there are compound emulsions, i.e. oil-in-water-in-oil type and water-in-oil-in-water type. Oil-in-water-in-oil (O/W/O type) compound emulsions will be generated from polymer flooding oil production

11.1 原油乳状液

11.1.1 原油乳状液的生成

（1）原油中含水并含有足够数量的天然乳化剂是生成原油乳状液的内在因素。

（2）石油生产中经常使用的缓蚀剂、杀菌剂、润湿剂和强化采油的各种化学剂等都是促使生成乳状液的乳化剂。

（3）各种强化采油方法会促使生成稳定的乳状液。

（4）井筒和地面集输系统内的压力骤降、伴生气析出、泵对油水增压、清管、油气混输等都会促使乳状液形成和稳定。

11.1.2 原油乳状液的分类

原油和水构成的乳状液主要有油包水型和水包油型两类乳状液。此外，还有复合乳状液，即油包水包油型、水包油包水型。聚合物驱采油常产生油包水包油型（O/W/O型）复合乳状液。

Crude oil demulsification and defoaming of foamed crude oil
原油破乳与起泡原油的消泡

第11章

frequently.

11.1.3 Harm of crude oil emulsion

(1) It will increase volume of liquid flow and reduce effective utilization of the equipment and piping.

(2) It will increase power consumption during transport.

(3) During process of heating, the fuel consumption will be increased under its influence.

(4) It will cause scaling and corrosion in metal pipes and equipment.

(5) It will adversely affect the normal refinery processing.

11.2 Crude oil demulsifier

Destruction of the emulsion is called demulsification. The demulsification process of crude oil emulsion consists of a series of steps including distance reduce of the dispersed water droplets, break of the interfacial film, combination and diameter increase of the water droplets, sedimentation and separation in oil phase, and so on. It is often referred as flocculation, coalescence and sedimentation of the water droplets. Interfacial film at external side of the dispersed phase is with high mechanical strength and could prevent combination and sedimentation of water droplets. How to destroy oil-water interfacial film becomes the key of crude oil demulsification since it could cause water droplets to coalesce and sediment.

The most frequent used agents for oilfield emulsion treatments are chemical demulsifiers. Although both of demulsifiers and emulsifiers are surfactants, effects of the two are the opposite. There were patents for demulsifier as early as the year of 1914. Hundreds of years passed, dosage of the demulsifier was dropped from around 1000 mg/L to not over several dozen mg/L. The technology has been rapidly developed, but research

11.1.3 原油乳状液的危害

（1）增大了液流的体积，降低了设备和管道的有效利用率。

（2）增加了输送过程中的动力消耗。

（3）增加了升温过程中的燃料消耗。

（4）引起金属管道、设备的结垢和腐蚀。

（5）对炼油厂加工过程的影响。

11.2 原油破乳剂

乳状液的破坏称破乳。原油乳状液的破乳过程是由分散水滴相互接近结合在一起、界面膜破裂、水滴合并粒径增大、在油相中沉降分离等一系列环节组成，常称为水滴的絮凝、聚结和沉降。由于分散相外围的界面膜有较高机械强度阻止水滴的合并沉降，所以原油乳状液的关键是破坏油水界面膜，使水滴聚结和沉降。

各油田乳状液处理中普遍使用化学破乳剂，破乳剂和乳化剂都是表面活性剂，但两者的作用却截然相反。1914年就有破乳剂专利，至今已有近百年历史，添加的剂量由早期1000mg/L左右降至现今的几到几十 mg/L，技术上有了迅速发展。但对破乳过程和破乳机理的研究仍处于较低水平。

on process and mechanism of demulsification is still at a low level.

11.2.1 Classification of crude oil demulsifier

Demulsifiers are often classified according to molecular structure, molecular weight, method of mosaic, polymerization sections, number of active hydrogen functional group in the starting agent, solubility, compound type, etc.

11.2.1.1 Classification according to number of polymerization sections

(1) Diblock copolymer: Such as AE8025, AE8051, and AE8031 from China.

The most widely used chemical demulsifier at home and abroad is nonionic polyoxyethylene polyoxypropylene ether presently. During synthesis of nonionic demulsifiers, the starting agent (containing active hydrogen) and a certain proportion of propylene oxide (PO) are first formulated into "lipophilic end" (this is the first section). Then a certain amount of ethylene oxide (EO) shall be polymized, which is the second section. That is why this product is called a diblock chemical demulsifier (in Chinese "diblock" means "two sections"), namely: lipophilic end (PO)m(EO)nH.

(2) Triblock copolymer.

Polymerize a propylene oxide section on diblock copolymer, a triblock copolymer demulsifier will be generated, i.e. lipophilic end (PO)m(EO)n(PO)zH, such as SP169, AP221, AP134 and AP3111 from China.

11.2.1.2 Classification according to solubility

According to solubility, the nonionic demulsifiers can be classified into three types: water-soluble, oil-soluble and partially water-soluble, as well as partially oil-soluble.

11.2.1 原油破乳剂的分类

破乳剂常按分子结构、分子量大小、镶嵌方式、聚合段落、起始剂具有活泼氢官能团的数量、溶解性能、化合物类别等进行分类。

11.2.1.1 按聚合段数划分

（1）二嵌段聚合物：如我国的 AE8025、AE8051、AE8031 等。

目前，国内外使用最多的化学破乳剂为非离子型聚氧乙烯聚氧丙烯醚。在非离子型破乳剂的合成过程中，将起始剂（含有活泼氢）与一定比例的环氧丙烷（PO）先配制成"亲油头"（此为第一段），然后接聚上一定数量的环氧乙烷（EO），此为第二段，这种产品就叫作二嵌段化学破乳剂，即：亲油头（PO）m（EO）nH。

（2）三嵌段聚合物。

在二嵌段聚合物的基础上再接聚一段环氧丙烷，即为三嵌段式破乳剂：油头（PO）m（EO）n（PO）zH。如我国的 SP169、AP221、AP134 和 AP3111 等。

11.2.1.2 按溶解性能划分

根据溶解性能，非离子型破乳剂可分为水溶性、油溶性和部分溶解于水、部分溶解于油三类。

11.2.1.3 Classification according to molecular weight

(1) Low molecular weight demulsifiers: Relative molecular mass<1000, such as inorganic acids, alkalis, salts, carbon disulfide, carbon tetrachloride, alcohols, phenols, ethers, etc.

(2) High molecular weight demulsifiers: Nonionic polyoxyethylene polyoxypropylene ether with relative molecular mass in range of 1,000~10,000, such as the famous demulsifiers of West Germany Dissolan4400, 4411, 4422, 4433 and so on. Chinese demulsifiers with model of AE, AP, BP, RA, etc. are also belonging to this category.

(3) Ultra high relative molecular weight demulsifiers: By taking the starting agents with multi reactive groups and cross-linking agents, or change catalysts, relative molecular mass of the polyether could reach to a range of tens of thousands to millions. As relative molecular mass increases, the dehydration effect will increase also.

11.2.2 Effect of demulsifier

During demulsification, the effects of demulsifiers are as follows:

(1) Demulsifiers are with higher activity than emulsifiers. It is believed by some literatures that activity of demulsifier should be 100-1000 times larger than emulsifier, so that the demulsifier can quickly pass through outer phase of the emulsion and disperse on oil-water interface, replace or neutralize the emulsifier, and reduce the interfacial tension and interfacial film strength of the emulsified water droplets.

(2) Demulsifier can eliminate electrostatic repulsion between water droplets and let them flocculate.

(3) It could help coalescence, which will destroy interfacial film around the emulsified water droplets and making them merge. In this case, the size of water droplets will be increased, and water droplets in crude oil will be

settled, causing separation of oil and water layers.

(4) It can wet solids and prevent the interfacial film composed of powder emulsifier from blocking condensation of the water droplets. Though solid particles such as clay, iron sulfide, drilling mud, etc. are hydrophilic, demulsifier can transfer these solid emulsifiers from oil-water interface into the crude oil.

Although mechanism of demulsifiers is not clear enough, there are two conclusions being drawn from long-term practice:① Only when molecular weight of the demulsifier greater than that of the natural emulsifier, the demulsification could be carried out effectively;② If a demulsifier is used as an emulsifier of oil-water mixture, a reverse phase emulsion, i.e., an O/W emulsion, will be generated. Figure 11-1 shows the dehydration process of demulsifier:

（4）能润湿固体，防止固体粉末乳化剂构成的界面膜阻碍水滴聚结。黏土、硫化铁、钻井泥浆等固体颗粒具有亲水性，破乳剂能把这些固体乳化剂从油水界面进入原油内。

尽管破乳剂的破乳机理尚不完善，但从长期实践中归纳出两点结论：①破乳剂的分子量大于天然乳化剂的分子量才能有效破乳；②若把破乳剂用作油水混合物的乳化剂，则生成反相乳状液，即 O/W 型乳状液。图 11-1 是破乳剂脱水过程：

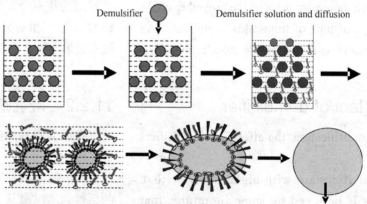

Figure 11-1　Dehydration process of demulsifier

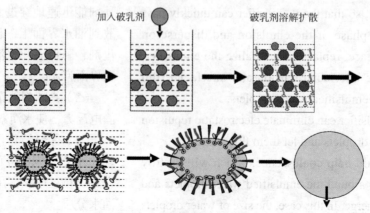

图11-1　破乳剂脱水过程

11.2.3 Selection of crude oil demulsifier

(1) Emulsion of paraffin-based crude oil (such as crude oil produced from Daqing oilfield) containing less colloid and less asphaltene is more likely to be destroyed, but sewage discharged after dehydration is often with high oil content. This is because the W/O emulsion is reversed to O/W type, indicating that the demulsifier is over hydrophilic. This problem will show up when diblock demulsifier such as 2070 and BP169 is used. To solve this problem, a PO section can be added to increase lipophilicity of the demulsifier. For example, a triblock demulsifier, such as SP169, AP221, etc. can be used directly, and oil content of sewage discharged will be reduced significantly.

(2) For crude oil with high colloid and asphaltene content, the emulsion is relatively stable. Thus it will be more difficult to destroy demulsification. However, the discharged sewage is with low oil content. Hence demulsifier used for crude oil of this type should be with good performance on demulsification. To obtain better demulsification effect, the demulsifier (such as AE type) with PO-EO diblock structure can be used. The demulsification speed is related with content of EO. If the content of colloid and asphaltenes increases, the EO content shall also be increased to make the dehydration speed faster.

(3) Dehydration rate of oil-soluble demulsifier is faster than that of water-soluble demulsifier. This is mainly because oil-soluble demulsifier will diffuse in form of molecular, hence the diffusion speed will be faster; for water-soluble demulsifier, the dehydration process will start from its entrance of oil phase from aqueous phase. It will redistribute in the oil phase, and then diffuse on surface of the emulsion water droplets. There will be two types of diffusion undergo: molecular and convection diffusion. Therefore, speed of dehydration rate for water-soluble demulsifier is slower than that of oil-soluble demulsifier.

(4) Under low temperature conditions, demulsifiers with low cloud point and phase transition temperature shall be chosen. These features will ensure the demulsification effect.

(5) Dehydration performance of compound demulsifier is better than that of the single demulsifier.

Mixing two or more types of demulsifiers in a certain ratio may be more effective than using any of them alone, this phenomenon is called a compounding effect (or synergistic effect). Generally, the demulsifiers with same ion type, or ionic and nonionic demulsifiers shall be compounded together. Since it may causing precipitate and fail, two anionic type demulsifiers are generally not compounded. However this is not always true because under certain conditions it is still workable. Some studies have shown that after adding a certain wetting agent or flocculant into nonionic demulsifier, the demulsification efficiency will be significantly improved. Type and proportion of compounding should be determined by experiments. Table 11-1 shows the dehydration experiment of compounded demulsifiers in Gaoshangpu oilfield and Shu IV district of Liaohe oilfield:

（4）低温破乳剂应选浊点较低的破乳剂。浊点低，相转变温度也低，易破乳。

（5）复配的破乳剂比单一破乳剂的脱水效果好。

将两种或两种以上的破乳剂按一定比例混合使用，其效力可能高于其中任何一种单独使用的效力，这种现象称为复配效应（或协同效应）。一般用相同离子类型的破乳剂复配，或离子、非离子型破乳剂复配，而阴、阴离子型破乳剂一般不能复配，会产生沉淀而失效，但在一定条件下仍可复配。一些研究表明，非离子型破乳剂中加入某种润湿剂或絮凝剂后，破乳效率明显提高，复配类型和比例应通过实验确定。表11-1是高尚堡和曙四联复配型破乳剂脱水实验：

Table 11-1 Compound demulsifier dehydration experiment

Oilfield	Demulsifier	Dosage/(mg/L)	Dehydration temp./℃	Sewage volume/mL				Sewage condition
				15min	30min	60min	90min	
Gaoshangpu	SP169	200	45	0	1	3	4	Clear
	TA1031			5	5	8	10	Turbid
	TA1031+SP169（3:1）			17	18	18	18	Clear
Shu IV district	SP169	300	75	1.5	3.5	4.5	5.0	Clear
	TA1031			3.0	9.0	9.0	9.0	Turbid
	TA1031+SP169（2:1）			12.5	13.5	13.5	13.5	Clear

表11-1 复配破乳剂脱水实验

原油产地	破乳剂	加量/(mg/L)	脱水温度/℃	脱水量 / mL				脱出水色
				15min	30min	60min	90min	
高尚堡	SP169	200	45	0	1	3	4	清
	TA1031			5	5	8	10	浑浊
	TA1031+SP169（3:1）			17	18	18	18	清
曙四联	SP169	300	75	1.5	3.5	4.5	5.0	清
	TA1031			3.0	9.0	9.0	9.0	浑浊
	TA1031+SP169（2:1）			12.5	13.5	13.5	13.5	清

11.2.4 Commonly used crude oil demulsifiers and their properties

(1) Alkyl phenolic resins-polyoxypropylene polyoxyethylene ether.

Such as polyoxyethylene polyoxypropylene alkyl phenol formaldehyde resin, the AR type demulsifier:

$$\left[Ar-CH_2 \right]_x \begin{matrix} O-(C_3H_6O)_m-(C_2H_4O)_nH \end{matrix}$$

(2) Polymethylphenyl silicone oil polyoxypropylene polyoxyethylene ether.

Such as polyoxyethylene polyoxypropylene methyl silicone oil:

$$H(C_2H_4O)_n-(C_3H_6O)_mO-\left(Si(CH_3)_2-O\right)_x-(C_3H_6O)_m-(C_2H_4O)_nH$$

(3) Polyphosphate.

Such as polyoxyethylene polyoxypropylene alkyl phosphate:

$$\begin{matrix} RO & O-(C_3H_6O)_m-(C_2H_4O)_nH \\ \;\;\;\;PH & \\ O & O-(C_3H_6O)_m-(C_2H_4O)_nH \end{matrix}$$

(4) Polyoxyethylene polyoxypropylene block copolymer and its modified products.

According to different initiators, demulsifier of this typed can be divided into several types, such as propylene glycol, glycerol, polyethene polyamine, phenolic resin, phenol-amine resin, etc.

①Propylene glycol block copolymer and its modified products, such as polyoxyethylene polyoxypropylene glycol ether, BE type demulsifier, polyoxyethylene polyoxypropylene propylene glycol ether rosinate, and chain extension products of polyoxyethylene polyoxypropylene propylene glycol ether dihydroxy acid.

②Polyethylene polyamineblock copolymer and its modified product, such as AE type polyoxyethylene polyoxypropylene ethylenediamine.

$$\text{CH}_2\text{N} \begin{cases} (C_3H_6O)_m-(C_2H_4O)_n-(C_3H_6O)pH \\ (C_3H_6O)_m-(C_2H_4O)_n-(C_3H_6O)pH \end{cases}$$
$$\text{CH}_2\text{N} \begin{cases} (C_3H_6O)_m-(C_2H_4O)_n-(C_3H_6O)pH \\ (C_3H_6O)_m-(C_2H_4O)_n-(C_3H_6O)pH \end{cases}$$

polyoxyethylene polyoxypropylene diethylene triamine, AP type polyoxyethylene polyoxypropylene pentaethylene hexamine and so on.

③Glycerol block copolymer, GP type.

聚氧乙烯聚氧丙烯二乙烯三胺、AP型聚氧乙烯聚氧丙烯五乙烯六胺等。

③丙三醇嵌段共聚物，GP型。

$$\text{CH}_3-\text{CH}-\text{O}-\text{O}-(C_3H_6O)_m-(C_2H_4O)_n$$
$$\text{CH}_3-\text{O}-(C_3H_6O)_m-(C_2H_4O)_n$$

④Phenol-amine resin type. polyoxyethylene polyoxypropylene phenol-amine resin, PFA type.

Properties of commonly used demulsifier are shown in Table 11-2.

④酚胺树脂类。聚氧乙烯聚氧丙烯酚胺树脂，PFA型。

常用破乳剂性质见表11-2。

Table 11-2 Properties of commonly used demulsifiers

Item		Main content	Application conditions
AE series	AE1910, AE1919, AE2040, AE4010, AE2010, etc.	Polyethyleneamine polyether	Oilfield, low temperature dehydration in refinery, viscosity reduction, suitable for crude oil with medium density and high wax content
SH series	SH9101, SH9105, SH9601, SH991, etc.	Polyoxyethylene polyoxypropylene ultra high molecular ether	Crude oil with high density, viscosity and acid value
GT series FC series	GT940, GT922, FC9301, FC961, etc.	Various types of polyoxypropylene polyoxyethylene ethers, oxypropylene ester compounded high molecular weight demulsifier	Suitable for high and low temperature demulsification of medium density, high salt and water containing crude oil, such as crude oil produced by Changqing, Shanbei, Oman and other oilfields
BP series	BP169, BP2040	Propylene glycol polyoxypropylene polyvinyl ether	Crude oil produced by Jianghan oilfield
ST series	ST-12, ST-13, ST-14	Phenol amine resin polyether	High density and viscosity crude oil
CD series	GD9901, MD01	Acrylic acid modified polyether	High density and viscosity heavy oil

表11-2 常用破乳剂性质

名称		主要成分	适应性
AE系列	AE1910、AE1919、AE2040、AE4010、AE2010, etc.	多亚乙基胺聚醚	油田、炼厂低温脱水，降黏，适合于中等密度、高含蜡原油
SH系列	SH9101、SH9105、SH9601、SH991, etc.	聚氧乙烯聚氧丙烯超高分子醚	高密度、高黏度、高酸值原油
GT系列 FC系列	GT940、GT922、FC9301、FC961, etc.	多种聚氧丙烯聚氧乙烯醚，氧丙烯酯复配高分子量破乳剂	适用于中等密度、高含盐含水原油，如长庆、陕北、阿曼等原油的高温、低温原油破乳
BP系列	BP169、BP2040	丙二醇聚氧丙烯聚乙烯醚	江汉原油
ST系列	ST-12、ST-13、ST-14	酚胺树脂聚醚	高密度、高黏度原油
CD系列	GD9901, MD01	丙烯酸改性聚醚	高密度、高黏度稠油

11.3 Defoaming of foamed crude oil

In liquid produced from oil well, there are associated gases or gases that could be used for flooding dissolved. Due to changes and disturbances of temperature and pressure in the separator, these gases will escape from the fluid. Strong foaming crude oil may form a stable foam layer in the separator. The foam layer will occupy large space in separator, affecting measurement of crude oil and liquid level control, causing gas with high liquid content. This will damage compressors and gas treatment facilities at downstream, and will cause tank overflow accidents if the condition is severe.

By standing, decompressing (vacuumizing), heating or pressurizing, the harmful foams can be eliminated. However, if the foams are needed to be removed in a short time, a defoamer will be a must.

11.3.1 Form mechanism of crude oil foam

The foaming problem of crude oil mainly occurs during process of oil and gas separation and crude oil stabilization. All these processes are related with release of natural gas (including hydrocarbons $C_1 \sim C_7$, mainly alkanes) from crude oil by reducing pressure and/or increasing temperature.

An oil-gas surface will be formed after natural gas released from crude oil. Surfactants (such as fatty acids, naphthenic acids, colloids, and asphaltenes) contained in crude oil will adsorb on the oil-gas surface, which can reduce surface tension of oil-gas surface or form a high-strength surface film, thereby affecting stability of the crude oil foams.

Size of bubbles in foams is not uniform. Since pressure in small bubbles is greater than of in large bubbles, gas in small bubbles can diffuse into large

11.3 起泡原油的消泡

油井采出液中溶解有伴生气或气驱用气,在分离器中由于温度压力变化与扰动等影响,气体会从采出液中逸出。对于发泡性强的原油,可能会在分离器内形成稳定的泡沫层,其存在会占据分离器内大量空间,影响原油计量及液位控制,造成气中带液现象,危害下游压缩机与气体处理设施,严重时还会发生冒罐事故。

消灭有害的泡沫可以利用静置、减压(抽真空)、加温或加压等办法,但是当需要在短时间内,迅速而有效地消除泡沫时,就需要借助于添加消泡剂。

11.3.1 原油泡沫形成机理

原油主要在油气分离和原油稳定过程中遇到起泡沫问题。这些过程都是通过降低压力和(或)升高温度,使天然气(包含$C_1 \sim C_7$的烃,主要是烷烃)从原油中释出的。

当天然气从原油中释出时,产生了油气表面。原油中含有的表面活性剂(如脂肪酸、环烷酸、胶质、沥青质)吸附在油气表面,能够降低油气表面张力或形成高强度的表面膜,从而影响原油泡沫的稳定性。

在泡沫中,气泡大小是不均匀的。由于小气泡内的压力大于大泡内的压力,因此小气泡中的

bubbles through liquid films. This will make the small bubbles gradually become smaller until disappear, and the large bubbles gradually become larger, thereby causing damage of the foams. This mechanism also exists in crude oil foams. Since surfactant adsorption film will suppress gas permeability of liquid film between the large and small bubbles, thereby improving stability of the crude oil foams.

11.3.2 Defoamer

Occurrence of foaming often brings difficulties to the chemical, pharmaceutical, water treatment and food processing industries, causing processing capacity reduction, affecting product quality and production operation. There are three main defoaming methods: physical, mechanical and chemical methods. Defoamers are often used in chemical defoaming method to inhibit or eliminate generation of the foams.

Defoamer is a general term of substances which are added to foaming liquid at a low concentration for preventing foam generation, reducing or even making the foams break. It is also called foam breaker, foam inhibitor, antifoaming agent or foam depressant. In order to suppress foams generation during usage, a defoamer shall be added to lubricating oil. The added defoamer can be considered as an additive and shall be kept at a certain concentration in the lubricating oil. To avoid adverse affect of the subsequent process, during refining process, the defoamer used shall not be remained in oil as much as possible. In this process, the defoamer is a kind of oil refining promoter.

11.3.2.1 Classification of crude oil defoamer

The chemical agent that can eliminate crude oil foams is called the crude oil defoamer. Crude oil defoamers can be divided into the following categories:

(1) Solvent crude oil defoamer.

Solvent crude oil defoamers refer to low molecular

Crude oil demulsification and defoaming of foamed crude oil

alcohols, ethers, alcohol ethers and esters which are commonly used as solvents. Since surface tension between the defoamers and both gas and oil are quite low, when these defoamers are sprayed on crude oil foams, they will rapidly causing the liquid film to be partially thinned, hence the foams will be destroyed.

The following solvents can be used as defoamers:

$C_5H_{11}OH$ (Pentanol)

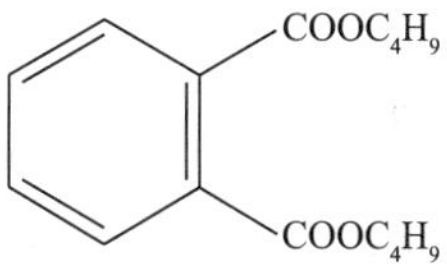

Dibutyl phthalate (with $COOC_4H_9$ groups)

(2) Surfactant crude oil defoamer.

Such defoamers refer to some of the certain surfactants that with branch structure. These defoamers could destroy foams because they will replace the surface active substances and generate unstable protective film.

Surfactants of the following structure can be used as defoamers:

$$H_2C-O-(C_3H_6O)_m-(C_2H_4O)_n-H$$
$$CH-O-(C_3H_6O)_m-(C_2H_4O)_n-H$$
$$H_2C-O-(C_3H_6O)_m-(C_2H_4O)_n-H$$

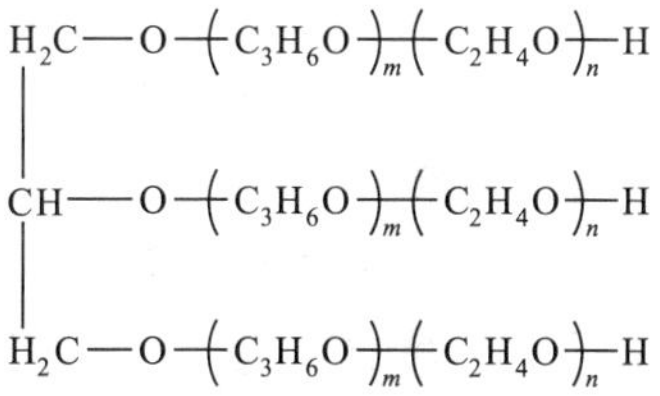

Polyoxyethylene polyoxypropylene glyceryl ether

(3) Polymer crude oil defoamer.

Defoamers of this type refer to polymers which have low surface tension with gas and low interfacial tension with oil. Their mechanism of defoaming is the same as that of solvent crude oil defoamer.

These defoamers mainly are polysiloxanes, such as:

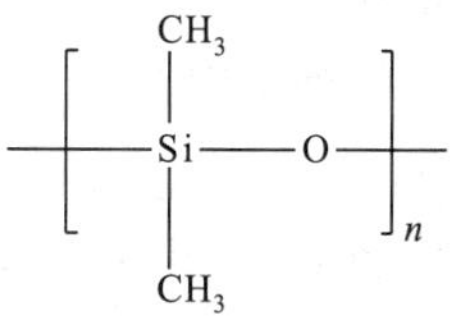

Polydimethylsiloxane

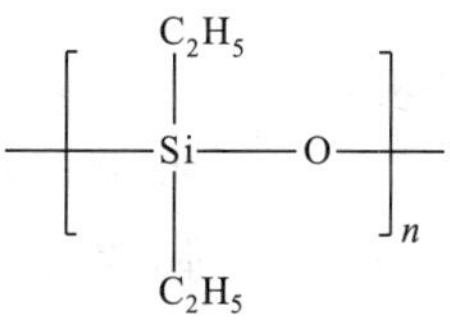

polydiethylsiloxane

这类消泡剂是指通常用作溶剂的低分子醇、醚、醇醚和酯。当将这些消泡剂喷洒在原油泡沫上时，由于它们与气的表面张力和与油的界面张力都低而迅速扩展，使液膜局部变薄而导致泡沫的破坏。

可用下列溶剂作消泡剂：

$C_5H_{11}OH$（戊醇）

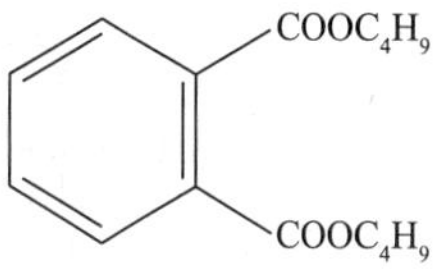

邻苯二甲酸二丁脂

（2）表面活性剂型原油消泡剂。

这类消泡剂是指一些有分支结构的表面活性剂。当将这些消泡剂喷洒在原油泡沫上时，由于它取代了原来稳定泡沫的表面活性物质后形成不稳定的保护膜，导致泡沫破坏。

可用下列结构的表面活性剂作消泡剂：

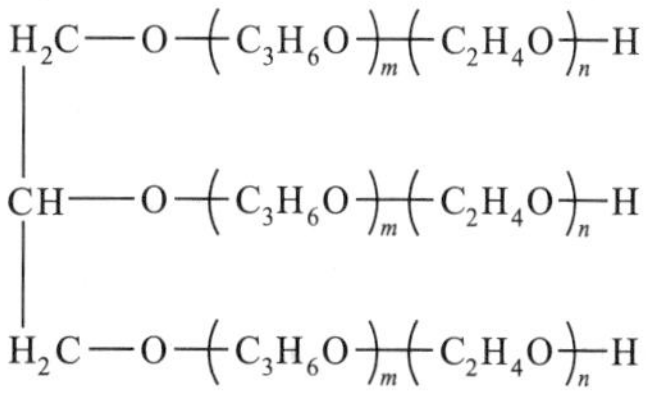

聚氧乙烯聚氧丙烯甘油醚

（3）聚合物型原油消泡剂。

这类消泡剂是指其与气的表面张力和与油的界面张力都低的聚合物。它的消泡机理与溶剂型原油消泡剂的消泡机理相同。

这类消泡剂主要有聚硅氧烷，如：

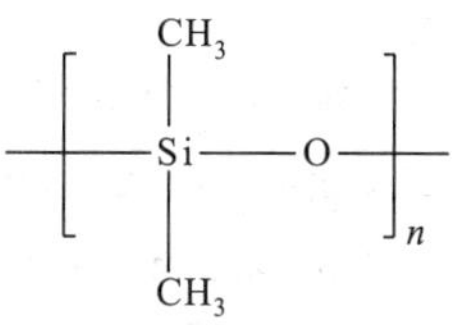

聚二甲基硅氧烷

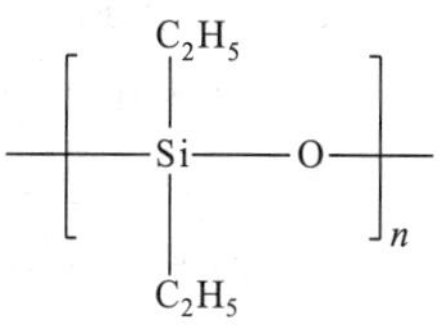

聚二乙基硅氧烷

Polysiloxane contains fluorine or modified by polyether can also be used, such as:

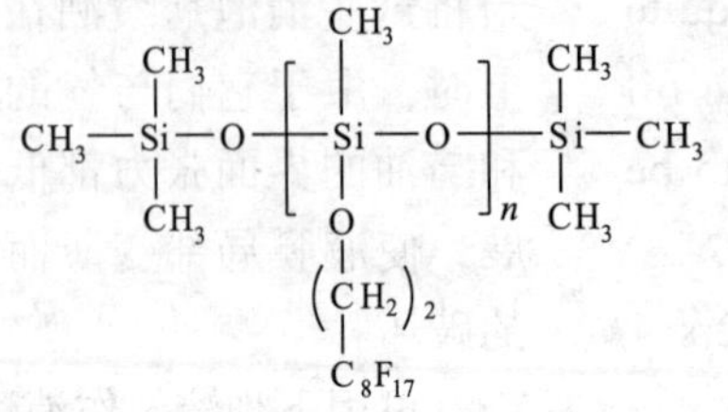

Polymethylfluoromethoxysiloxane

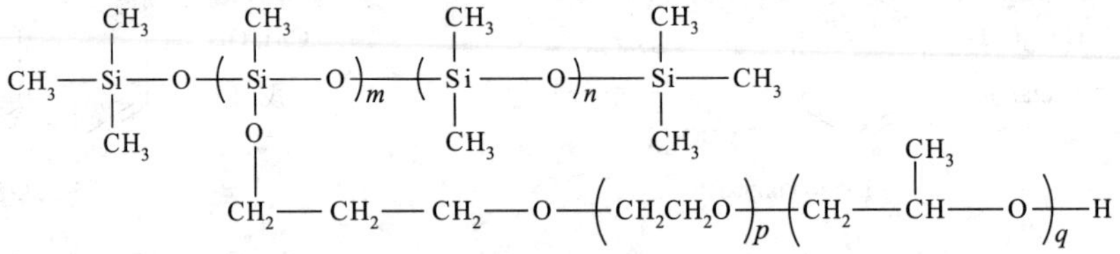

Polysiloxane modified by polyether

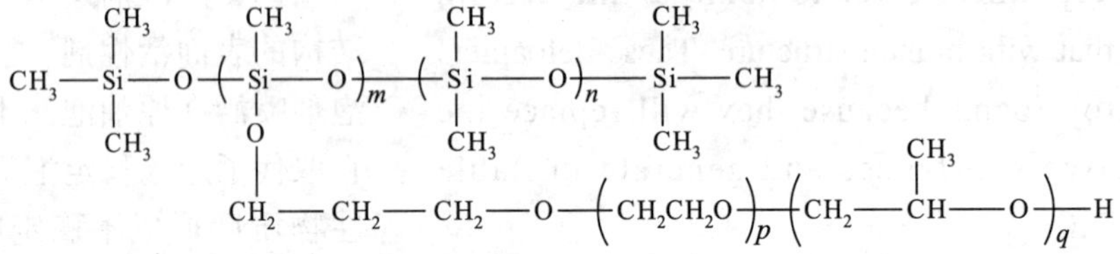

也可用含氟聚硅氧烷和聚醚改性的聚硅氧烷，如：

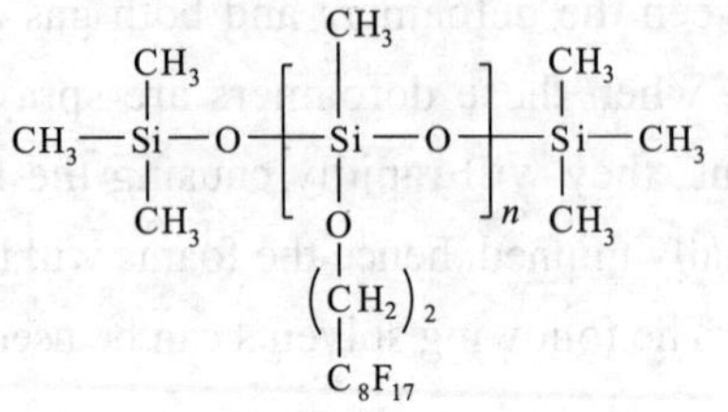

聚甲基氟代葵氧基硅氧烷

聚硅氧烷的聚醚改性产物

11.3.2.2 Mechanism of defoamer

(1) Viewpoint of reducing the partial surface tension (σ).

From this point of view, σ of the defoamer is smaller than that of the foaming liquid. Defoamer will first adsorb on film of the bubbles and then invades into it, making σ of the invaded parts significantly reduced. However, rest of the film surface still maintains the original large σ, getting a tension difference on film of the bubbles. This tension difference will cause a strong tension that pulling the weak part of bubble film and make the bubbles break, as shown in Figure 11-2.

(2) Viewpoint of expansion.

This mechanism can be illustrated by Figure 11-3. First, the defoamer droplet D intrudes into the bubble film F, making itself a part of the film; then it

11.3.2.2 消泡剂作用机理

（1）降低部分表面张力（σ）观点。

这种观点认为，消泡剂的 σ 比发泡液的小。当消泡剂与泡沫接触后，吸附于泡膜上，继而侵入膜内，使该部分的 σ 显著降低，而膜面其余部分仍然保持着原来较大的 σ，这种在泡膜上的张力差异，使较强张力牵引着张力较弱的部分，从而使泡破裂，如图 11-2 所示。

（2）扩张观点。

这种抗泡机理可用图 11-3 说明。首先是消泡剂小滴 D 侵入泡膜 F 内，使消泡剂小滴成为膜的

Crude oil demulsification and defoaming of foamed crude oil
原油破乳与起泡原油的消泡 第11章

will expand on the film. As it expands, the part of film it entered initially will become thinner, which will finally cause break of the bubbles. This view holds that effect of defoamer is related to free energy of the system.

(3) Viewpoint of penetration.

This view holds that defoamer could increase permeability of bubble film to air; thereby it will accelerate bubble combination, reduce strength and elasticity of the bubble film, and break the bubbles finally. If surface viscosity of the foam liquid film is high, strength of the liquid film will be increased.

一部分，然后在膜上扩张，随着消泡剂的扩张，消泡剂进入部分最初开始变薄，最后破裂。该观点认为消泡剂的作用与体系的自由能相关联。

（3）渗透观点。

这种观点认为，消泡剂的作用是增加气泡壁对空气的渗透性，从而加速泡沫合并，减小泡膜壁的强度和弹性，达到消泡目的。泡沫液膜的表面黏度高会增加液膜的强度。

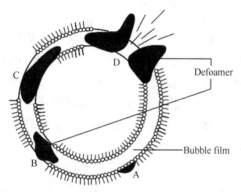

Figure 11-2 Defoamer reduces σ of liquid film partially and breaks the foam

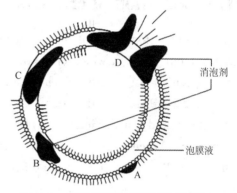

图11-2 消泡剂降低局部液膜σ而破泡

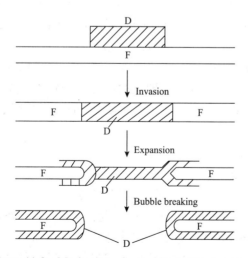

Figure 11-3 Mechanism of expansion bubble breaking

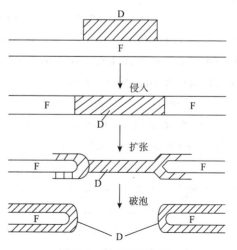

图11-3 扩张破泡机理

Pour point depression and drag reduction transportation of crude oil
第12章 原油降凝与减阻输送

12.1 Pour point depression transportation of crude oil

Condensation point of crude oil refers to the highest temperature at which crude oil loses fluidity under specified test conditions. There are two reasons for crude oil loses its liquidity: One is because viscosity of crude oil will increase with the decrease of temperature. When viscosity rises to a certain extent, crude oil will lose its fluidity; the other is caused by paraffin (wax) in crude oil. As the temperature getting lower, it will reach to wax precipitation temperature first and causing wax crystals precipitate. If it continuously getting lower, number of wax crystals will increase, which will resulting them to grow and coalesce until forming a network structure. This structure will fix the crude oil and making it lose its fluidity.

Transportation of crude oil that being treated by pour point depressing method in long distance pipelines is called pour point depressed transportation of crude oil. Methods of crude oil pour point depressing including:

(1) Physical pour point depression method.

Physical pour point depression method is a heat treatment method. By this method, the crude oil shall be heated firstly to the optimum heat treatment temperature, and then be cooled at a certain rate, so condensation point of the crude oil will be lowered.

12.1 原油的降凝输送

原油的凝点是指在规定的试验条件下原油失去流动性的最高温度。原油失去流动性有两个原因：一个是原油的黏度随温度的降低而升高，当黏度升高到一定程度时，原油即失去流动性；另一个是由原油中的蜡引起的，当温度降低至原油的析蜡温度时，蜡晶析出，随着温度的进一步降低，蜡晶数量增多，并长大、聚结，直到形成遍及整个原油的结构网，原油即失去流动性。

原油的降凝输送是指用降凝法处理过的原油在长输管道中的输送。原油降凝法有下列几种：

（1）物理降凝法。

物理降凝法是一种热处理方法。该法首先将原油加热至最佳的热处理温度，然后以一定的速率降温，达到降低原油凝点的目的。

Pour point depression and drag reduction transportation of crude oil

(2) Chemical pour point method.

Chemical pour point depression method is a method that achieved goal of pour point depression by adding pour point depressants to crude oil. The chemical agent that could lower condensation point of crude oil is called a crude oil pour point depressant.

There are two main types of chemical crude oil pour point depressants: One is a surfactant crude oil pour point depressant, such as petroleum sulfonate and polyoxyethylene alkyl amine. This type of pour point depressant will adsorb on surface of paraffin (wax) crystals to prevent generation of the network structure. The other is polymeric crude oil pour point depressants. On main chain and (or) branch of these pour point depressants, there are non-polar parts that could be cocrystallized (eutectic) with paraffin (wax) molecules as well as polar parts that can cause distortion of paraffin (wax) crystals.

(3) Chemical-physical pour point depression method.

This is a comprehensive pour point method. Operation of this method requires heat treatment of the crude oil after addition of pour point depressant.

According to table 12-1, condensation points of crude oil being treated by chemical-physical pour point depression method are lower than condensation points of those being treated by heat. Besides condensation point, crude oil treated by this comprehensive method shows better low temperature fluidity than heat treated crude oil. This is mainly because it is not only with natural crude oil pour point depressant (colloidal, asphaltene) but also contains the added polymer pour point depressant.

Table 12-1 Effect on condensation point of crude oil after heat treatment and comprehensive treatment

Oilfields	Condensation point before treatment /℃	Condensation point after heat treatment /℃	Condensation point after comprehensive treatment /℃
Daqing	32.5	17.0	12.3
Jianghan	26.0	14.0	6.0
Renqiu	34.0	17.0	13.5
Hongjingzi	17.0	8.0	1.5

表12-1　热处理与综合处理对原油凝点的影响

原油产地	处理前凝点/℃	热处理后凝点/℃	综合处理后凝点/℃
大庆油田	32.5	17.0	12.3
江汉油田	26.0	14.0	6.0
任丘油田	34.0	17.0	13.5
红井子油田	17.0	8.0	1.5

12.1.1 Classification of pour point depressant

Pour point depressant is an oil soluble organic high macromolecule compound or a polymer. Addition of appropriate amount of pour point depressant to crude oil containing paraffin (wax) will change morphology and structure of the paraffin (wax) crystal in crude oil under certain conditions. Thereby it will make condensation point of the crude oil significantly lower and improve the low temperature fluidity of the waxy crude oil. Hence the pour point depressant is also known as flow improver.

Presently in oilfields at home and abroad, the most frequently used crude oil pour point depressants are comb-shaped polymers.

12.1.1 降凝剂的种类

降凝剂是一种油溶性高分子有机化合物或聚合物。在含蜡原油中添加适量的降凝剂，在一定条件下就能改变原油中蜡晶形态和结构，从而显著地降低含蜡原油的凝点。改善含蜡原油的低温流动性能，因此，降凝剂又称流动改进剂。

目前，国内外各油田应用的原油降凝剂多为梳形聚合物。

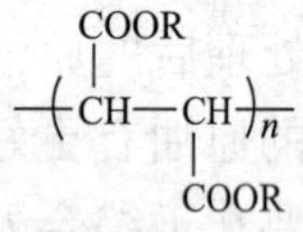

Poly (dialkyl fumarate)

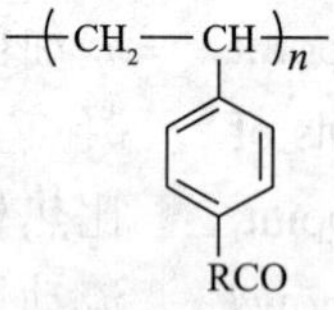

Polyacylstyrene

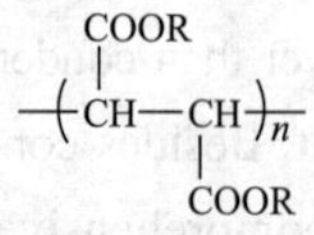

聚富马酸二烷基酯

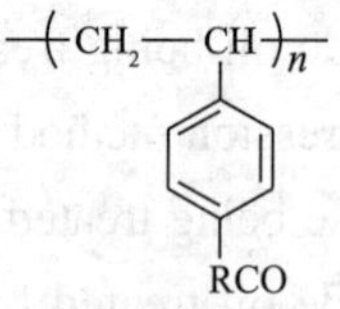

聚酰基苯乙烯

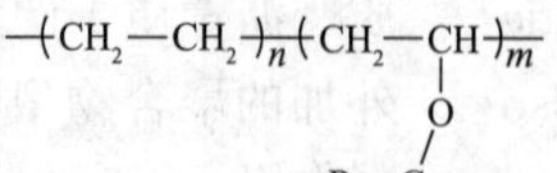

Ethylene-vinyl carboxylate copolymer

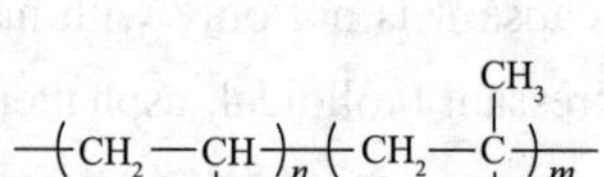

Ethylene-methacrylate copolymer

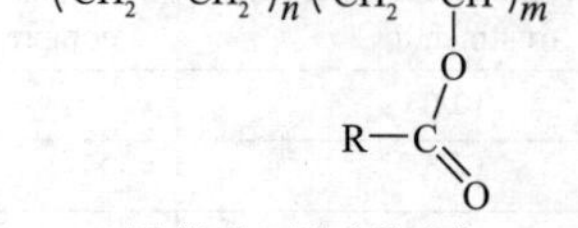

乙烯-羧酸乙烯酯共聚物

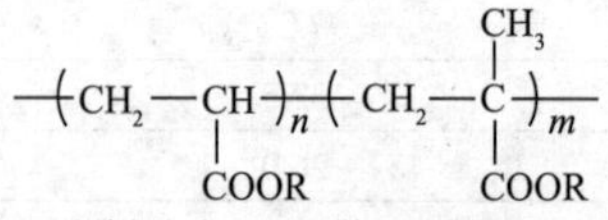

丙烯酸酯-甲基丙烯酸酯共聚物

R of the above compounds shall be C_{18}-C_{24}. In case that length of alkyl in side chain of these pour point depressants are similar to length of alkane in the most concentrated paraffin (wax) of the crude oil, the pour point depressants will be with good performance, and effect of pour point depression and viscosity reduction is remarkable. This type of pour point depressants had been used in Shengli, Zhongyuan, Qinghai and other oilfields, and range of their pour point depression is 16-25℃.

12.1.2 Mechanism of pour point depressant

The crude oil pour point depressant is a polymer or condensation polymer. It is with a polar side chain and an alkane main chain similar to hydrocarbon molecule of wax. Theoretically, the ideal pour point depressant should affect the whole process of crystalline core generation as well as growth and adhesion of paraffin (wax) crystals. By doing this, it can prevent paraffin (wax) to form a three dimensional network structure, reduce condensation point of waxy crude oil and improve low temperature fluidity of the crude oil. According to different paraffin (wax) deposition stages that will be affected by pour point depressants, there are three main mechanisms currently:

(1) Effect of crystalline core: Similar with colloid and asphaltene, during stage of crystalline core generation, the pour point depressant will be crystallized first at a temperature higher than that of the wax in crude oil. In this case it will become the center of paraffin (wax) crystal development, and making number of small wax crystals in oil increase, hence large paraffin (wax) clusters will not be easily formed.

(2) Effect of adsorption: In stage of paraffin (wax) crystal growth, the pour point depressant molecules will be adsorbed around paraffin (wax) crystals more easily. Thereby it can prevent the new formed paraffin (wax) crystals from attaching, inhibit generation of the three dimensional network

structures and reduce condensation point of the crude oil.

(3) Effect of eutectic: During growth of paraffin (wax) crystals, the non-polar group on molecule of pour point depressant and hydrocarbon chain that similar to paraffin molecule will form eutectics, and the polar group on pour point depressant molecule will stop further growth and adhesion of paraffin (wax) crystals by electrostatic repulsion and steric hindrance (Figure 12-1、Figure 12-2).

（3）共晶作用：在蜡晶的生长过程中，降凝剂分子的非极性基团与石蜡分子相似的烃链共晶，而降凝剂分子的极性基团则通过静电斥力和空间位阻等作用阻碍蜡晶进一步长大和相互连接（图12-1、图 12-2）。

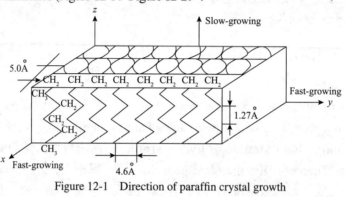

Figure 12-1 Direction of paraffin crystal growth

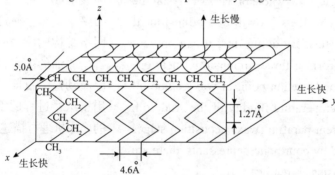

图12-1 石蜡结晶生长方向

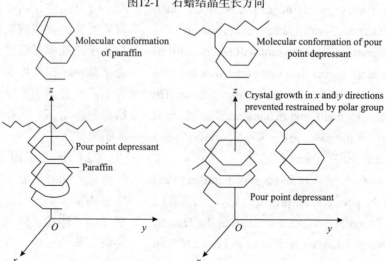

Figure 12-2 Molecular conformation and eutectic of paraffin and pour point depressant at low temperature

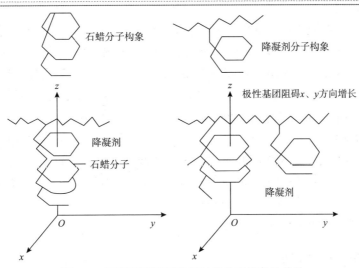

图12-2 石蜡及降凝剂低温时的分子构象及共晶

(4)Theory of paraffin (wax) solubility improvement.

According to this theory, the pour point depressant will work as a surfactant. By adding the pour point depressant, solubility of paraffin (wax) in the oil will be increased. This will reduce generation and increase dispersion of paraffin (wax). After the paraffin (wax) is dispersed, charges on its surface will cause repulsive force between the paraffin (wax) crystals, letting them not easy to form a three dimensional network structure. Hence condensation point is lowered. According to crystallography, solubility improvement of the solute caused by pour point depressant will decrease supersaturation of the solution, thereby reducing the apparent growth rate and hindering growth of the crystal.

Effect of pour point depressants in modification process of crude oil additives is related to its type and chemical structure. During different stages of paraffin (wax) crystal growth, there may be one or more mechanisms working at the same time. Study on the working mechanism shall be carried out according to the specific situation.

12.2 Drag reduction transportation of crude oil

With enlargement of oil pipeline network scale and the continuous increase oil products demand, pipeline

（4）改善蜡的溶解性理论。

改善蜡的溶解性理论认为，降凝剂如同表面活性剂，加降凝剂后，增加了蜡在油品中的溶解度，使析蜡量减少，同时又增加了蜡的分散度，且由于蜡分散后的表面电荷的影响，蜡晶之间相互排斥，不容易形成三维网状结构，而降低凝点。结晶学还认为，如果添加降凝剂改善了溶质的溶解性，会使溶液的过饱和度下降，从而降低表观成长速率，阻碍晶体的生长。

降凝剂在原油加剂改性过程中所起的作用与降凝剂的种类及其本身的化学结构有关，而且在蜡晶生长的不同阶段，可能只有一种机理起主要作用，也可能几种机理同时存在，还需根据具体情况具体分析。

12.2 原油的减阻输送

随着输油管网规模的增大和油品需求量的不断增加，管道输

transportation is getting more and more important in gathering and transportation system. In order to increase economic benefits and control the elastic transport capacity of the pipelines effectively, studies on drag reduction and transportation flow rate increase becomes particularly necessary.

Comparing with pipeline modification, use a drag reducer is a better method for transportation flow rate increase because it will show effect quickly and with good economic efficiency and operability.

12.2.1　Drag reducer

Drag reducer, as the name suggests, is a group of agents that could reduce the drag produced through transportation of fluids. It is divided into two major categories of water soluble and oil soluble. Between these two categories, the oil soluble drag reducers have attracted much attention. It contains two types: One is surfactant compounds, and the other is a group of highly flexible linear polymers with ultra high molecular weight ($M>10^6$). However, the usage, dosage, advantages and disadvantages of these two types of drag reducers are quite different.

Surfactant drag reducer works by forming rod-like and network microscopic micelles inside the fluid solution. Although its shear resistance is good, it only works when the concentration reaches to critical concentration. Besides, since it needs to be evenly mixed with the fluid, the total costs will be increased to solve problem of adding methods and mixing equipment. It may also easy to cause pollution to oil products due to the large addition concentration (currently in the state of experimental research).

Polymer drag reducer is usually used in small dosage and its performance on drag reduction and flow rate increase is good. However, due to its long molecular chain, high shearing effect of the turbulent fluids will break the molecular chains and resulting in a sudden decrease in drag reduction effect. This is the biggest

送在集输系统中所占的比重日益增大，如何实现输油管道减阻增输，在实现提高经济效益的同时有效控制管道弹性输送能力显得尤为重要。

相比管线改造方法，使用减阻剂进行管道增输，见效快速、经济效益好且可操作性强。

12.2.1　减阻剂

减阻剂，顾名思义是一种能够减小流体输送过程中所受阻力的试剂。分为水溶性和油溶性两大类。备受关注的油溶性减阻剂有两类：一类是表面活性剂化合物，另一类是具有超高分子量（$M>10^6$）的高柔性线型高分子，同样具有减阻作用，但这两类减阻剂用法、用量及优缺点大相径庭。

表面活性剂减阻剂是通过在流体溶液内部形成棒状及网状的微观胶束实现减阻的效果，虽然其抗剪切性能良好，但只有当浓度达到临界浓度时才有减阻作用。并且需要与流体混合均匀，这又存在一个添加方法和混合设备的问题，经济性不好。同时添加浓度大，易对油品产生污染（目前主要处于实验研究状态）。

高分子聚合物减阻剂通常用量小，减阻增输效果好。但是，由于分子链较长，湍流流体的高剪切作用会使其分子链断裂，导致减阻效果骤降，且这种破坏是不可逆的，一旦遭到剪切，减阻

Pour point depression and drag reduction transportation of crude oil

defect of polymer drag reducer (widely used in the field) because the damage is irreversible. Once being sheared, the drag reduction effect will not be restored.

12.2.2 Commonly used crude oil drag reducers

Crude oil drag reducer is generally an oil soluble polymer with its molecular weight of 10^6 or more. It shall be with excellent solubility, shear resistance, and oxidation resistance, etc. Generally, the larger the relative molecular mass is, the worse the shear resistance will become. If there are short side chains on the molecule, shear resistance performance will be increased. However the side chain should not be too long or too many because it will make molecule flexibility worse and deduct the resistance reduction performance of the drag reducer.

The drag reducers currently used at home and abroad (Table 12-2) are:

效果将不再恢复，这是高分子聚合物减阻剂最大的缺陷（现场应用广泛）。

12.2.2 常用原油减阻剂

作为原油减阻剂，一般是相对分子质量大于等于 10^6 的油溶性聚合物。要求具有优良的溶解性、抗剪切性、抗氧化性等。通常相对分子质量越大，抗剪切性越差，分子中含有短侧链，抗剪切性增强，但侧链不宜过长过多，这使得分子的柔顺性变差，减阻性降低。

目前国内外使用的减阻剂（表12-2）有：

Table 12-2 Commonly used crude oil drag reducers

Drag reducer		Model	Relative molecular mass	Performance and application
$-(CH_2-\underset{\underset{COOR}{\|}}{C}\underset{CH_3}{\|})_n-$ (Polymethacrylate)		PDR—I	10^6	Crude oil drag reducer
		PDR—II		
		PDR—III		
$-(CH_2-\underset{\underset{R}{\|}}{CH})_n-$ (Poly-α-olefin)		CDR—102	10^6-10^7	Dosage is 0.75-3μg/g
1—copolymer of butane and α olefin		FLO	10^6-10^7	Dosage is 0.48-1.8μg/g
		TM—1003		

表12-2 常用原油减阻剂

名 称		代 号	相对分子质量	性能及用途
$-(CH_2-\underset{\underset{COOR}{\|}}{C}\underset{CH_3}{\|})_n-$ （聚甲基丙烯酸酯）		PDR—I	10^6	原油减阻剂
		PDR—II		
		PDR—III		
$-(CH_2-\underset{\underset{R}{\|}}{CH})_n-$ （聚α-烯烃）		CDR—102	10^6~10^7	用量为0.75~3μg/g
1—丁烯与α烯烃共聚物		FLO	10^6~10^7	用量为0.48~1.8μg/g
		TM—1003		

12.2.3 Mechanism of drag reducer

The chemical agent that can reduce drag of crude oil pipelines is called a crude oil drag reducer. Crude oil drag reducer only works in turbulent flow field. In the turbulent flow field, velocity and direction of the fluid particles will change randomly and forming large and small eddy currents. By absorbing energy from the fluid, large scale vortexes can be deformed and broken into small scale vortexes. In addition to moving in the axial direction, the vortex also moves in radial direction randomly. These useless movements will consume a large amount of energy. The eddy currents are weakened, calmed and disappeared under effect of new stagnation force and pipeline shear stress. The energy carried by vortexes is converted into heat energy and lost, and at the same time, new eddy currents are generated, and the whole process will repeat again.

Crude oil drag reducer is an oil soluble linear polymer compound, and its drag reducing effect is a purely physical effect. Molecules of the drag reducer will not interact with molecules of the oil and will also not affect chemical properties of the oil. The drag reducing effect only related to flow characteristics of the drag reducer closely. After entering to the oil flow, the macromolecular chain of drag reducer will naturally stretch along with flow direction by its own unique new elasticity. This stretch will affect movement of the fluid particles. Radial force of the fluid particles will cause molecules of the drag reducer distorting and reelingly deforming. Relying on mutual attraction between the molecules, the drag reducer molecules can resist the force of the fluid particles, changing direction and size of the fluid particles. This will convert part of the radial force into axial force in direction of the currents, thereby reducing the undesirable energy consumption. Macroscopically, friction loss of the fluid is reduced, which means effect of the drag reduction is shown.

12.2.3 减阻剂作用机理

能降低原油管输阻力的化学剂称为原油减阻剂。原油减阻剂只在湍流场中起作用。在湍流场中，流体质点的运动速率和方向都随机变化，形成大大小小的涡流，大尺度的旋涡从流体中吸收能量瞬间可发生形变、破碎而转化成小尺度旋涡。旋涡除了沿轴向运动外，还随机地径向运动，这些无用功消耗了较大的能量。湍流在新滞力和管壁剪切应力作用下被减弱、平息、消失，旋涡所携带的能量转化为热能而散失，同时又会有新的涡流产生，周而复始。

原油减阻剂都是油溶性线型高分子化合物，减阻剂的减阻作用是一个纯物理作用，减阻剂分子与油品的分子不发生作用，也不影响油品的化学性质，只与其流动特性密切相关。减阻剂加入油流中，依靠本身特有的新弹性，大分子链顺流动方向自然拉伸取向，这种取向会影响流体质点的运动。流体质点的径向作用力使减阻剂分子发生扭曲，旋转变形。而减阻剂分子依靠分子间相互引力抵抗流体质点的作用力，改变流体质点的作用方向和大小，使一部分做无用功的径向力转化为顺流向的轴向力，从而减少了无用功的消耗，宏观上表现为减少了流体的摩阻损失，即起到减阻作用。

在层流中，流体受新滞力作用，没有像湍流那样的旋涡耗散，

Due to effect of the new stagnation forces, there is no vortex dissipation like turbulence in fluid of the laminar flows. Therefore, addition of drag reducer will not reduce the friction. As turbulence caused by Reynolds number increase appears, the drag reducer works. The larger the Reynolds number is, the more obvious the drag reduction effect will become. However, if Reynolds number is very large, shear stress of the fluid will degrade molecule of the drag reducer, and the drag reduction effect will continuously decrease until completely disappear.

According to factors affecting drag reduction, the requirements for a drag reducer are:

(1) Solubility. The drag reducer must be dissolved in the pumping fluid.

(2) Shear stability. The drag reducer must be with sufficient stability so it will not degrade in turbulence.

(3) No downstream effects. The drag reducer should not adversely affect equipment and oil quality of the downstream refinery, i. e. it shall not affect engine performance of the oil products.

(4) Low dosage and high economic efficiency.

因此，加入减阻剂也不会降低摩阻。随着雷诺数增大进入湍流，减阻剂就显露出减阻作用。雷诺数越大，减阻效果越明显。但是，当雷诺数相当大时，流体的剪切应力会使减阻剂分子降解，减阻效果反而下降，甚至完全失去减阻作用。

根据影响减阻的因素等考虑，对减阻剂的要求为：

（1）可溶性。减阻剂必须溶于泵输送流体中。

（2）剪切稳定性。减阻剂必须有足够的稳定性，在湍流中不降解。

（3）无下游效应。减阻剂不应对下游炼厂的设备和油品质量有不良的影响，也就是不影响使用成品油的发动机的性能。

（4）用量少，经济效益高。

第13章 Metal corrosion and chemical protection 金属的腐蚀与化学防护

13.1 Metal corrosion

Metal corrosion refers to the phenomenon that a metal surface is destroyed by chemical or electrochemical reaction with a surrounding medium.

Metal corrosion is a serious problem. There are 10% of metals (95% of which is iron and steel) corroded each year.

According to characteristics of the corrosion process, metal corrosion can be divided into chemical corrosion and electrochemical corrosion. The former is without corrosion current and will not form battery; the latter is carried out by corrosion currents and will form battery. Corrosion generated in oil recovery is mainly electrochemical corrosion. For example, the corrosion generated during oil well acidification is electrochemical corrosion. The acid will help to form many micro batteries on uneven structure on steel surface, causing corrosion of steel.

According to characteristics of surface damage, metal corrosion can be divided into uniform corrosion and localized corrosion. From point of macroscopic view, the former is incapable of distinguishing anode region (the position where anode reaction occurs) and cathode region (the position where cathode reaction occurs); while the latter is capable of doing that. Localized corrosion is more harmful than uniform

13.1 金属腐蚀

金属腐蚀是指金属表面与周围介质发生化学或电化学反应而受到破坏的现象。

金属腐蚀是一个严重的问题，每年共10%金属（其中95%钢铁）损耗于腐蚀。

若按腐蚀过程的特点，金属腐蚀可分为化学腐蚀和电化学腐蚀。前者不构成电池，是没有腐蚀电流的腐蚀；后者构成电池，是有腐蚀电流的腐蚀。采油中遇到的腐蚀主要为电化学腐蚀。如油井酸化时所产生的腐蚀是电化学腐蚀，因酸的存在，使结构不均匀的钢铁表面形成许多微电池，引起钢铁腐蚀。

若按表面破坏的特点，金属腐蚀可分为均匀腐蚀和局部腐蚀。前者是无法在宏观上区分阳极区（阳极反应发生的位置）和阴极区（阴极反应发生的位置）的腐蚀；后者是可以在宏观上区分阳极区和阴极区的腐蚀。局部腐蚀比均匀腐蚀的危害更大。从上面的电极反应可以看到，阳极电流在数

corrosion. It can be seen from the electrode reaction above that value of the anode current must be equal to that of the cathode current. When localized corrosion occurs, the anode reaction will be carried out on a small area, which will greatly reduce time of corrosion perforation on equipment and pipelines.

According to characteristics of corrosive medium, metal corrosion can be divided into acidic medium corrosion and neutral medium corrosion. Acidic medium refers to a medium with its pH less than 5. Various acids of different concentration used in oil recovery, such as hydrochloric acid, hydrofluoric acid, phosphoric acid, sulfuric acid, formic acid, acetic acid, fluoboric acid, and their mixtures all belongs to this type. Water containing acid gases such as H_2S and CO_2 can also be divided into this medium type. H^+ is the main depolarizer for this corrosive medium type. Neutral medium refers to a medium with its pH in range of 5–9. Cooling water, boiler water, washing water, well killing water, water well injection water, oil well produced water and aqueous solution of various salts (such as sodium chloride, potassium chloride, ammonium chloride, calcium chloride, calcium bromide, zinc bromide) belongs to this type. O_2 is the main depolarizer for this corrosive medium type.

13.2 Prevention method of metal corrosion

There are many methods to prevent metal corrosion. The main methods include: changing nature of the metal, separating the protected metal from corrosive medium, conducting metal surface treatment, improving the corrosion environment or implementing e lectrochemical protection.

Among them, improving the corrosion environment is of great significance on corrosion reduction and prevention. For example, metal corrosion could be reduced and prevented by adding a substance capable of reducing the corrosion rate (i. e. the corrosion inhibitor). Corrosion inhibitor is a chemical material which could significantly reduce the rate of metal corrosion. Since it is used in small dosage, and the operation is simple and economical, the corrosion inhibitor is a commonly used method for anti-corrosion.

种化学物质，将它少量地加入腐蚀介质中，就可显著地减小金属腐蚀的速率。由于缓蚀剂用量少，简便而且经济，故是一种常用的防腐手段。

13.2.1　Mechanism of corrosion inhibitor

13.2.1　缓蚀剂的作用机理

Mechanism of corrosion inhibitor is quite complex. There are two viewpoints for mechanisms of corrosion inhibitor currently: electrochemical and chemical. In viewpoint of chemical, mechanism of inorganic and organic corrosion inhibitors are different. In petrochemical industry, the most widely and commonly used corrosion inhibitors are organic corrosion inhibitors. Their mechanisms will be discussed in detail below.

缓蚀剂作用机理比较复杂，目前存在以电化学和化学观点来解释的缓蚀作用机理，在化学观点中无机缓蚀剂和有机缓蚀剂作用机理又不同。在石化行业，使用最多的、最普遍的是有机缓蚀剂，下面将详细讨论它们的作用机理。

13.2.1.1　Adsorption mechanism

13.2.1.1　吸附机理

There are many kinds of organic compounds which can be used as corrosion inhibitors. Structures of these compounds are much in common. Most of them are with polar groups centered on atoms with large electronegativity such as O, N, and S, and non-polar groups centered on C and H. Polar groups are hydrophilic, they can form an adsorption layer on the active point or the entire metal surface by physical and chemical adsorption. Bonding strength between adsorption layer and the metal depends on adsorption property and bonding strength between corrosion inhibitor and the metal. The non-polar groups are located at the end that deviates from the metal. By varying degrees of coverage and wetting, it can affect the corrosion inhibition effect. Organic corrosion inhibitor can stabilize the energy state of metal surface by adsorption. It could also form a hydrophobic film

可以作缓蚀剂的有机化合物种类很多，有缓蚀性能的化合物在结构上多有共同之处，含有电负性大的O、N、S等原子为中心的极性基团，和以C、H为中心的非极性基团。极性基团亲水，它可以通过物理吸附和化学吸附，在金属表面的活性点或整个表面形成吸附层，吸附层和金属之间的结合强度取决于缓蚀剂和金属之间的吸附性质、键合强度。非极性基团位于离开金属的方向，它通过不同程度的覆盖和润湿作用，来影响缓蚀效果。有机缓蚀剂通过吸附，使金属表面的能量状态趋向稳定，并由非极性基构成憎水性膜，使电荷转移和腐蚀

with the non-polar group, which will suppress charge transfer and corrosive diffusion, reducing degree of metal corrosion.

Corrosion inhibition efficiency of the organic corrosion inhibitor is related with the following aspects: geometry of corrosion inhibitor molecule, cross section area, electron density of N atom, τ bond between heteroatom and metal atom, ionic potential and electron affinity, dissociation constant, nature of the functional group, spatial effect, molecular structure and dipole moment and so on.

(1) Mechanism of physical adsorption: Cations of organic amine (RnNX) can electrostatically adsorb on a negatively charged metal surface. By juxtaposing on surface of the metal, the amine cations can form a barrier layer. In the amine cations, the larger the C atoms number in alkyl group R is, the better the corrosion inhibition effect will be. Organic compounds that react with H^+ to form cations can also conduct physically adsorption. The stronger the alkaline become, the stronger the tendency to form cations will be. And the easier adsorption carries out, the higher the corrosion inhibition efficiency will become.

(2) Mechanism of chemical adsorption: Chemical adsorption is formed by electron sharing of empty d-orbital of the metal and the following electrons: unpaired electron from the polar groups of corrosion inhibitor, double bond, triple bond and the τ electron on benzene ring. The corrosion inhibition effect could be improved by changing the electron density of unpaired electrons. Rate of chemical adsorption is slower than physical adsorption, and effect of chemical adsorption is stronger. It is an irreversible adsorption process. This kind of adsorption does not need cooperation of metal surface charges, and the neutral molecules can also be adsorbed. Its corrosion inhibition performance is better than physical adsorption. Since corrosion inhibitor provides electrons to adsorb with the metal, different

剂的扩散都受到抑制，降低金属的腐蚀。

有机缓蚀剂的缓蚀效率和缓蚀剂分子的几何尺寸、截面积、N原子的电子密度、杂原子与金属原子之间的τ键、离子势和电子亲和力、离解常数、起作用基团的本质、空间效应、分子结构和偶极距等有关。

（1）物理吸附机理：有机胺阳离子（RnNX），在负电荷的金属表面静电吸附，由于胺阳离子在金属表面并列形成阻挡层，胺阳离子中的烷基R的C原子数越大，缓蚀效果就越好。与H^+反应形成阳离子的有机化合物也可以进行物理吸附，碱性愈强，形成阳离子的趋势愈强，愈易于吸附，缓蚀效率就愈高。

（2）化学吸附机理：化学吸附是缓蚀剂极性基团的未配对电子、双键、三键和苯环上的τ电子，与金属的空$d-$轨道形成电子共享完成的，改变未配对电子的电子密度可改进缓蚀效果。化学吸附比物理吸附慢，作用较强，是不可逆吸附过程，这种吸附不依靠金属表面电荷，中性分子同样可以吸附。缓蚀作用较物理吸附好一些。由于缓蚀剂提供电子与金属形成吸附，因此化合物中不同的取代基团影响极性基的电子密度，影响缓蚀效果。

substituent groups in the compound will affect electron density of the polar group, which will further affects the corrosion inhibition effect.

13.2.1.2 Hard–soft–acid–base (HSAB) theory

In chemical adsorption, as the electron donor, the polar group is the L base; while as the electron acceptor, the metal is the L acid. In some researches, acid and base are also divided into soft and hard categories. The general low valent metal cations are soft acids, and the ones with high valent are hard acids; Compounds with large electronegativity such as compounds of N, O and aniline are hard bases, while compounds with relatively low electronegativity such as compounds of S, P, Br, I or anion are soft bases. To achieve high stability, hard acid shall be combined with hard base, and soft acid shall combine with soft base. Hard acids are electron acceptors, which are with small volume, high positive charge, and low polarizability. Its outer electrons are combined firmly. Soft acids are with big volume. Their positive charges are low or no positive charges. Polarization of the soft acids is high, and they have easily excited d-electrons. The outer electrons of the soft acids are combined loosely. Hard bases are with low polarization and high electronegativity. They are difficult to oxidize and their outer electrons are tightly combined, the electrons are hard to lose. Soft bases are with high polarity and low electronegativity. They are easy to be oxidized, and their outer electrons are loosely combined, making them as good electron donors.

13.2.1.3 Molecular structure influence of corrosion inhibitor

When considering adsorption of the polar groups on corrosion inhibitors, influence of the corrosion inhibitor molecular structure should also be considered. There are two aspects that shall be considered mainly: shadowing effect and steric hindrance.

(1)Shadowing effect: The polar groups in corrosion

13.2.1.2 软硬酸碱原理

在化学吸附中，极性基团是电子给予体，是L碱；金属是电子接受体，是L酸。在研究中有人又把酸、碱分成软、硬两大类，低价的一般金属阳离子是软酸，高价的则是硬酸；电负性大的化合物如N、O的化合物和苯胺是硬碱，电负性较小的化合物如S、P、Br、I的化合物或阴离子是软碱。硬酸与硬碱结合、软酸与软碱结合，稳定性高。硬酸是受电子体，体积小，正电荷高，极化性低，其外层电子抓得牢。软酸体积大，正电荷低或等于零，极化性高，并有易激发的d-电子，外层电子抓得松。软碱极化性低，负电性高，难氧化，外层电子抓得紧，电子难失去。软碱极性高，电负性低，易氧化，外层电子抓得松，是好的电子给予体。

13.2.1.3 缓蚀剂分子结构的影响

在考虑缓蚀剂极性基团的吸附作用时，还要考虑缓蚀剂分子结构的影响，主要有遮蔽效应和空间障碍。

（1）遮蔽效应：缓蚀剂分子中极性基团吸附在金属表面，非

inhibitor molecule will adsorb on the metal surface, and the non-polar groups will cover the metal surface. In this case movement of the substance will be inhibited and corrosion inhibition effects will be achieved. For corrosion inhibitors that mainly working by effect of adsorption, their shadowing effect is related with cross section area of the molecule and the effective covering area. When both of them are large, the corrosion inhibition effect is strong.

(2)Steric hindrance: Steric hindrance refers to reduction of the corrosion inhibition efficiency of non-polar corrosion inhibitors with side chains caused by adverse effect on adsorption of molecules on metal surface due to the side chains. For example, in 2-tert-butyl sulfide, although the electronegativity of the S atom is very large, the effect of corrosion inhibition could not be shown due to steric hindrance.

Due to influence of corrosion inhibitor structure and law of polar group adsorption, the corrosion inhibition efficiency will also decrease with the decrease of adsorption force of polar groups. They are following an order like this: acid > ammonium > alcohol > ether.

13.2.1.4 Theory of film formed by interfacial reaction

In addition to corrosion inhibition effect on metal interface, corrosion inhibitors can also form an interface film by reacting, softening and integrating with the metal, consolidating the corrosion inhibition effect.

Under specific conditions, some corrosion inhibitors can be converted into another substance that interacts strongly with metal surface and inhibit corrosion. For example, dibenzyl sulfide can work as a corrosion inhibitor directly in acid, however, aldehyde can only be working as a corrosion inhibitor in acid after forming alcohol aldehyde and acetaldehyde and mixing with the croton condensate. Another type of interfacial reaction is to form a film of polymer or

极性基团覆盖在金属表面，对物质的运动起抑制作用，而起到缓蚀作用。对吸附能力相当的缓蚀剂，遮蔽效应的遮蔽范围与有效遮盖面积和分子的截面积有关，当两者都大时，缓蚀效果就强。

（2）空间障碍：空间障碍就是有支链的非极性基缓蚀剂，由于其支链的缘故，影响了其他分子与金属表面的吸附，从而降低了缓蚀效率。如2-叔丁基硫化物，S原子电负性很大，但由于空间分支障碍，导致并无缓蚀效果。

缓蚀剂结构影响，还有极性基团吸附规律影响，随着极性基团吸附力减少，缓蚀效率也减少，它们有如下顺序：酸＞铵＞醇＞醚。

13.2.1.4 界面反应成膜理论

缓蚀剂除与金属界面吸附发生缓蚀作用外，还可以通过与金属结合处发生反应和转化软化、整合等作用，形成相界膜，起到缓蚀作用。

有的缓蚀剂，在金属表面的特定条件下，转化成另外一种与金属表面作用较强的物质，起到缓蚀作用，如二苄基硫化物，在酸中起缓蚀作用的是二苄基硫化物；醛在酸中形成了醇醛和乙醛的巴豆缩合物混合物后，才起缓蚀作用。另一类界面反应的类型是形成聚合或缩合物膜，这类药

condensation compound. This type of agent could help to form polymer or condensation compound corrosion inhibitor, or form a film at the interface by reacting with a corrosion inhibiting component that could form polymer. Corrosion inhibitors of this type are with high corrosion inhibition efficiency and are suitable for a wide temperature range. Such substances include amines, condensation and polymerization products of aldehydes, polyvinyl pyridine, polyvinyl piperazine, polyvinylamine, oxy formaldehyde resin, polyacetylene polyalkylamine, triazine polymer, etc. They could form stable complex with metal ions which has a strong protective effect, inhibiting the corrosion.

Some corrosion inhibitors can react with metals or ions to form a complex film, which is an important type of phase interface corrosion inhibitor. Some of corrosion inhibitors of this type could react with metal surface and form a complex protective layer, which could inhibit corrosion. For example, mercaptobenzothiazole, benzotriazole and tannin can all form stable complexes with metals. There is also a type of corrosion inhibitor which could form a film by coordination integration products that making from metal surface ions and polar groups of the molecule, such as —HN_2, —OH, —SH, —COOH, etc. When this film is with low solubility and dense enough, it will show good metal surface adhesion, and protect the metal from being corroded.

13.2.2 Classification of corrosion inhibitor

(1) Classified by electrochemistry: According to inhibition effect on electrode process, the corrosion inhibitor can be divided into anode type (mitigating the anode reaction), cathode type (mitigating the cathode reaction), and mixed type (mitigating both of the reactions).

(2) Classified by chemistry: According to chemistry properties, the corrosion inhibitors can be divided into inorganic and organic type. This classification is

剂是以聚合物或缩合物作为缓蚀剂，或者以可以形成聚合物的缓蚀组分在界面的反应形成膜，起到缓蚀作用。这类缓蚀剂缓蚀效率较高，适合的温度范围宽。这类物质如胺、醛类缩聚产物、聚乙烯吡啶、聚乙烯呱嗪、聚乙烯胺、氧基甲醛树脂等，还有聚乙炔聚烷基胺、三嗪高聚物等，它们同金属离子形成稳定的络合物，具有强烈的保护作用，起到缓蚀作用。

有的缓蚀剂能与金属或离子反应形成络合物膜，是相界缓蚀剂的又一重要类型。它们有的与金属表面反应形成络合物保护层，起到缓蚀作用，如巯基苯并噻唑、苯并三氮唑和单宁等，都是与金属形成了稳定的络合物。还有一类缓蚀剂，通过分子中的极性基团，如—HN_2、—OH、—SH、—COOH 等与金属表面离子形成配位整合物膜，当这类膜溶解度低、致密时，才能有良好的金属表面附着性，从而起到保护作用。

13.2.2 缓蚀剂分类

（1）按电化学分类：根据缓蚀剂对电极过程的抑制作用，可分为阳极类（减缓阳极反应）、阴极类（减缓阴极反应）和混合类（同时减缓两种的反应）。

（2）按化学分类：分为无机缓蚀剂和有机缓蚀剂两大类，这种分类在研究缓蚀作用机理方面很重要。

very important for studying of the corrosion inhibition mechanism.

(3) Classified by effect of electrode reaction: According to effect of electrode reaction, the corrosion inhibitor could be divided into corrosion inhibitor with interface corrosion inhibiting effect, corrosion inhibitor with electrolyte layer corrosion inhibiting effect, corrosion inhibitor with film corrosion inhibiting effect, and corrosion inhibitor with passivation corrosion inhibiting effect.

(4) Classified by form of usage: According to form of usage, the corrosion inhibitor could be divided into oil soluble, water soluble and volatile corrosion inhibitor.

(5) Classified by type of corrosion inhibitor film: According to type of corrosion inhibitor film, the corrosion inhibitor could be divided into oxidation film type, adsorption film type, precipitation film type and reaction conversion film type corrosion inhibitor. The properties of various films types are shown in Table 13-1.

（3）按对电极反应的作用分类：分为界面缓蚀作用缓蚀剂、电解质层缓蚀作用缓蚀剂、膜缓蚀作用缓蚀剂和钝化作用缓蚀剂。

（4）按使用形式分类：分为油溶性缓蚀剂、水溶性缓蚀剂和挥发性缓蚀剂。

（5）按缓蚀剂膜的种类分类：分为氧化型膜缓蚀剂、吸附型膜缓蚀剂、沉淀型膜缓蚀剂和反应转化型膜缓蚀剂。各类膜的性质见表13-1。

Table 13-1　Properties of various surface films of corrosion inhibitors

Type of film		Typical corrosion inhibitors	Properties of the film
Oxidation film		Chromate, nitrite, molybdate	It is a dense and thin film （0.003-0.02 μm) with strong adhesion to matrix metal and good corrosion inhibiting performance
Adsorption film		Polar group containing organic compounds (such as amines), aldehydes and heterocyclic compounds	It is a film that could be formed smoothly in acid and nonaqueous solution. The film is ultra thin （Single or multi-molecular layer) and with bad stability
Precipitation film	Ion in water type	Polymeric phosphate, zinc salt, silicate	The film is porous and thick, and has poor adhesion to the matrix metal
	Metal ion type	Mercaptobenzothiazole （MBT) and some certain chelating agents	The film is dense, relatively thin, and has good adhesion to the matrix metal
Reaction conversion film		Alkyne derivatives (such as alkyne acetone), polycondensates and polymers	The film is porous, thick, and with the best corrosion resistance and stability comparing with films above

表13-1　缓蚀剂各类表面膜的性质

膜的种类		典型缓蚀剂	膜的特性
氧化性膜		铬酸盐、亚硝酸盐、钼酸盐	致密，膜薄（0.003~0.02 μm），与基本金属附着力强，防腐蚀性能优良
吸附性膜		含极性基团有机物（如胺类）、醛和杂环化合物	在酸和非水溶液中形成良好的膜，膜极薄（单或多分子层），膜的稳定性较差
沉淀膜型	水中离子型	聚合磷酸盐、锌盐、硅酸盐	膜多孔且较厚，与基体金属附着性较差
	金属离子型	巯基苯并噻唑（MBT）、某些螯合剂	膜致密、较薄，与基体金属附着性较好
反应转化型膜		炔类衍生物（如炔丙酮）、聚缩物和聚合物	膜多孔、较厚，防腐蚀性和稳定性最好

Oxidation film corrosion inhibitor will directly or indirectly oxidizes the metal to be protected by forming a metal oxidation film on its surface, preventing the corrosion reaction from proceeding. Generally, these agents have good protective effect on transition metals of Fe group that could be passivated, but little effect on non-transition metals. Precipitation film corrosion inhibitors are water soluble, but it could form a precipitation film after interacting with other ions in corrosive medium, which is poorly soluble or insoluble in water, protecting the metal from corrosion. This film is porous, and its effect is worse than that of the oxidation film. Its bonding strength with metal surface is also poor. Since there are polar groups on molecule of adsorption film corrosion inhibitors which can adsorb on surface of the metal and form a film. At the same time the non-polar groups of their molecules in water base will hinder contact of water or corrosion medium and the metal surface, protecting the metal from corrosion. The reaction conversion film is formed through interfacial reaction or transformation of the corrosion inhibitor, the corrosive medium, and the metal surface.

13.2.3　Commonly used corrosion inhibitor

The commonly used corrosion inhibitors are shown in Table 13-2:

Table 13-2　List of commonly used corrosion inhibitors

Item	Abbreviation	Molecular formula	Performance and application
Ethylene diamine tetramethylene phosphonate	EDTMP	$(HO)_2P(O)-CH_2-N(CH_2P(O)(OH)_2)-CH_2CH_2-N(CH_2P(O)(OH)_2)-CH_2-P(O)(OH)_2$	Brown viscous oily liquid, low toxicity, no pollution, good stability, can be used as slow release scale inhibitor
Ethylene diamine tetramethylene phosphonate	AMP	$(MO)_2P(O)-H_2C-N(CH_2P(O)(OM)_2)-CH_2CH_2-N(CH_2P(O)(OM)_2)-CH_2-P(O)(OM)_2$	Brown viscous oily liquid, low toxicity, no pollution, good stability, can be used as slow release scale inhibitor

续表

Item	Abbreviation	Molecular formula	Performance and application
Aminotrimethylp hosphonic acid	ATMP (NTMP)	—	Yellow viscous liquid or white crystal, low toxicity, good thermal stability, scale inhibition rate ≥85%. It is usually used in boilers and water pipes, and could be used as corrosion inhibitor also
Amino phosphate	JC—841	Aminophosphoric acid	Light yellow liquid or white solid, with effect of corrosion inhibition and strong scale inhibition performance
Scale and corrosion inhibitor	CT_1—31	Organic phosphate	Light brown liquid, low toxicity, chemically stable. It is suitable for high temperature, high hardness base water treatment, and is with effect of anti-scaling
Corrosion fouling inhibitor	CT_4—21B	Organic phosphate	White powder. It is suitable for corrosion inhibition of water injection system
Scale and corrosion inhibitor	CT_4—21A	—	Light brown red liquid. Its performance and application is same with CT_4—31
Corrosion inhibitor	CT_1—10	Organic amines and amides are its main agents.	Brownish yellow liquid, low toxicity, chemically stable. It is suitable for high salinity water and water injection systems
Corrosion inhibitor	CT_2—7	Organic amines and amides are its main agents.	Brownish red liquid. Its performance and application is same as above. This corrosion inhibitor is with strong pitting corrosion resistance
Corrosion inhibitor	SL—1	Compounding agent taking polyoxyethylene amine as main component.	Brownish yellow liquid. Performance and application information is same as above
Scale and corrosion inhibitor	8185	NTMP, HPAM, 7571 compounding agent	Dark brown liquid. It is water soluble and suitable for hot water washing system
Corrosion inhibitor	SQ	—	Brownish red liquid
Corrosion inhibitor	CT_2—1	Organic polyamine compound	Dark brown viscous liquid. It is suitable for corrosion inhibition of pipelines that transporting oil and gas containing H_2S and CO_2
Corrosion inhibitor	CT_2—2	Organic polyamine compound	Brown liquid. Application is same as above

表13-2 常用缓蚀剂一览表

名　称	代　号	分子式	性能及用途
乙二胺四甲基磷酸	EDTMP	$(HO)_2P(O)CH_2\!-\!\!\underset{\underset{CH_2P(O)(OH)_2}{\big|}}{N}\!CH_2CH_2\!N\!\underset{\underset{CH_2P(O)(OH)_2}{\big|}}{\big|}CH_2P(O)(OH)_2$	棕色黏稠油状液体，低毒，无污染、稳定性好，可用于缓释阻垢剂
乙二胺四甲基磷酸盐	AMP	$(MO)_2P(O)CH_2\!-\!\!\underset{\underset{CH_2P(O)(OM)_2}{\big|}}{N}\!CH_2CH_2\!N\!\underset{\underset{CH_2P(O)(OM)_2}{\big|}}{\big|}CH_2P(O)(OM)_2$	棕色黏稠油状液体，低毒，无污染、稳定性好，可用于缓释阻垢剂

续表

名　称	代　号	分子式	性能及用途
胺基三甲基膦酸	ATMP (NTMP)	—	黄色黏稠液体或白色结晶，低毒，热稳定性好，阻垢率≥85%，用于锅炉及水管，兼缓蚀作用
胺基磷酸盐	JC—841	胺基磷酸	淡黄色液体或白色固体，兼缓蚀作用，阻垢力强
防垢缓蚀剂	CT_1—31	有机磷酸盐	浅棕色液体，低毒，化学稳定，适合高温，高硬度碱性水处理，兼防垢作用
缓蚀缓垢剂	CT_4—21B	有机磷酸盐	白色粉末，适合注水系统缓蚀
防垢缓蚀剂	CT_4—21A	—	浅棕红色液体，性能、用途同CT_4—31
缓蚀剂	CT_1—10	有机胺和酰胺为主剂	棕黄色液体，低毒，化学稳定，适合高矿化度水和注水系统
	CT_2—7	有机胺和酰胺为主剂	棕红色液体，性能用途同上，防点蚀性强
	SL—1	聚氧乙烯胺为主的复配剂	棕黄色液体，其余同上
防垢缓蚀剂	8185	NTMP，HPAM，7571复配剂	黑褐色液体，可溶于水，适合热水热洗系统
缓蚀剂	SQ	—	棕红色液体
	CT_2—1	有机多胺化合物	棕褐色黏稠液体，适合含H_2S、CO_2油气的输送管道缓蚀
	CT_2—2	有机多胺化合物	棕色液体，用途同上

Oilfield wastewater treatment 第14章
油田含油污水处理

14.1 Overview of oilfield wastewater

14.1.1 Source of oilfield wastewater

Oilfield wastewater is a kind of sewage produced by dehydration and separation of crude oil after production, hence is also called oilfield produced water. Its main source is the crude oil dehydration station. Water remained at bottom of various crude oil storage tanks, sewage produced after cleaning the crude oil with high salt content and the well cleanout wastewater entering the sewage treatment station are also source of the wastewater in oilfield.

Normally there is not much wastewater produced in early development stage of oilfields. However, most of China's oilfields have entered the middle and late stages of development currently and relying on secondary and tertiary oil recovery technologies to improve oil recovery. Secondary oil recovery includes development methods such as water injection and gas injection, in which water injection is more commonly used. According to this method, high pressure water is injected into the formation to replace crude oil out from the oil well. This result the water content of crude oil continuously rising, the figure has reached 80% in many oilfields, and some even reached more than 90%. High water content of the production fluid will increase

14.1 油田含油污水概述

14.1.1 油田含油污水来源

油田含油污水是伴随原油采出后，经原油脱水分离而产生的一类污水，又称油田采出水。它的主要来源是原油脱水站，其次是各种原油储罐的罐底水，清洗含盐量较高的原油后的污水以及进入污水处理站的洗井废水等。

油田含油污水在油田早期开发过程中往往较少，然而目前，我国的大多数油田已经进入开发的中后期阶段，依靠二次采油和三次采油技术提高采收率。二次采油包括注水和注气等开发方式，其中注水开发较为常用，即向地层中注入高压水驱动原油使其从油井中开采出来。原油的含水率不断上升，很多油田的采出液含水率都达到了80%，有些甚至达到了90%以上。采出液高含水增加了原油集输过程中的燃料动力消耗并且会引起管道和设备的结垢腐蚀。原油脱水分离出的油田

fuel and power consumption during gathering and transportation process of the crude oil and cause fouling and corrosion of pipes and the equipment. With proper treatment, the wastewater after crude oil dehydration and separation is the main source of injection water.

14.1.2 Properties of oilfield wastewater

Although due to different oil production methods, crude oil characteristics, geological and other conditions, the properties of water produced from oilfield are vary, there are still commonalities. Generally, it contains various components such as petroleum, suspended solid, dispersed oil, suspended oil, and chemical agent. Produced water in oilfield is with the following characteristics normally:

(1) High oil content. Generally, produced wastewater contains 1000 – 2000 mg/L crude oil, and some can even reach to more than 5000 mg/L. About 90% of the crude oil is floating oil and disperse oil, and 10% is emulsified oil.

(2) Containing suspended solid particles. The particle size is generally in range of 1-100μm, mainly including clay minerals, microorganisms, corrosion scale products and organic matter. The common microorganisms are mainly iron bacteria, saprophyte and sulfate-reducing bacteria, which are monofilamentous or filaments with side chains that formed by many cells.

(3) High salt content. Generally, oilfield produced wastewater is with high content of inorganic salts ranging from several thousand to tens of thousands, or even hundreds of thousands mg/L, mainly including Ca^{2+}, Mg^{2+}, HCO_3^-, Ba^{2+}, Sr^{2+}, Cl^- and other ions.

(4) Contains dissolved gases, mainly including H_2S, CO_2, O_2 and so on.

(5) Some oilfield wastewater contains polymers. This is mainly found in tertiary polymer flooding oilfield. The polymers mainly include polyacrylamides and biopolymers. Sewage of this kind contains a large

含油污水经适当处理后是回注水的主要来源。

14.1.2 油田含油污水性质

由于采油方法、原油特性、地质等条件不同，油田采出水的性质各异，但又存在共性。一般含有石油类、固体悬浮物、分散油及悬浮油、化学药剂等多种成分。一般来说，油田的采出水有以下特点：

（1）含油量高。一般采油污水含有 1 000 ~2000 mg/L 的原油，有些含油量可达 5 000 mg/L 以上。原油中约 90% 为浮油和分散油，10% 为乳化油。

（2）含有悬浮固体颗粒。颗粒粒径一般为 1~100μm，主要包括黏土矿物、微生物、腐蚀结垢产物和有机物等。常见的微生物主要是铁细菌、腐生菌和硫酸盐还原菌，这些细菌是由多个细胞连成的单丝状或具有侧枝的丝状体。

（3）高含盐量。油田采油污水一般无机盐含量很高，从几千到几万甚至十几万 mg/L，主要包括 Ca^{2+}、Mg^{2+}、HCO_3^-、Ba^{2+}、Sr^{2+} 和 Cl^- 等离子。

（4）含溶解气体。主要包括 H_2S、CO_2、O_2 等。

（5）部分油田含油污水含聚合物。主要存在于三次采油聚合物驱油田，聚合物主要包括聚丙烯酰胺类和生物聚合物。含聚污

amount of residual polymer. It is with high viscosity, small oil bead particle diameter and high degree of emulsification.

During process of produced water injection, crude oil and suspended solids will plug the formation. With the high salinity, these pluggings will accelerate corrosion and scaling, and the organic matter will promote growth of microorganism, causing corrosion and plug. If the water is discharged, petroleum hydrocarbons and heavy metals will affect respiratory, nervous and reproductive systems of human and animals, which are carcinogenic and teratogenic. What's more, the toxicity can be enriched and accumulated step by step in animal and plant through the food chain, causing ecological environment pollution.

14.1.3 Reuse and discharge standards for oilfield wastewater

Reinjection of produced wastewater is with the following advantages: Save water resources while reducing environmental pollution caused by wastewater discharge; Produced wastewater contains surfactant substances, which can improve displacement efficiency; Since it is with high salinity, produced water will not cause clay particles expansion and reduces permeability; It could mix well with miscible phase of the oil reservoir and will not cause precipitation. However, the produced water must be processed to meet the relevant standards before reinjection.

The industry standard SY/T 5329—2012, namely *Water quality standard and practice for analysis of oilfield injecting water in clastic reservoirs* promulgated by National Energy Board, is the water quality standard for injection water. Water used for injection in reservoirs generally requires stable water quality and good fluid compatibility with oil reservoirs; clay mineral will not form hydration expansion or suspension after water is injected into oil reservoir; large amounts of suspended

水中含有大量残余聚合物、黏度大、油珠粒径小、乳化程度高。

采出水回注时原油和悬浮物会堵塞地层，高矿化度会加速腐蚀、结垢，有机物会促进微生物繁殖，造成腐蚀和堵塞。采出水外排时，石油烃和重金属能够影响生物的呼吸系统、神经系统和生殖系统，具有致癌、致畸生物毒性，并能通过食物链在动植物体内逐级富集和放大，对生态环境造成污染。

14.1.3 油田含油污水回用及排放标准

采出水用于回注有以下优点：可以节约水资源，同时减少因污水排放造成的环境污染；采出水含有表面活性剂物质，能提高洗油能力；采出水矿化度高，不会引起黏土颗粒膨胀而降低渗透率；采出水与油层混相配伍性良好，不会产生沉淀。但采出水需经处理达到相关标准后才可以回注。

注水水质标准参见国家能源局颁布的《碎屑岩油藏注水水质指标及分析方法》SY/T 5329—2012 行业标准。油藏注水一般要求水质稳定，与油层流体配伍性好；水注入油层后黏土矿物不产生水化膨胀或悬浊；水中不能含有大量悬浮物、有机淤泥和油；对设备、管线腐蚀性小；当进行两种及以上水源混注时，应首先

solids, organic sludge and oil shall not be contained in the water; the water for injection shall not etching the pipelines and equipment much; Before mixing two or more water sources, an laboratory experiment should be carried out first to confirm they are with good compatibility and will not cause damage to the oil reservoir. The main control indexes for water quality are shown in Table 14-1.

进行室内实验，证实其配伍性好，对油层无伤害。推荐水质主要控制指标见表14-1。

Table 14-1　The main control indexes for water quality

Average air permeability of injection layer/μm^2	≤0.01	>0.01-≤0.05	>0.05-≤0.5	>0.5-≤1.5	>1.5
Suspended solids content/(mg/L)	≤1.0	≤2.0	≤5.0	≤10.0	≤30.0
Median diameter of suspended solids/μm	≤1.0	≤1.5	≤3.0	≤4.0	≤5.0
Oil content/(mg/L)	≤5.0	≤6.0	≤15.0	≤30.0	≤50.0
Average corrosion rate/(mm/a)	<0.076				
SRB/(number/mL)	≤10	≤10	≤25	≤25	≤25
Iron bacteria/(number/mL)	$n \times 10^2$	$n \times 10^2$	$n \times 10^3$	$n \times 10^4$	$n \times 10^4$
Saprophyte/(number/mL)	$n \times 10^2$	$n \times 10^2$	$n \times 10^3$	$n \times 10^4$	$n \times 10^4$

表14-1　推荐水质主要控制指标

注入层平均空气渗透率/μm^2	<0.10			0.1~0.6			>0.6		
标准分级	A1	A2	A3	B1	B2	B3	C1	C2	C3
悬浮固体含量/（mg/L）	≤1.0	≤2.0	≤3.0	≤3.0	≤4.0	≤5.0	≤5.0	≤7.0	≤10.0
悬浮物颗粒直径中值/μm	≤1.0	≤1.5	≤2.0	≤2.0	≤2.5	≤3.0	≤3.0	≤3.5	≤4.0
含油量/（mg/L）	≤5.0	≤6.0	≤8.0	≤8.0	≤10.0	≤15.0	≤15.0	≤20	≤30
平均腐蚀率/（mm/a）	<0.076								
点腐蚀	A1、B1、C1级：试片各面都无点腐蚀；A2、B2、C2级：试片有轻微点腐蚀；A1、B1、C1级：试片有明显点腐蚀								
SBR菌/（个/毫升）	0	<10	<25	0	<10	<25	0	<10	<25
铁细菌/（个/毫升）	$n \times 10^2$			$n \times 10^3$			$n \times 10^4$		
腐生菌/（个/毫升）	$n \times 10^2$			$n \times 10^3$			$n \times 10^4$		

As water content in production fluid rises, the amount of produced water increases rapidly, and part of the produced water needs to be treated and discharged. Quality of water discharged shall be in accordance with GB 8978—1996, i.e. *Integrated Wastewater Discharge Standard*. The maximum allowable concentration of some pollutants for discharge is shown in Table 14-2.

随着采出液综合含水上升，采出水量快速增加，一部分采出水需要处理后外排。外排水质标准参见《污水综合排放标准》GB 8978—1996，部分污染物最高允许排放浓度见表14-2。

Table 14-2   Maximum allowable concentration of some pollutants for discharge (excerpt) (for constructions building after January 1, 1998)

S.N.	Pollutant	Source of pollutants	Standard of primarytreatment/(mg/L)	Standard of secondary treatment/(mg/L)	Standard of tertiarytreatment /(mg/L)
1	pH value	All factories and facilities that produce pollutants	6–9	6–9	6–9
2	Chromaticity (dilution factor)	All factories and facilities that produce pollutants except dye industry	50	80	—
3	Suspended solids(SS)	Municipal secondary wastewater treatment plants	20	30	—
		Other factories and facilities that produce pollutants	70	200	400
4	Chemical oxygen demand (COD)	Petrochemical industry (including refining of petroleum)	100	150	500
5	Petroleum	All factories and facilities that produce pollutants	5	10	20
6	Volatile phenol	All factories and facilities that produce pollutants	0.5	0.5	2.0
7	Sulfides	All factories and facilities that produce pollutants	1.0	1.0	1.0
8	Ammonia nitrogen	Pharmaceutical raw materials, dyes, petrochemical industry	15	50	—

表14-2   部分污染物最高允许排放浓度（摘录）（1998年1月1日后建设的单位）

序号	污染物	适用范围	一级标准/（mg/L）	二级标准/（mg/L）	三级标准/（mg/L）
1	pH值	一切排污单位	6~9	6~9	6~9
2	色度（稀释倍数）	除染料工业外其他排污单位	50	80	—
3	悬浮物（SS）	城镇二级污水处理厂	20	30	—
		其他排污单位	70	200	400
4	化学需氧量（COD）	石油化工工业（包括石油炼制）	100	150	500
5	石油类	一切排污单位	5	10	20
6	挥发酚	一切排污单位	0.5	0.5	2.0
7	硫化物	一切排污单位	1.0	1.0	1.0
8	氨氮	医药原料药、染料、石油化工工业	15	50	—

14.2 Technology and treatment agent of oilfield wastewater treatment

Water quality of wastewater from different oilfields is quite different, and purification treatment requirements are also different. Thus their treatment processes are diverse. Take the gravity wastewater treatment process as an example (Figure 14-1). Oil from

14.2 油田含油污水处理技术及处理剂

各油田含油污水水质差异较大，净化处理要求不同，处理流程多样。以重力式污水处理流程为例（图14-1），污水经除油处

the wastewater will be removed first, and the coagulant is added for coagulation treatment. Then made it flocculated and separate the flocs. After that, it shall go through buffering, lifting and filtering. Finally the bactericide will be added, which will be the last process of water treatment process. Purified water made through this process could be used for reinjection. Wastewater produced from back washing filtration containers will be recycled into other wastewater for treatment, and recovered oil from the wastewater will be sent to the crude oil gathering system or used as fuel. It can be seen that various treatment technologies and agents are involved in the wastewater treatment process.

理后投加混凝剂进行混凝处理，沉降分离絮体后得到混凝出水，再经缓冲、提升、过滤，过滤出水加杀菌剂，得到净化水，用于回注。滤罐反冲洗排水回收加入污水原水进行处理，回收的油送至原油集输系统或作为燃料。可以看出，油田含油污水处理过程中涉及多种处理技术及处理剂。

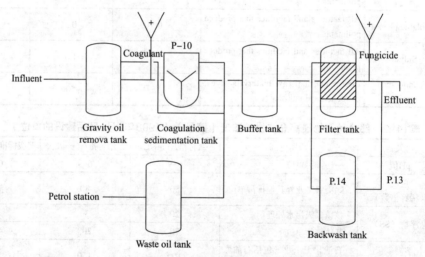

Figure 14-1　Schematic diagram of the gravity wastewater treatment process

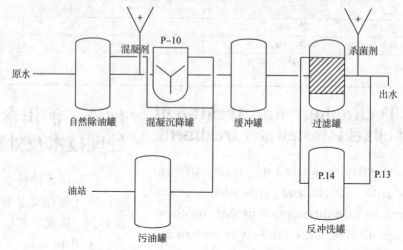

图14-1　重力式污水处理示意图

· 310 ·

14.2.1 Oil removal

Methods of oil removal are normally physical, it including gravity oil removal, inclined plate technology, coalescence (coalescent) technology, hydrocyclone, and gas flotation, etc.

Efficiency of oil removal can be improved by using some chemical reagents. Although oil in wastewater is in form of oil beads, deoiling agent can make them aggregate and easier to float in the oil removal equipment. In this case the oil will be removed quickly. The deoiling agents are:

(1) Cationic polymers. Such as hydroxy aluminum, hydroxy iron, etc. These multi-nuclear hydroxy bridge complex ions can neutralize electronegativity of the oil beads surface and bridging the oil beads, making them aggregate and float.

(2) Surfactants, such as alkyl trimethyl ammonium chloride, polyoxyethylene polyoxypropylene glycol ether, and etc. Surfactant can replace the original adsorption film on surface of the oil beads, greatly weakening their protective effect. Hence the oil beads are easy to aggregate and float.

14.2.2 Coagulation filtration

Coagulation treatment refers to remove most of the colloids, suspended solid and some dissolved organic matters in the sewage by chemical reagents. The agent used in coagulation is called a coagulant or a flocculant.

14.2.2.1 Classification of flocculants

Flocculants including: Inorganic polymer flocculant, organic polymer flocculant, bio-flocculant and composite flocculant.

Structure of inorganic polymer flocculant is easily arranged into regular form of microcrystalline, which will further compose chains and branches, such as basic aluminum chloride (BAC), polyaluminum chloride

14.2.1 除油

除油一般使用物理方法，包括重力除油、斜板除油、粗粒化（聚结）除油、旋流除油、气浮除油等技术。

除油过程中使用一些化学试剂可以提高除油效率。污水中的油是以油珠的形态存在于水中，除油剂可以使它们易于在除油设备中聚集、上浮得以去除。除油剂包括：

（1）阳离子型聚合物。如羟基铝、羟基铁等，这些多核羟桥络离子可中和油珠表面负电性和桥接油珠，使油珠聚集上浮。

（2）表面活性剂。如烷基三甲基氯化铵、聚氧乙烯聚氧丙烯丙二醇醚等表面活性剂可以取代油珠表面原有的吸附膜，大大削弱保护作用，从而使油珠易于聚集上浮。

14.2.2 混凝过滤

混凝处理是利用化学药剂去除污水中的大部分胶体、固体悬浮物和部分溶解性有机物。混凝中使用的药剂称为混凝剂或絮凝剂。

14.2.2.1 絮凝剂分类

絮凝剂包括：无机高分子絮凝剂、有机高分子絮凝剂、生物絮凝剂和复合絮凝剂。

无机高分子絮凝剂结构易排列成有规则的微晶型，进而组成链状和分枝状，如碱式氯化铝（BAC）、聚合氯化铝（PAC）、聚

(PAC), polyaluminum sulfate (PAS), polyferric sulfate (PFS) and polyferric chloride (PFC).

According to the source, organic polymer flocculants can be divided into two major categories, natural and synthetic. Flocculants made from extracting and modifying high molecular substances such as starch, cellulose, vegetable gums and chitin from nature, animals or plants are called natural polymer flocculants. Synthetic polymer flocculant is a polymer flocculant with different chain length and functional groups artificially prepared according to different needs. It can be divided into four major types: cationic type, anionic type, zwitterionic type and nonionic type. Polyacrylamide (PAM) and its derivatives are the most widely used synthetic organic polymer flocculants since they are accounting for more than 55% of the total amount.

Microbial flocculants include metabolites produced by microorganisms, microbial cell wall extracts, and microbial cells. With flocculating activity, these substances will help aggregation of suspended solids and destabilize the colloids. Currently, the research on raw materials, operations and equipment of microbial flocculant is not mature enough and the production cost is high. Hence it still can't compete with inorganic and organic flocculants.

Inorganic, organic and microbial flocculants are with their own advantages and disadvantages. For example, inorganic polymeric flocculants are with relatively low cost, but their degree of polymerization and flocculation performance in water treatment is generally lower than organic flocculants. For complex and stable sewage, treatment effect of a single flocculant is often not ideal. Hence more and more researches are carrying out on composite flocculants. Composite flocculant can be divided into two types: compound flocculant and composite flocculant. According to the specific sewage quality and water conditions, two

合硫酸铝（PAS）、聚合硫酸铁（PFS）和聚合氯化铁（PFC）等。

有机高分子絮凝剂按照来源不同可以分为天然和人工合成两大类。将存在于自然界或动植物体内的高分子物质例如淀粉、纤维素、植物胶类和甲壳素等提取后进行改性，制备的产品称为天然高分子絮凝剂。人工合成高分子絮凝剂，是根据不同需求人工制备的具有不同链长度和官能团的高分子絮凝剂，可以分为阳离子型、阴离子型、两性型和非离子型四大类型。在合成类有机高分子絮凝剂中，聚丙烯酰胺（PAM）及其衍生物应用最为广泛，占聚合型高分子絮凝剂总量的55%以上。

微生物絮凝剂包括微生物产生的代谢产物、微生物细胞壁提取物和微生物细胞，这些物质具有絮凝活性，能使固体悬浮物连接在一起，使胶体脱稳。目前，微生物絮凝剂原料、操作和设备研究还不够成熟，生产成本较高。仍然不能与无机、有机絮凝剂竞争。

无机、有机和微生物絮凝剂各有优点和不足。例如，无机聚合絮凝剂价格较低廉，但是其在水处理中聚合度及絮凝效果通常仍低于有机絮凝剂。对于复杂稳定的废水，单一絮凝剂的处理效果往往不太理想。对于复合絮凝剂的研究越来越多。复合絮凝剂可以分为复配絮凝剂和复合使用絮凝剂。根据具体废水水质和水况不同，使用两种或两种以上絮

or more flocculants shall be used in combination. Their synergistic effect will improve the flocculation efficiency and reduce the treatment cost.

14.2.2.2 Mechanism of coagulation

The coagulation process of wastewater is complicated, and it is generally believed that there are two processes in it: "agglutination" and "flocculation". The process of colloids destabilize and form fine aggregates is called "aggregation", and the process of aggregates aggregating to large flocs is called "flocculation." In actual coagulation, these two processes can be regarded as carrying out simultaneously and cannot be separated. Coagulation mechanism may vary according to different coagulant type and water quality. Generally speaking, there are four kinds of mechanism: electric neutralization, compression of the electrical double layer, adsorption bridging, and sweep net capturing. The colloidal ions are charged. Electric neutralization refers a process that partial charges of the colloidal particles are neutralized by ions with the opposite charge. This will reduce electrostatic repulsion, making ions easily collide and aggregate with other particles.

The micelle has an electric double layer structure, that is, a compact layer and a diffusion layer. Concentration of the counter ions on surface of colloidal particles is the largest, and it will gradually decrease on direction from surface of colloidal particles to the solvent. Compressed electric double layer refers a process that adding electrolyte to the solution and making the ion concentration increase. By doing this thickness of the diffusion layer is reduced, thus when two colloidal particles are getting close, the collision distance will be reduced, and the mutual repulsive force will be reduced also. Then the colloidal particles will be easy to collide and aggregate.

Adsorption bridging refers that since the polymer

coagulants are all with linear structures and each linear molecular chain can adsorb many colloidal particles like the bridges. And this is helpful to form large flocs. Sweep net capturing means that colloidal particles can be trapped by the formed flocs or precipitated to form larger flocs, speeding the precipitation up.

14.2.2.3 Settlement

After the wastewater is coagulated, the sewage with flocs (alum flocs) will enter coagulation sedimentation tanks for sedimentation and separation. Since oil and some of the suspended substances are with small density, they could be removed by floating. Flocs with large density will sink to bottom of the tanks and removed. After sedimentation and separation, there are still fine flocs, suspended solids, colloidal substances, emulsified oil and microorganisms, which can be further filtered. During filtration process, the wastewater will slowly flow through a filter bed/tank with a certain thickness. The filter bed/tank is normally composed of a porous medium. The impurities will be adsorbed on surface of the porous filter material. As the filtration proceeds, the impurities will deposit on surface of the filter and form a film. Hence in order to improve filtration effect and increase filtration resistance, the filter material should be backwashed regularly. The filter material is required to have sufficient mechanical strength, chemical stability, reasonable price, certain porosity, spherical shape and large specific surface area as well as extensive supply sources. Commonly used filter materials include quartz sand, walnut shell, activated carbon, anthracite, ceramsite, polyvinyl chloride particles, etc. Filtration efficiency is related to material, pore size and particle size of the filter material. To conduct treatment of different water quality, different filter materials can be combined with different sizes into a double-layer or three-layer filter tank. In order to prevent loss of the filter material, a support layer is also required at the bottom.

14.2.2.3 沉降

含油污水经混凝处理后，带有絮体（矾花）的污水进入混凝沉降罐进行沉降分离，油和部分悬浮物密度较小，上浮去除。絮体密度较大，下沉至罐底被除去。经沉降分离后，出水仍含有细小的絮体、悬浮物、胶体物质、乳化油和微生物等，可以进一步进行过滤处理。过滤过程中，污水缓慢流经多孔介质组成的具有一定厚度的过滤床/池，杂质被吸附在多孔滤料表面，随着过滤进行，杂质储集在滤料表面形成一层膜。提高过滤效果，同时提升过滤阻力，需定期对滤料进行反冲洗清理。滤料需具有足够的机械强度、化学稳定性、来源广泛、价格合理、具有一定的孔隙度、外形为球状、具有较大的比表面积。常用的滤料包括石英砂、核桃壳、活性炭、无烟煤、陶粒、聚氯乙烯颗粒等，过滤效率与滤料的材质、孔径、粒径等相关。针对处理水质不同，可由不同材质、粒级的滤料组合成双层、三层滤料池，为防止滤料流失，底部还需设置承托层。

14.2.3 Corrosion inhibition

Since it is with high salinity and contains corrosive gases and microorganisms, oilfield wastewater may cause chemical and electrochemical corrosion, damaging conveyers and processing equipment. In order to prevent corrosion of metal equipment, corrosion inhibitors can be used. Corrosion inhibitor refers to a material that could prevent or slow metal corrosion with a small dosage in corrosive medium. According to inhibiting effect of corrosion inhibitor on the electrode process, it can be divided into three categories:

(1) Anodic corrosion inhibitor. It is with ability of oxidization, can passivate the metal surface and inhibit metal corrosion. This type of inhibitors includes dichromate, molybdate, nitrite, salicylate and the like.

(2) Cathodic corrosion inhibitor. It can reduce or eliminate depolarizers or increase cathode reaction over potentials. This type of inhibitors includes hydrazine, hydrazine, sodium sulfite, arsenic salts, mercury salts, and the like.

(3) Mixed corrosion inhibitor. It can slow down reaction rate of anodic and cathodic reactions simultaneously, also known as masking corrosion inhibitor. It can directly adsorb or attach to metal surface, or form insoluble protective film by a secondary reaction to separate the metal from corrosive medium. This type of inhibitors includes zinc sulfate, barium chloride and the like.

There are many kinds of corrosion inhibitors, and the mechanism is complicated. It mainly includes electrochemical theory, adsorption theory and film formation theory (see the corrosion inhibition chapter for details).Corrosion inhibitors with good effects in oilfields include nitrogen containing organic compounds, fatty amines and their salts, amides and imidazolines. Different corrosion inhibitors shall be chosen according to nature of the corrosive medium. For example, fatty

14.2.3 防腐缓蚀

油田含油污水矿化度高、含有腐蚀性气体和微生物,与输送设备、处理设备接触时,会产生化学、电化学腐蚀,破坏设备。为防止金属设备腐蚀可以投加缓蚀剂。凡是在腐蚀介质中添加少量物质就能防止或减缓金属的腐蚀,这类物质就成为缓蚀剂。根据缓蚀剂对电极过程的抑制作用可以将其分为三类:

(1)阳极型缓蚀剂。具有氧化性,能使金属表面钝化而抑制金属溶蚀,如重铬酸盐、钼酸盐、亚硝酸盐、水杨酸盐等。

(2)阴极型缓蚀剂。能减少或消除去极化剂或增加阴极反应过电位。如肼、联胺、亚硫酸钠、砷盐、汞盐等。

(3)混合型缓蚀剂。可以同时减缓阴、阳极反应速度,也称为掩蔽型缓蚀剂。能直接吸附或附着在金属表面,或由次生反应形成的不溶性保护膜使金属与腐蚀介质隔离。如硫酸锌、氯化铍等。

缓蚀剂种类繁多,机理错综复杂,主要包括电化学理论、吸附理论和成膜理论(详见缓蚀章节)。油田中使用效果较好的缓蚀剂包括含氮有机化合物、脂肪胺及其盐类、酰胺及咪唑啉类等。针对腐蚀介质的性质不同,选择不同的缓蚀剂。例如,抑制硫化氢腐蚀可以用脂肪胺类缓蚀剂,抑制二氧化碳腐蚀可以用咪唑啉类缓蚀剂。在进行现场投加之前,要先进行室内挂片实验,确定合

amine corrosion inhibitor can be used to suppress hydrogen sulfide corrosion, and imidazoline corrosion inhibitor can be used for inhibiting carbon dioxide corrosion. Before onsite application, an indoor coupon test should be carried out to determine the appropriate corrosion inhibitor type, dosage and dosing method.

14.2.4　Scale removal and prevention

The oilfield wastewater contains a high concentration of carbonate, sulfate and chloride. When the temperature, pressure, dissolved gas and other conditions change, the ion balance will be broke and scales such as calcium carbonate and calcium sulfate will be formed and mixing with organic matter, corrosion products, and etc., these scales will block the pipeline, increase friction and energy consumption, as well as affecting normal production.

The common scales in oilfield water are:

(1) Calcium carbonate scale. Formed by combination of calcium ion and carbonate ion or bicarbonate ion, the reaction is as follows:

$$Ca^{2+}+CO_3^{2-}=\!=\!=CaCO_3\downarrow$$

$$Ca^{2+}+2HCO_3^-=\!=\!=CaCO_3\downarrow +CO_2\uparrow +H_2O$$

(2) Magnesium carbonate scale. Magnesium carbonate will hydrolyze in water and form magnesium hydroxide, which is with low solubility:

$$MgCO_3+H_2O \longrightarrow Mg(OH)_2+CO_2$$

(3) Calcium sulfate scale.

$$Ca^{2+}+SO_4^{2-}=\!=\!=CaSO_4$$

(4) Barium sulfate scale.

$$Ba^{2+}+SO_4^{2-}=\!=\!=BaSO_4$$

(5) Iron precipitate.

The common scale inhibitors of oilfields include:

14.2.4　除垢防垢

油田含油污水含有较高浓度的碳酸盐、硫酸盐、氯化物，当温度、压力、溶解气等条件改变，打破离子平衡就会形成碳酸钙、硫酸钙等水垢，与污水中有机质、腐蚀产物等一起混合堵塞管线，增大摩阻，增加能耗，影响正常生产。

油田水常见的水垢：

（1）碳酸钙垢。由钙离子与碳酸根或碳酸氢根结合生成，反应如下：

（2）碳酸镁垢。碳酸镁在水中水解成溶解度很小的氢氧化镁：

（3）硫酸钙垢。

（4）硫酸钡垢。

（5）铁沉淀物。

油田常用的防垢剂包括：

(1) Phosphates.

Phosphates include inorganic phosphates and organic phosphates (Figure 14-2). Inorganic phosphates, such as trisodium phosphate, tetrasodium pyrophosphate, etc. can effectively prevent calcium carbonate scale, but are easy to rapidly hydrolyze with calcium ions and forming insoluble calcium phosphate. Their applicable temperature is not higher than 80 ℃. Organic phosphates, such as amino trimethylene phosphonic acid (ATMP), hydroxyethylidene diphosphonic acid (HEDP), ethylenediamine tetramethylene phosphoric acid (EDTMP), etc. are not easily hydrolyzed. Since they are with excellent scaling prevention effect and compatibility, they are widely used.

（1）磷酸盐类。

包括无机磷酸盐和有机磷酸盐（图 14-2），无机磷酸盐如磷酸三钠、焦磷酸四钠等可以有效防碳酸钙垢，但是易于随温度升高快速水解与钙离子形成难溶的磷酸钙，适用温度不高于80℃。有机磷酸盐如氨基三甲基膦酸（ATMP）、羟基亚乙基二膦酸（HEDP）、乙二胺四甲叉磷酸（EDTMP）等不易水解，防垢效果和配伍性好，使用广泛。

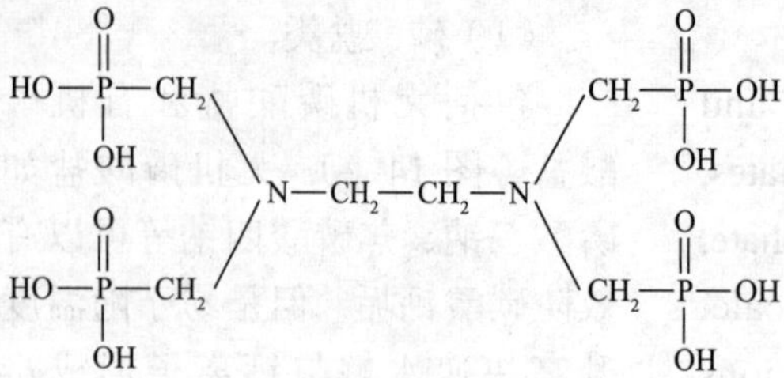

Figure 14-2   Phosphate scale inhibitors

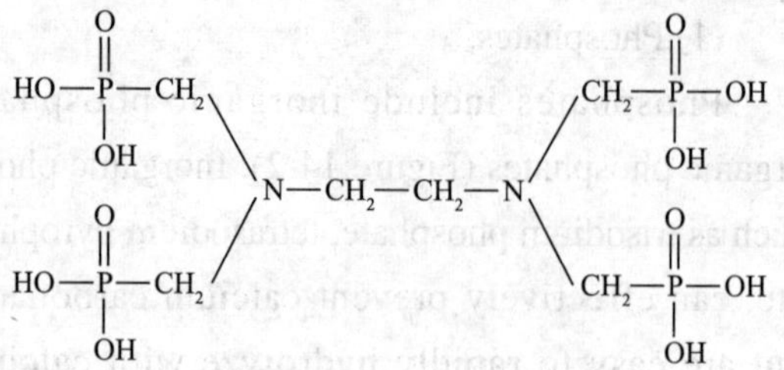

图14-2   磷酸盐类防垢剂

(2) Polymers.

As a polymer scale inhibitors, polymaleic anhydride (HPMA) could prevent scales of calcium carbonate and barium sulfate. Other scale inhibitors, such as polyacrylic acid (PAA) and polyacrylamide (PMA), are also belonging to polymer scale inhibitors (Figure 14-3).

（2）聚合物类。

聚合物防垢剂中聚马来酸酐（HPMA）具有防碳酸钙、硫酸钡的能力，此外还有聚丙烯酸（PAA）、聚丙烯酰胺（PMA）等聚合物防垢剂（图14-3）。

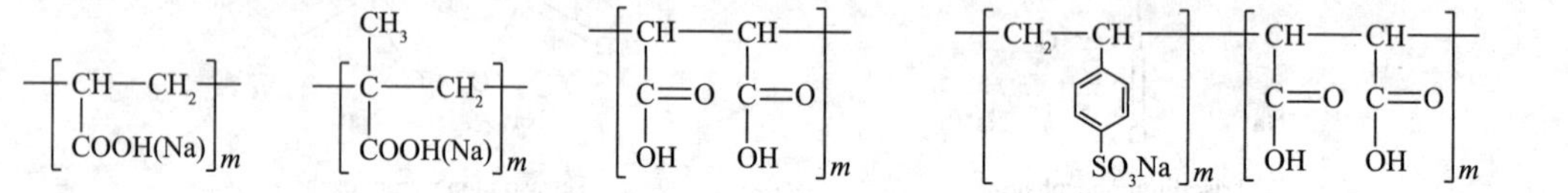

Sodium polyacrylate   Sodium polymethacrylate   Polymaleic acid   Copolymer of styrene sulfonic acid-polymaleic anhydride

Figure 14-3   Polymer scale inhibitors

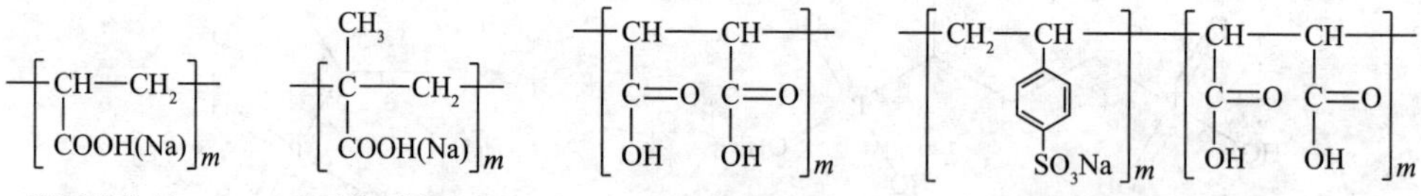

聚丙烯酸(钠)   聚甲基丙烯酸(钠)   聚马来酸   苯乙烯磺酸-马来酸酐共聚物

图14-3   聚合物类防垢剂

Two or more scale inhibitors can be used in combination for synergy effect.

Mechanism of scale inhibitor includes:

(1) Dispersion. Polymer scale inhibitor is with high charge density, and copolymer also with surface activity, crystalline core of the scales are dispersed and stable. In this case forming of scale is prevented.

(2) Masking. Combination of chelation or complexation scale inhibitors can form precipitated metal ions, inhibiting formation of precipitates by metal ions and anions.

两种或两种以上防垢剂可复配使用，以期协同增效。

防垢剂的作用机理包括：

（1）分散作用。聚合物防垢剂具有较高的电荷密度，共聚物还具有表面活性，成垢晶核分散稳定，起到防垢的效果。

（2）掩蔽作用。螯合或络合防垢剂结合能形成沉淀的金属离子，抑制金属离子与阴离子形成沉淀；

(3) Flocculation. The long molecular chain of coagulant will adsorb crystalline core of the scales and form flocs, and the flocs will be removed later. Thereby forming of scale is prevented.

(4) Change the crystal form. Scale inhibitor will enter the crystal structure and change the original crystal form, so the crystal cannot continue to grow. Thereby formation of the crystal scale is prevented.

14.2.5 Sterilization

Oilfield wastewater contains a large amount of organic pollutants, which will provide conditions for bacterial growth. Bacteria and their metabolites may cause corrosion and blockage in processing equipment and pipelines. Common harmful bacteria include sulfate-reducing bacteria (SRB), saprophytic bacteria (TGB), iron bacteria (FB) and the like. Sulfate-reducing bacteria will reduce inorganic sulfate in water to hydrogen sulfide under anaerobic conditions, which will corrode equipment and pipelines, and the corrosion product ferrous sulfide will cause blockage. Saprophyte and its mucus will mix with other impurities and forming bacterial coherent sludges. These sludges will adhere and block the pipeline and equipment.

Commonly used bactericides include:

(1) Oxidation type bactericide. By strong oxidation in water, such bactericides will damage the bacterial cell structures, completing sterilization.

①Chlorine gas. Chlorine gas is a commonly used oxidation type bactericide in oilfields. Chlorine will produce hypochlorous acid in water. The hypochlorous acid will then decompose and produce atomic oxygen, which could enter cells of bacteria, oxidize and destroy their metabolic system.

$$Cl_2+H_2O \Longrightarrow HCl+HClO(次氯酸/Hypochlorous\ acid)$$

$$HClO \longrightarrow [O]+HCl$$

②Chlorine dioxide. Chlorine dioxide is a broad spectrum bactericide that could sterilize pathogenic microorganisms, heterotrophic bacteria, sulfate reducing bacteria and fungi effectively. Chlorine dioxide also produces atomic oxygen in water, which is with 2.5 times oxidizing ability than chlorine. Its diffusion speed and penetration ability in water is also faster than chlorine, making better sterilization effect and application prospect.

②二氧化氯。二氧化氯是一种广谱型杀菌剂，对病原微生物、异养菌、硫酸盐还原菌、真菌等均有较高的杀灭作用。二氧化氯在水中也可产生原子态氧，其氧化能力是氯气的2.5倍，在水中的扩散速度和渗透能力也比氯快，具有较好的杀菌效果和应用前景。

$$2ClO_2+H_2O \longrightarrow 2HCl+5[O]$$

③Ozone. Ozone (O_3) is an allotrope of oxygen. It consists of an oxygen molecule carrying an oxygen atom [O].Ozone is unstable and with strong oxidative ability.

③臭氧。臭氧（O_3）是氧气的同素异形体，由一个氧分子携带一个氧原子[O]组成，具有不稳定性和很强的氧化能力。

$$O_3 \longrightarrow O_2+[O]$$

In addition to above, potassium ferrate, potassium permanganate, ammonium chloride, sodium hypochlorite can also be used for sterilization according to mechanism of oxidation. The oxidation type bactericides are with good bactericidal effect, low price and extensive source of supply, but their effects are short and stability is not good.

(2) Non-oxidization type bactericide. Bactericide of this type can be further divided into non-ionic type and ionic type. Non-ionic type bactericides such as organic aldehydes, cyanide-containing compounds, heterocyclic compounds, etc. (Figsure. 14-4) are mainly conducting sterilization by infiltrating into bacterial cells or complexing with bacteria components.

此外，高铁酸钾、高锰酸钾、氯铵、次氯酸钠等也可以通过氧化作用杀菌。氧化型杀菌剂杀菌效果好、价格低、来源广，但其效果维持时间短、稳定性较差。

（2）非氧化型杀菌剂。可分为非离子型和离子型。非离子型杀菌剂如有机醛类、含氰类化合物、杂环化合物等（图14–4），主要是靠渗透到细菌细胞内或与细菌的组分络合来杀菌。

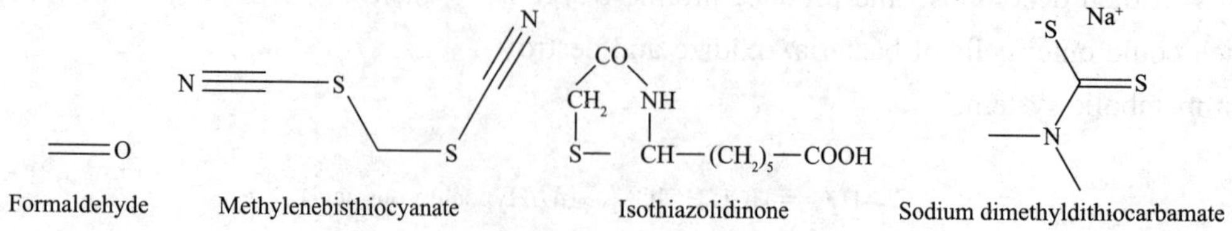

Figure 14-4 Commonlyused non-ionic bactericides

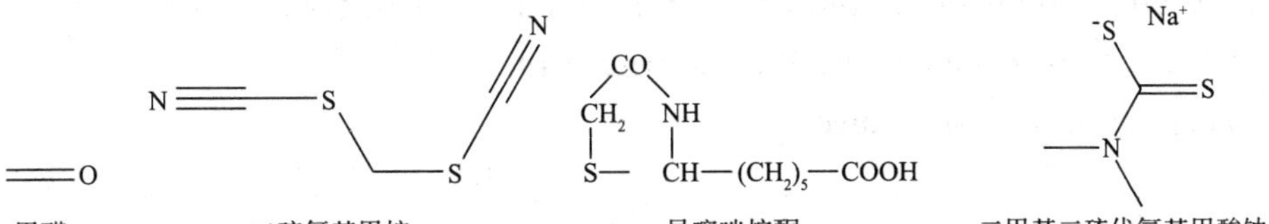

Oilfield wastewater treatment
油田含油污水处理 第14章

图14-4 常见的非离子型杀菌剂

Ionic bactericides include quaternary ammonium salts, quaternary phosphonium salts, alkyl guanidine and the like (Figure. 14-5). Among them, the quaternary ammonium salts are one of the most extensively used cationic bactericide type since they could not only be used for sterilization but also could strip coherent sludges.

离子型杀菌剂包括季铵盐类、季磷盐、烷基胍等（图14-5）。其中，季铵盐是目前使用广泛、有效的阳离子杀菌剂之一，不仅具有杀菌效果，对黏泥也有很好的剥离效果。

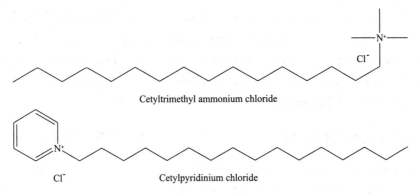

Cetyltrimethyl ammonium chloride

Cetylpyridinium chloride

Figure 14-5 Commonly used ionic bactericides

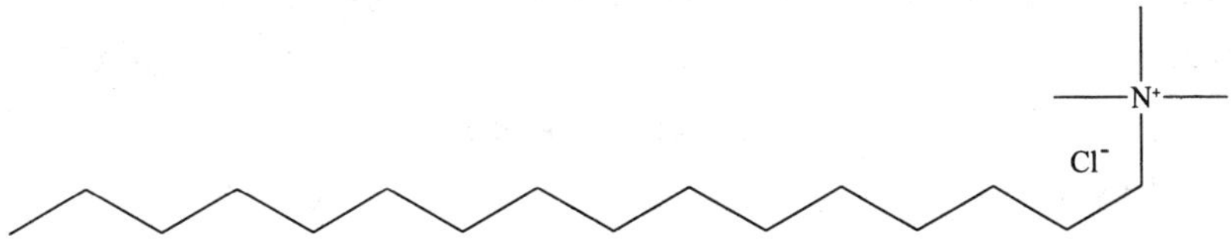

十六烷基三甲基氯化铵

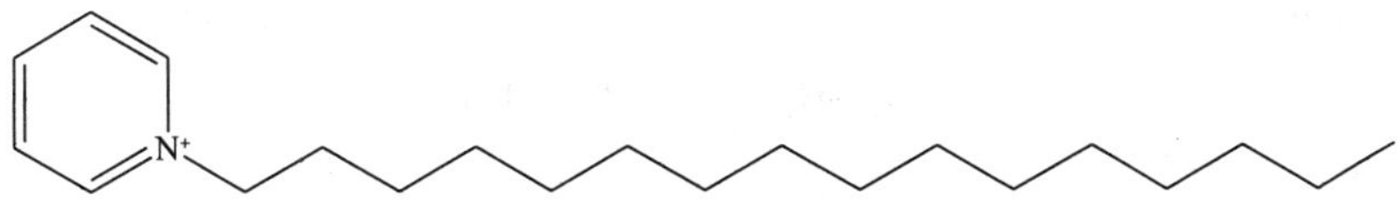

十六烷基氯化吡啶

图14-5 常见的离子型杀菌剂

Cationic polymer bactericides such as bis-quaternary ammonium salts or poly-quaternary ammonium salts could be obtained from modifying the quaternary ammonium salts. For example, the bis-

在季铵盐的基础上对其进行改性可以得到双季铵盐、聚季铵盐等阳离子聚合物杀菌剂。如长链叔胺与1,3-二溴丙烷在溶剂乙

quaternary ammonium salts could be synthesized by reaction of long chain tertiary amine and 1,3-dibromopropane in ethanol solvent.

醇中反应合成双季铵盐。

$$C_{12}H_{25}N(CH_3)_2 + BrCH_2CH_2CH_2Br \longrightarrow [C_{12}H_{25}-\underset{CH_3}{\overset{CH_3}{N}}-(CH_2)_2-\underset{CH_3}{\overset{CH_3}{N}}-C_{12}H_{25}]^{2+} 2Br^-$$

Under catalytic conditions, poly-quaternary ammonium salts could be synthesized by copolymerization of dichloroethyl ether and tetramethylethylenediamine.

在催化条件下，二氯乙醚与四甲基乙二胺可发生共聚反应合成聚季铵盐类。

$$n\underset{CH_3}{\overset{CH_3}{N}}-CH_2CH_2-\underset{CH_3}{\overset{CH_3}{N}} + nCl-CH_2CH_2-O-CH_2CH_2-Cl \xrightarrow[\Delta]{\text{催化剂/Catalyst}} \left[-\underset{CH_3}{\overset{CH_3}{N^+}}-CH_2CH_2-\underset{CH_3}{\overset{CH_3}{N^+}}-CH_2CH_2-O-CH_2CH_2-\right]_n 2nCl^-$$

Bactericide mechanism includes affecting the bacterial enzyme activity, inhibiting protein synthesis, destroying cell wall of bacteria, and inhibiting synthesis of nucleic acid, etc. To a compound bactericide obtained by compounding two or more bactericides and surfactants, since each agent can work synergistically, the bactericidal ability will be increased.

杀菌剂作用机理包括影响菌体酶活性、抑制蛋白质合成、破坏细菌细胞壁、阻碍核酸合成等。对两种及两种以上杀菌剂与表面活性剂复配得到的复合型杀菌剂，各单剂之间协同增效，提高杀菌能力。

14.2.6 Oxygen removal

Dissolved oxygen in sewage will aggravate corrosion of the metal and should be removed. The commonly used oxygen scavengers and their mechanisms are as follows:

(1) Sodium sulfite.

$$2Na_2SO_3 + O_2 \longrightarrow 2Na_2SO_4$$

(2) Formaldehyde.

$$2CH_2O + O_2 \longrightarrow 2HCOOH$$

(3) Hydrazine.

$$NH_2-NH_2 + O_2 \longrightarrow N_2 + 2H_2O$$

(4) Thiourea.

$$4NH_2-\underset{}{\overset{S}{C}}-NH_2 + O_2 \longrightarrow 2\ \underset{H_2N}{\overset{HN}{\underset{}{}}}C-S-S-C\underset{NH_2}{\overset{NH}{\underset{}{}}} + 2H_2O$$

(5) Isoascorbic acid.

14.2.6 除氧

污水中的溶解氧会加剧金属腐蚀，应将其去除，常用的除氧剂及机理如下：

（1）亚硫酸钠。

（2）甲醛。

（3）联氨。

（4）硫脲。

（5）异抗坏血酸。

$$2\ \begin{array}{c}HO-C-CH-CH-CH_2OH\\ HO-C\ \ \ O\ \ \ OH\\ \|\\ C\\ \|\\ C\end{array}\ \text{(异抗坏血酸)} + O_2 \longrightarrow 2\ \begin{array}{c}O=C-CH-CH-CH_2OH\\ O=C\ \ \ O\ \ \ OH\\ \|\\ C\\ \|\\ C\end{array} + 2H_2O$$

参 考 文 献

[1] 杨昭,李岳祥.油田化学[M].哈尔滨:哈尔滨工业大学出版社,2016.
[2] 陈大均,陈馥.油气田应用化学(第二版)[M].北京:石油工业出版社,2015.
[3] 张玉亭.胶体与界面化学[M].北京:中国纺织出版社,2008.
[4] Birdi K S. Surface and Colloid Chemistry:Principles and Applications[M]. Bosa Roca:Taylor & Francis Inc,2009.
[5] 张玉平.油田基础化学[M].天津:天津大学出版社,2006.
[6] 李永太.提高采收率原理与方法[M].北京:石油工业出版社,2008.
[7] 赵福麟.油田化学[M].东营:中国石油大学出版社,2007.
[8] 赵福麟.油田化学常用述语.中华人民共和国石油天然气行业标准.SY/T 5510.
[9] Pål Skalle. Drilling Fluid Engineering[M]. TELLURIDE:Ventus Publishing,2011.
[10] Faïza Bergaya,Benny K.G. Theng and Gerhard Lagaly(Eds.). Developments in Clay Science 1 Handbook of Clay Science[M]. Amsterdam:Academic Press,Elsevier,2006.
[11] Johannes Karl Fink. Petroleum Engineer's Guide to Oil Field Chemicals and Fluids,Second Edition[M]. Houston:Gulf Professional Publishing,2015.
[12] Ryen Caenn,H.C.H. Darley,George R. Gray. Composition and Properties of Drilling and Completion Fluids(Seventh Edition)[M]. Houston:Gulf Professional Publishing,2017.
[13] Alexandre Lavrov. Lost Circulation:Mechanisms and Solutions[M]. Houston:Gulf Professional Publishing,2016.
[14] Johannes Karl Fink. Water-Based Chemicals and Technology for Drilling,Completion,and Workover Fluids[M]. Houston:Gulf Professional Publishing,2015.
[15] 张明昌.固井工艺技术[M].北京:中国石化出版社,2016.
[16] 丁保刚,王忠福.固井技术基础[M].北京:石油工业出版社,2006.
[17] 《钻井手册》编写组.钻井手册(上)[M].北京:石油工业出版社,2013.
[18] Editing group of Drilling handbook. Drilling handbook. Volume 1[M]. Beijing:Petroleum Industry Press,2013.
[19] 齐国强,王忠福.固井技术基础(第2版)[M].北京:石油工业出版社,2016.
[20] 张德润,张旭.固井液设计及应用[M].北京:石油工业出版社,2002.
[21] ZHANG Derun,ZHANG Xu. Design and application of cementing fluid[M]. Beijing:Petroleum Industry Press,2002.
[22] 周金葵,李效新.钻井工程[M].北京:石油工业出版社,2007.
[23] 孙明光.钻井、完井工程基础知识手册[M].北京:石油工业出版社,2002.
[24] 孙富全,侯薇,靳建洲,等.超低密度水泥浆体系设计和研究[J].钻井液与完井液,2007,24

(3): 31-34.
- [25] 姚志翔. 超高密度水泥浆体系的研究与应用 [J]. 钻井液与完井液, 2015, 32 (1): 69-72.
- [26] 靳建洲, 孙富全, 侯薇, 等. 胶乳水泥浆体系研究及应用 [J]. 钻井液与完井液, 2006, 23 (2): 37-39.
- [27] 卢海川, 李洋, 宋元洪, 等. 新型固井触变水泥浆体系 [J]. 钻井液与完井液, 2016, 33 (6): 73-78.
- [28] 肖京男, 刘建, 桑来玉, 等. 充气泡沫水泥浆固井技术在焦页9井的应用 [J]. 断块油气田, 2016, 23 (6): 835-837.
- [29] 杜凯, 黄凤兴, 伊卓, 等. 页岩气滑溜水压裂用降阻剂研究与应用进展 [J]. 中国科学: 化学, 2014, 44 (1): 1696-1704.
- [30] 车航, 崔会杰, 邱兵, 等. 压裂液延迟交联与快速破胶技术 [J]. 钻井液与完井液, 2003, 20 (4): 24-26.
- [31] 严志虎, 戴彩丽, 赵明伟, 等. 清洁压裂液的研究与应用进展 [J]. 油田化学, 2015, 32 (1): 141-145.
- [32] YAN Zhihu, DAI Caili, ZHAO Mingwei, et al. Application and research progress on clean fracturing fluid [J]. Oilfield chemistry, 2015, 32 (1): 141-145.
- [33] 中华人民共和国石油天然气行业标准 SY/T 6376—2008, 压裂液通用技术条件.
- [34] Mumallah N.A. Factors Influencing the Reaction Rate of Hydrochloric Acid and Carbonate Rock, SPE 21036.
- [35] Li Y, Sullivan R.B. An Overview of Current Acid Fracturing Technology With Recent Implications for Emulsified Acids. SPE26581.
- [36] Templeton C.C. and Richardson E.A., Karnes, G.T. and Lybarger, J.H.: Self-Generating Mud Acid, SPE5153.
- [37] Lund K., Fogler, H.S. and McCune C.C.: On predicting the flow and Reaction of HCl/HF Acid Mixtures in Porous Sandstone Cores, Soc. Pet. Eng. J. (1976) 248-260; Trans., AIME, 261.
- [38] 闫治淘, 许新华, 涂勇, 等. 国外酸化技术研究进展 [J]. 油气地质与采收率, 2002, 9 (2): 86-87.
- [39] 关富佳, 姚光庆, 刘建民. 泡沫酸性能影响因素及其应用 [J]. 西南石油学院学报, 2004, 26 (1): 65-67.
- [40] 陈赓良, 黄瑛. 酸化工作液缓速作用的理论与实践 [J]. 钻井液与完井液, 2004, 21 (1): 50-54.
- [41] 关富佳, 姚光庆, 向蓉. 乳化酸的优越性能及油层酸化应用研究 [J]. 新疆石油学院学报, 2003, 15 (2): 50-52.
- [42] 付美龙. 油田化学原理 [M]. 北京: 石油工业出版社, 2015.
- [43] 于涛, 丁伟, 曲广淼. 石油高等院校特色教材, 油田化学剂(第二版)[M]. 北京: 石油工业出版社, 2008.
- [44] 岳湘安, 王尤富, 王克亮. 提高石油采收率基础 [M]. 北京: 石油工业出版社, 2007.
- [45] 侯吉瑞. 提高原油采收率新进展: Advances in enhanced oil recovery [M]. 北京: 化学工业出版社, 2015.

[46] 李永太，孔柏岭，李辰．全过程调剖技术与三元复合驱协同效应的动态特征［J］．石油学报，2018，39（6）．

[47] 杜吉国．新型水基清蜡剂的研制［D］．西安：西安石油大学，2017．

[48] 吴自勤，王兵，孙霞．薄膜生长［M］．第2版．北京：科学出版社，2017．

[49] 尚兴隆，白博峰．壁面吸附复合液滴的变形与运动特性［J］．工程热物理学报，2017，38（12）：2636-2640．

[50] 王伯君．井下固体防蜡防垢技术的研究［D］．大庆石油学院，2006．

[51] 刘爱华，章结斌，陈创前．油田清防蜡技术发展现状［J］．石油化工腐蚀与防护，2009，26（1）：1-4．

[52] 潘一，杨尚羽，杨双春，等．化学防砂剂的研究进展［J］．油田化学，2015，32（3）：449-454．

[53] Antus Mahardhini, Izzad Abidiy, Hugues Poitrenaud, et al. Chemical Sand Consolidation as a Failed Gravel Pack Sand-Control Remediation on Handil Field, Indonesia［C］．SPE-174240-MS，2015．

[54] Dana Aytkhozhina, David Mason, Raul Marulanda, et al. Chemical Sand Consolidation – Developing a Strategic Capability Across a Wide Portfolio［C］．SPE/IADC-173092-MS，2015．

[55] F Haavind, SS Bekkelund, A Moen, et al. Experience With Chemical Sand Consolidation as a Remedial Sand-Control Option on the Heidrun Field［C］．SPE 112397，2008．

[56] Castor T P, Somerton W H, Kelly J F. Recovery mechanisms of alkaline flooding［M］//Surface phenomena in enhanced oil recovery. Springer, Boston, MA, 1981：249-291. Mayer E H, Weinbrandt R M, Irani M R, et al. ALKALINE WATERFLOODING ITS THEORY, APPLICATION AND STATUS［J］．Enhanced Oil Recovery，1982：191．

[57] 赵福麟,张艳玉,徐英霞．有中间相存在的微乳状液相图［J］．华东石油学院学报（自然科学版），1985，4：005．

[58] Buckley J S, Liu Y, Monsterleet S. Mechanisms of wetting alteration by crude oils［J］．SPE journal，1998，3（01）：54-61．

[59] Doll T E. An update of the polymer-augmented alkaline flood at the Isenhour Unit, Sublette County, Wyoming［J］．SPE reservoir engineering，1988，3（02）：604-608．

[60] Bock J, Valint P L, Pace S J. Enhanced oil recovery with hydrophobically associating polymers containing sulfonate functionality：U.S. Patent 4，702，319［P］．1987-10-27．

[61] Taylor K C, Hawkins B F, Islam M R. Dynamic interfacial tension in surfactant enhanced alkaline flooding［J］．Journal of Canadian Petroleum Technology，1990，29（01）．

[62] Wyatt K, Pitts M J, Surkalo H, et al. Alkaline-Surfactant-Polymer Technology Potential of the Minnelusa Trend, Powder River Basin［C］//Low Permeability Reservoirs Symposium. Society of Petroleum Engineers，1995．

[63] 冯叔初，郭揆常．油气集输与矿场加工［M］．东营：中国石油大学出版社，2006．

[64] 任卓琳，牟英华，崔付义．原油破乳剂机理与发展趋势［J］．油气田地面工程，2005，24（7）：16-17．

[65] 王学会，朱春梅，胡华玮，等．原油破乳剂研究发展综述［J］．油田化学，2002，19（4）：379-381．

[66] 杜玉海,康仕芳.优秀原油破乳剂所具备的性能初探[J].高分子通报,2006(11):92-95.

[67] 李美蓉,冯刚,娄来勇,等.原油破乳剂筛选及破乳效果研究[J].精细石油化工进展,2006,7(11):14-18.

[68] 周恒,邢晓凯,国旭慧,等.原油发泡问题研究进展[J].石油化工高等学校学报,2018,31(1):8-12.

[69] 寇杰,董培林,刘广友,等.油气集输技术数据手册[M].北京:中国石化出版社,2013.

[70] 代晓东,贾子麒,李国平,等.原油降凝剂作用机理及其研究进展[J].油气储运,2011,30(2):86-89.

[71] 刘诚,张宁,代晓东,等.新型油品降凝剂合成研究进展[J].油气储运,2011,30(12):884-888.

[72] 杨飞,李传宪,林名桢,等.含蜡原油降凝剂与石蜡作用机理的研究进展与探讨[J].高分子通报,2009(8):24-31.

[73] 赵书华,刘飞飞,王树立,等.含蜡原油降凝剂的研究进展及其应用[J].常州大学学报(自然科学版),2015,27(3):45-50.

[74] 王哲,马贵阳,赵状,等.管输含蜡原油降凝技术研究进展[J].当代化工,2014(12):2588-2590.

[75] Wu C, Zhang J L, Li W, et al. Molecular dynamics simulation guiding the improvement of EVA-type pour point depressant[J]. Fuel, 2005, 84(16): 2039-2047.

[76] 黄志强,马亚超,李琴,等.天然气管输减阻剂减阻效果现场评价方法研究[J].西南石油大学学报(自然科学版),2016,38(4):157-165.

[77] 曹云.天然气减阻剂室内性能评价及现场应用试验[D].青岛:中国石油大学,2011.

[78] 叶天旭,王铭浩,李芳,等.天然气管输减阻剂的研究现状[J].应用化工,2010,39(1):104-106.

[79] 关中原,李春漫,尹国栋,等.EP系列减阻剂的研制与应用[J].油气储运,2001,20(8):32-34.

[80] 李国平,杨睿.国内外减阻剂研制及生产新进展[J].油气储运,2000,19(1):3-7.

[81] 潘一,孙林,杨双春,等.国内外管道腐蚀与防护研究进展[J].腐蚀科学与防护技术,2014,26(1):77-80.

[82] 刘红.基于盐酸酸洗的含氮有机缓蚀剂制备、缓蚀性能及机理研究[D].成都:西南石油大学,2015.

[83] 胡鹏飞,文九巴,李全安.国内外油气管道腐蚀及防护技术研究现状及进展[J].河南科技大学学报:自然科学版,2003,24(2):100-103.

[84] 张大全,高立新,周国定.国内外缓蚀剂研究开发与展望[J].腐蚀与防护,2009,30(9):604-610.

[85] 赵永生,庞正智.2-己基咪唑作为铜的盐酸酸洗缓蚀剂作用机理的研究[J].北京化工大学学报(自然科学版),2003,30(1):36-39.

[86] 张金钟,毛学强,刘万元,等.评价缓蚀剂性能方法的研究进展[J].化学工程与装备,2011(12):139-140.

[87] 张文. 油田污水处理技术现状及发展趋势[J]. 油气地质与采收率, 2010, 17 (2): 108-110.

[88] Ottaviano J G, Cai J, Murphy R S. Assessing the decontamination efficiency of a three-component flocculating system in the treatment of oilfield-produced water [J]. Water Research, 2014, 52 (3): 122-130.

[89] Fakhru'Lrazi A, Pendashteh A, Abdullah L C, et al. Review of technologies for oil and gas produced water treatment. [J]. Journal of Hazardous Materials, 2009, 170 (2): 530-551.

[90] Torres L, Yadav O P, Khan E. A review on risk assessment techniques for hydraulic fracturing water and produced water management implemented in onshore unconventional oil and gas production. [J]. Science of the Total Environment, 2015, 539: 478-493.

[91] 赵林, 徐玉霞, 马超, 等. 油田污水处理高分子絮凝剂的发展现状[J]. 石油地质与工程, 2005, 19 (3): 67-70.

[92] 王海峰, 王增林, 张建. 国内外油田污水处理技术发展概况[J]. 油气田环境保护, 2011, 21 (2): 34-37.

[93] 张贵才. 油田污水防垢与缓蚀技术研究[D]. 成都: 西南石油学院, 2005.

[94] 毕秀范. 水处理剂在油田污水回注系统的应用[J]. 工业水处理, 2005, 25 (2): 75-76.